SILICON SYSTEMS FOR WIRELESS LAN

ADVANCED SERIES IN ELECTRICAL AND COMPUTER ENGINEERING

The purpose of this series is to publish work of high quality by authors who are experts in their respective areas of electrical and computer engineering. Each volume contains the state-of-the-art coverage of a particular area, with emphasis throughout on practical applications. Sufficient introductory materials will ensure that a graduate and a professional engineer with some basic knowledge can benefit from it.

Published:

Vol. 22: *Silicon Systems for Wireless LAN*
edited by Zoran Stamenković, Gildas Leger and Alberto Bosio

Vol. 21: *Application of Omics, AI and Blockchain in Bioinformatics Research*
edited by Jeffrey J. P. Tsai and Ka-Lok Ng

Vol. 20: *Computational Methods with Applications in Bioinformatics Analysis*
edited by Jeffrey J. P. Tsai and Ka-Lok Ng

Vol. 18: *Broadband Matching: Theory and Implementations (Third Edition)*
by Wai-Kai Chen

Vol. 17: *Practical Signal Processing and Its Applications: With Solved Homework Problems*
by Sharad R. Laxpati and Vladimir Goncharoff

Vol. 16: *Design Techniques for Integrated CMOS Class-D Audio Amplifiers*
by Adrian I. Colli-Menchi, Miguel A. Rojas-Gonzalez and Edgar Sanchez-Sinencio

Vol. 15: *Active Network Analysis: Feedback Amplifier Theory (Second Edition)*
by Wai-Kai Chen

Vol. 14: *Linear Parameter-Varying System Identification: New Developments and Trends*
by Paulo Lopes dos Santos, Teresa Paula Azevedo Perdicoúlis, Carlo Novara, Jose A. Ramos and Daniel E. Rivera

Vol. 13: *Semiconductor Manufacturing Technology*
by C. S. Yoo

Vol. 12: *Protocol Conformance Testing Using Unique Input/Output Sequences*
by X. Sun, C. Feng, Y. Shen and F. Lombardi

Vol. 11: *Systems and Control: An Introduction to Linear, Sampled and Nonlinear Systems*
by T. Dougherty

Vol. 10: *Introduction to High Power Pulse Technology*
by S. T. Pai and Q. Zhang

For the complete list of titles in this series, please visit
http://www.worldscientific.com/series/asece

Advanced Series in Electrical and Computer Engineering – Vol. 22

SILICON SYSTEMS FOR WIRELESS LAN

Editors

Zoran Stamenković
IHP, Leibniz-Institut für innovative Mikroelektronik, Germany

Gildas Leger
CSIC, Spain & IMSE-CNM, Spain

Alberto Bosio
École Centrale de Lyon, France

World Scientific

NEW JERSEY • LONDON • SINGAPORE • BEIJING • SHANGHAI • HONG KONG • TAIPEI • CHENNAI • TOKYO

Published by

World Scientific Publishing Co. Pte. Ltd.
5 Toh Tuck Link, Singapore 596224
USA office: 27 Warren Street, Suite 401-402, Hackensack, NJ 07601
UK office: 57 Shelton Street, Covent Garden, London WC2H 9HE

British Library Cataloguing-in-Publication Data
A catalogue record for this book is available from the British Library.

Advanced Series in Electrical and Computer Engineering — Vol. 22
SILICON SYSTEMS FOR WIRELESS LAN

ISBN 978-981-121-071-6 (hardcover)
ISBN 978-981-121-072-3 (ebook for institutions)
ISBN 978-981-121-073-0 (ebook for individuals)

For any available supplementary material, please visit
https://www.worldscientific.com/worldscibooks/10.1142/11564#t=suppl

https://doi.org/10.1142/9789811210723_fmatter

Foreword

Up to date, there are very few books covering all the steps (from the system-level to the transistor-level) necessary to design, model, verify, implement, integrate, and test a silicon system. Our book targets this empty space and intents to bridge the gap between system and circuit designers, technologists, and physicist. It is extremely important nowadays (and will be more important in the future) for system and circuit designers to understand the physical implications of system and circuit solutions based on hardware/software co-design as well as for technologists and physicists to cope with the system and circuit requirements in terms of energy, speed, and data throughput.

Today's integrated silicon circuits and systems for wireless communications are of a huge complexity. The problem is tackled using configurable general-purpose processors and designing custom hardware accelerators, mixed-signal, and analog circuits. The design flow starts with a functional model at system level, where major functional blocks are defined and no timing information is given. Afterwards, all the functionalities described before are mapped to hardware and will be defined down to circuit level (register-transfer level synthesis). At that point, a cycle accurate model of the digital part is ready for synthesis using standard CMOS logic gates (technology mapping). Once the circuit is synthesized, the physical view (layout) of hardware is created using transistors and interconnections. Functional correctness of the design has to be verified at all three (system, circuit, and transistor) levels. Next step is the integration of digital, mixed-signal, and analog parts into a silicon system (manufacturing). Finally, the system must be tested against random and systematic faults. Therefore, the book addresses all the steps of design, modeling, verification, implementation, integration, and test of advance silicon systems for wireless local-area networks (WLAN).

The book is aimed to help new device, circuit, and system designers to function knowledgeably in the area. It can serve as a comprehensive handbook and a user guide, at the same time, to readers who have already acquired basic knowledge of microelectronics.

Part 1 (Design) consists of three chapters. Chapter 1 (System Architecture) introduces wireless communication systems as a topic and describes WLAN transceiver architecture. Chapter 2 (Digital Baseband) briefly presents transmitter (input buffer, scrambler, signal field generator, encoder, interleaver, mapper, inverse fast Fourier transform, pilot insertion, cyclic-prefix insertion, and preamble insertion) and receiver (synchronizer, fast Fourier transform, channel estimator, demapper, deinterleaver, decoder, and descrambler) components of an Orthogonal Frequency Division Multiplexing (OFDM) WLAN digital baseband. Chapter 3 (Analog Front-End) introduces basic concepts and architecture of an OFDM WLAN analog front-end.

Part 2 (Modeling) encompasses four chapters. Chapter 4 (System Models) considers system models and partitioning the system into hardware and software components doing hardware/software co-design. Chapter 5 (Digital Circuit Models) describes the main digital components and circuits necessary to implement an OFDM WLAN transceiver (configurable general-purpose processors, digital-signal processors, debug-support units, buses, controllers, and peripherals). Chapter 6 (AMS Circuit Budgets) provides an overview of the design methodology of both transmitter and receiver part of an OFDM analog front-end. Chapter 7 (Embedded Memories) presents the main aspects of memory architecture, design, and technology.

Part 3 (Implementation and Integration) comprises four chapters. Chapter 8 (Implementation Methodologies) describes methodologies used to implement an OFDM WLAN transceiver based on Multiple-Input Multiple-Output (MIMO) communication and signal combining in the analog front-end. Chapter 9 (Semi-Custom Implementation of Digital Circuits) considers implementation details of MAC and digital baseband processors of an RF-MIMO OFDM WLAN transceiver. Chapter 10 (Full-Custom Implementation of Analog and Mixed-Signal Circuits) presents implementation aspects (including the package and PCB) of the aforementioned transceiver's analog front-end. Chapter 11 (Integration) briefly reviews package and antenna types and provides integration details of the complete transceiver.

Part 4 (Verification and Testing) is also composed of four chapters. Chapter 12 (Verification) introduces the main concepts, techniques, and coverage metrics of design verification (including a unique standard called Universal Verification Methodology). Chapter 13 (Digital Testing) considers testing of digital integrated circuits as a process of detecting faults (and defects) introduced by manufacturing. Chapter 14 (Analog, Mixed-Signal, and RF Circuits Test) describes performance-oriented and defect-oriented testing of analog, mixed-signal, and RF circuits.

Chapter 15 (Embedded Memory Test) discusses failure modes and mechanisms specific to SRAM and non-volatile memories.

As editors, we thank our co-authors (Antonio Gines, Bojan Dimitrijevic, Ernesto Sanchez, Hassen Aziza, Mile Stojcev, and Milos Krstic) for the great work. We also wish to thank many colleagues (Ignacio Santamaria, Javier Via, Jesus Gutierrez, Jesus Ibanez, Jesus Perez, Klaus Tittelbach-Helmrich, Laura Gonzalez, Michael Wickert, Ralf Eickhoff, Rolf Kraemer, Susana Ruiz, Uwe Mayer, and Victor Elvira) for their exceptional scientific and technical contributions.

Enjoy the book,

Zoran Stamenkovic
Gildas Leger
Alberto Bosio

https://doi.org/10.1142/9789811210723_fmatter

Contents

Part-2 Modeling

Part-3 Implementation and Integration

Part 1

Design

https://doi.org/10.1142/9789811210723_0001

Chapter 1

System Architecture

Zoran Stamenković
IHP - Leibniz-Institut für innovative Mikroelektronik
Frankfurt (Oder), Germany

Mile Stojčev
University of Niš
Niš, Serbia

1.1 Introduction

Electronic devices can freely communicate in different ways but the Open Systems Interconnection (OSI) model shown in Figure 1.1 has been used as a reference for decades [1]. It is a conceptual framework that uses layers to visually describe the functions of a networking system. This can help understanding physical issues, network problems, and application challenges, as well as working across the layer stack. The following seven layers (going down from top to bottom) make this communication stack:

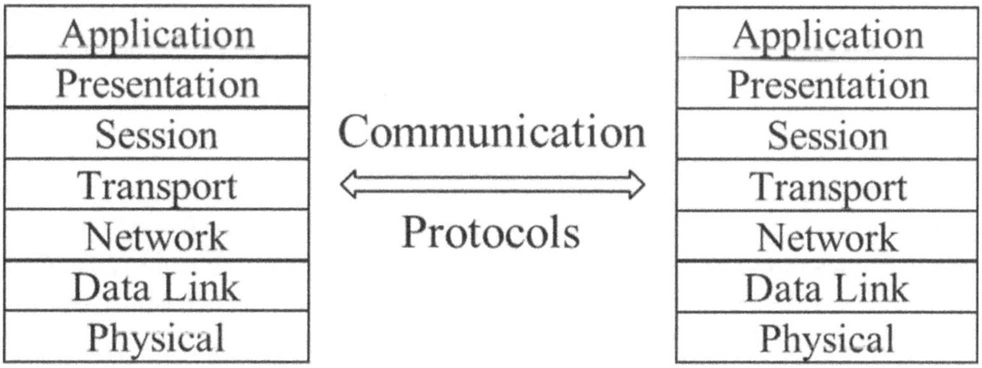

Fig. 1.1. Communication layers of OSI model

Application Layer

The Application Layer is at the top, visible and closest to the end user. Users directly interact with applications that work at this layer. Internet browsers, e-mail tools, and text/picture processors are examples of such applications.

Presentation Layer

The Presentation Layer prepares or translates an application format to a network format or vice versa. In other words, the layer presents data for the application or the network. A good example of this is cryptography: encryption and decryption of data for secure transmission and reception.

Session Layer

Electronic devices (computers, lap-tops, phones, printers, sensor nodes, etc.) need to create a session if they want to communicate. This is done at the Session Layer, which is responsible for setup, configuration (e.g. how long a system waits for a response), and termination of the applications at each end of the session.

Transport Layer

The Transport Layer deals with coordination of the data transfer between clients and hosts: how much data to send, how to split large amounts of data into smaller packets, select a proper data rate, define where it goes, etc. A typical example of the transport layer is the Transmission Control Protocol (TCP), which is built on top of the Internet Protocol (IP), commonly known as TCP/IP.

Network Layer

The main functionality of the Network Layer is routing. This layer is responsible for data packet forwarding and routing through different routers. They establish the communication between two distant electronic devices (a computer and a server or a base station and a mobile client) selecting the most efficient or fastest path/connection.

Data Link Layer

The Data Link Layer provides the transfer of data between two directly connected nodes. It obeys and maintains the sharing rules on the physical medium when several transmitting instances want to access it. Normally, the Data Link Layer also corrects the errors of the physical layer. It comprises two sublayers: the Medium Access Control (MAC) and the Logical Link Control (LLC). Network switches operate at this layer.

Physical Layer

The Physical Layer represents the electrical and physical features of the system. This includes different communication media (copper wire, optical fiber, directed electromagnetic beam, and radio link) and corresponding physical requirements (physical and logical interfaces, power supply, regulations on logic levels, output power level, etc.). The Physical Layer also specifies the signal modulation and data coding schemes, and provides channel assessment techniques (channel clear or busy).

1.2 Wireless Communication Systems

Wireless communication can be used to connect small handheld devices, telephones, printers, laptops, and desktop computers anywhere within their communication range. Wireless communication systems (Figure 1.2) include mobile cellular networks, wireless metropolitan area networks (WMAN), wireless local area networks (WLAN), and wireless personal area networks (WPAN). Depending on the coverage, these networks can be established in buildings, halls, campuses, and cities, as well as regionally and globally.

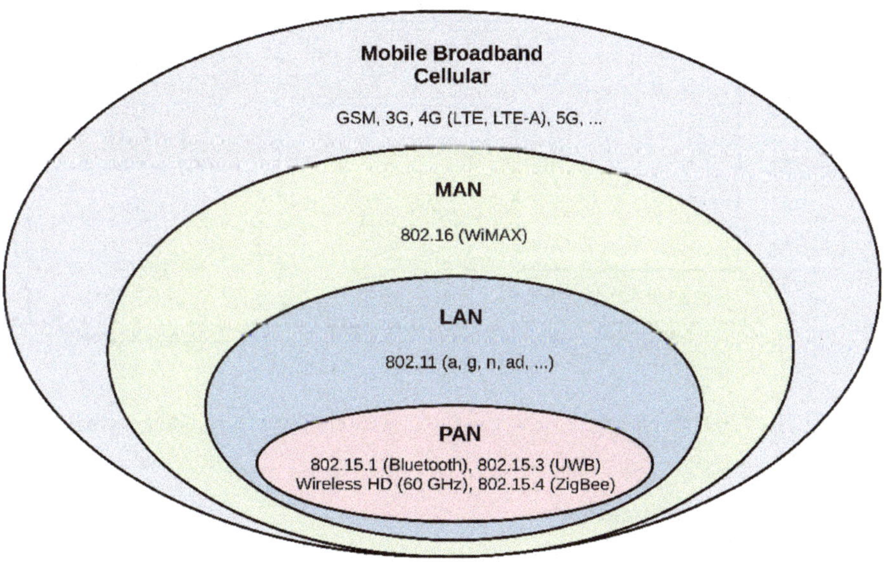

Fig. 1.2. Wireless communication networks

These various networks and coverage ranges have forced researchers and engineers to propose, design, and implement many different wireless products, standards, and services (Figure 1.3). Moreover, different applications and markets impose different system requirements. Audio systems do not require high data rates and can tolerate a high bit-error probability but cannot tolerate a significant delay. On the other hand, data transmission requires much higher data rates and a very small bit-error rate but does not have a severe delay requirement. Real-time video systems require high data rates coupled with the same delay constraints as audio systems, while paging and short messaging have very low data rate requirements and no delay constraints. Diverse requirements of different applications make it difficult to build a single wireless system that can satisfy all these requirements simultaneously. Therefore, wireless communication systems and protocols are (and will stay) fragmented in order to support different applications.

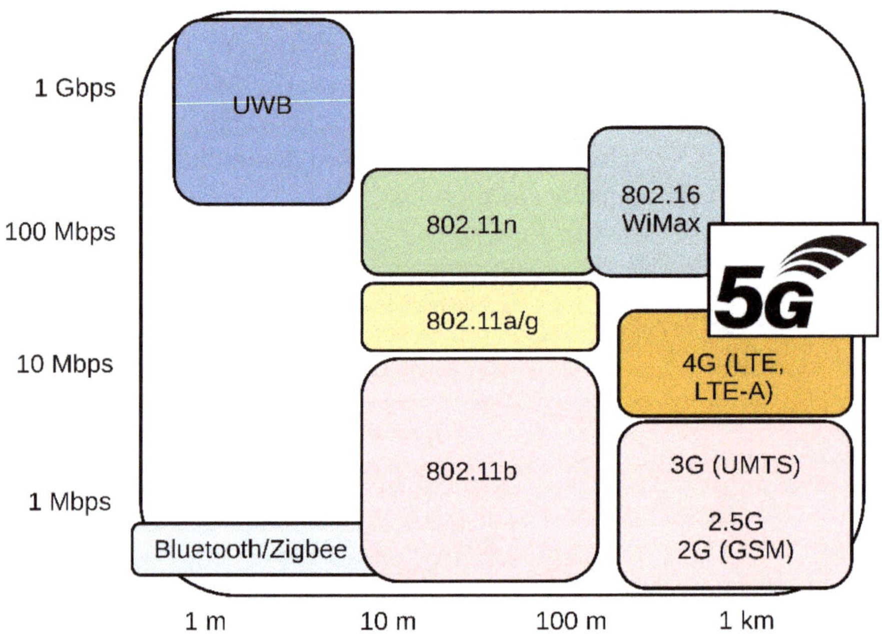

Fig. 1.3. Data rate and coverage range of wireless communication systems

1.2.1 *Mobile communication*

In mobile cellular networks, many users communicate across a single frequency band through mobile telephones. Mobile telephone systems are designed to provide a two-way audio communication with regional or national coverage. Mobile systems support lightweight handheld terminals operating inside and outside buildings at both pedestrian and vehicle speeds. They reuse the same frequency spectrum at spatially separated locations exploiting path loss. The coverage area of a cellular system is divided into non-overlapping cells where a set of channels is assigned to each cell. This same channel set is used in another distant cell. Operation within a cell is controlled by a centralized base station.

The base stations are placed on tall buildings or hills and transmit at high power with the cell coverage of a few kilometers. These large cells are called macrocells. Signals propagate out from the base stations uniformly in all directions, so a mobile client moving in a circle around the base station would have approximately constant received power. The need for increased capacity in areas with increased user density and the reduced size and cost of base station electronics have made feasible more advanced mobile systems that have smaller cells with the base stations close to street level or inside buildings transmitting at much lower power. These smaller cells are called microcells, picocells, or femtocells, depending on the size.

The first generation of mobile cellular networks (1G-NMT) was purely analogue. Then digital mobile systems (Figure 1.3) started from the second (2G-GSM), the third (3G-UMTS), and the fourth (4G-LTE) to the current fifth generation (5G-NR). They are described in numerous international standards [2].

1.2.2 *Long-range communication*

Wireless MAN offers an alternative to cabled access networks, such as fiber optic links, coaxial cables, and digital subscriber lines. It has the capacity to address broad areas without the costly infrastructure required in deploying cable links. WMAN technology brings the broadband wireless access to a building and users inside the building can connect to it with conventional networks such as Ethernet or wireless LAN. Moreover, it can support nomadic and mobile users (for instance, a stationary or slow-moving vehicle).

The IEEE Standard 802.16 [3] defines a set of air interfaces based on a common MAC protocol but with physical layer specifications dependent on the spectrum of use and the associated regulations. This standard offers the opportunity to reach many more customers less expensively (although at lower data rates) and targets at users in individual homes or small and medium enterprises.

1.2.3 *Mid-range communication*

The mid-range and low-power wireless communication can be used to network electronic devices such as smart telephones, laptops, printers, and TV sets. These wireless local area networks allow users to wirelessly connect (within close proximity) to a transceiver. It is necessary to implement a user authentication technique, otherwise the transceiver (router) can be freely accessed.

Wireless LAN (defined by the IEEE Standard 802.11 [4]) provides high-speed data within a small region where users move from place to place. Wireless devices that access these networks are typically stationary or moving at pedestrian speeds. All wireless LANs use one of the ISM frequency bands [5]. A license is not required to operate in these bands. WLAN can have either a star architecture, with the wireless access points placed throughout the coverage region, or a peer-to-peer architecture, where the wireless terminals self-configure into a network. These networks can reach a coverage range of a few hundred meters (Figure 1.3).

1.2.4 *Short-range communication*

The short-range communication technology permits wireless connection of various electronic devices to a system for the data transfer. Smart telephones and laptops can be connected to a hands free earphone, mouse, and keyboard. The IEEE Standard 802.15 [6] defines wireless personal area networks (WPANs) as a perfect fit for these applications. They offer a coverage range of 1 m to 100 m and a wide variety of data rates. However, since the devices connected in WPANs should be mobile and lightweight, energy consumption is the main concern.

There are three different approaches to the wireless PAN technology (Figure 1.3). The first is Bluetooth which offers data rates of up to 3 Mbps and ranges of up to 100 m at transmit power of 1 mW. The second of these technologies, ZigBee, offers data rates of up to 250 kbps and can support links with a very low duty cycle. Hence, it is suitable for deployment in battery-powered devices that must survive a few years after charging. As the last one, ultra-wide band (UWB) technology provides low-power pulses that deliver data rates comparable to Ethernet.

1.2.5 *Why do we focus on WLAN only?*

We focus on WLAN only since it is one of the most important technologies of computer networking, connects users to the Internet, and can be used for free in industry halls, homes, restaurants, airports, and many other places. WLAN is a single technology that can help make all connections seamless and provide a high level of security and privacy.

1.3 WLAN Transceiver Architecture

WLAN is a network in which mobile devices can communicate wirelessly (over a radio link). Each mobile device is equipped with a transmitter and a receiver (in a single word, transceiver) to provide communication at mid-range distances. The IEEE Standard 802.11 [4] specifies the two lowest OSI model layers of WLAN: Medium Access Control (MAC) sublayer and Physical Layer (PHY). The MAC is a set of rules determining how to access the medium and send the data, while the details of transmission and reception are left to the PHY.

1.3.1 *WLAN MAC sublayer*

Most WLANs operate in the infrastructure Basic Service Set (BSS) mode. In this case, all WLAN stations communicate via single wireless hops with a central entity denoted as the Access Point (AP). Normally, the AP acts as a gateway to the wired communication infrastructure (Internet), so that WLAN stations can connect to peer nodes all over the world. Alternatively, WLAN nodes can form an *ad-hoc* network (independent basic service set, IBSS). This mode operates without a central controller (controlling the access to the network cell, the traffic in the cell, and so on) and normally without a gateway to the infrastructure.

WLAN stations sense the wireless medium before transmitting a frame using two Listen-Before-Talk (LBT) mechanisms: a Physical Carrier Sense (P-CS) and a Virtual Carrier Sense (V-CS). The first is a part of the physical layer, while the second is a MAC sublayer function. Figure 1.4 shows the hierarchical position of these mechanisms. If either of them indicates busy medium conditions, the station does not attempt to transmit. Once the wireless medium is idle, a station may send a frame. To avoid multiple frames being transmitted simultaneously, WLAN implements the Collision Avoidance (CA): stations need to wait for a certain random period of time before accessing the wireless medium.

P-CS is a means of the Clear Channel Assessment (CCA). It detects transmissions of similar systems, while the Energy Detection (ED) informs on medium usage by dissimilar physical layers or pure interference. ED indicates medium usage independent of the modulation and shape of a signal. For any valid Physical Layer Convergence Protocol (PLCP) header received, P-CS indicates a busy wireless medium for the duration of a frame.

V-CS informs stations about ongoing or planned transmissions, e.g. expected acknowledgement frames or more fragments belonging to the same frame. All stations that are not in power-save mode, constantly monitor the wireless medium and retrieve reservation information from any frame they can decode. Frames provide the reservation information in the MAC header. If the duration field is present, stations set a Network Allocation Vector (NAV) to the respective value. The NAV works as a count-down timer. As long as the timer sets a value different from zero, V-CS indicates a busy medium.

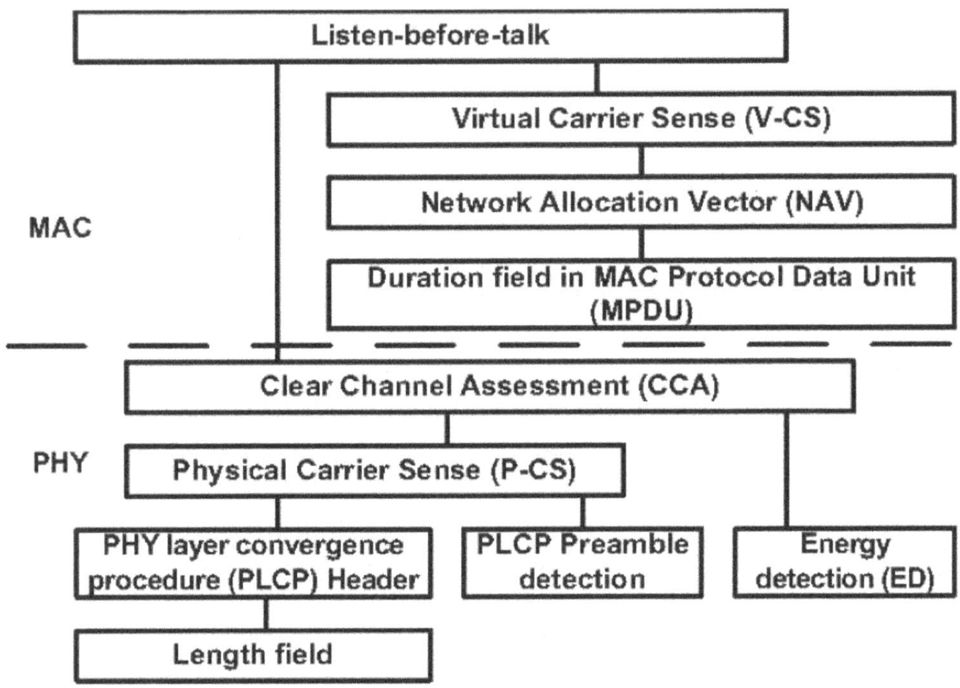

Fig. 1.4. Task partitioning between MAC and PHY layers [4]

Since the WLAN transceiver transmits and receives at the same RF channel, but the received wireless signal strength is much smaller than the emitted one, it cannot transmit and receive concurrently and cannot detect collisions. It has to rely on a collision avoidance (CA) procedure waiting for a random time period before transmitting. Two basic parameters are defined (their durations depend on the PHY layer technology):

- Short Inter-frame Space (SIFS), the minimum time gap between two transmissions on the medium.
- Time slot which defines the period needed for ED and result indication to the MAC sublayer.

The SIFS time is the amount of time to elapse between tightly coupled frame transfers, in particular between a frame and the corresponding acknowledgement frame, but also between several fragments of the same (logical) frame. The SIFS time provides sufficient time for change of the transceiver mode (the receiver of a frame switches to the transmit mode and the transmitter turns to the receive mode). For a new frame transfer, the transmitting station can access the wireless medium only if the medium remains idle for at least a duration of SIFS+EDSlot. This duration is known as the Point Coordination Function Inter-frame Space (PIFS). To provide a central coordination entity or Point Coordinator (PC) with the highest priority, other stations must wait longer. They access the wireless medium at earliest after the SIFS+2xEDSlot period of time which is called Distributed Coordination Function Inter-frame Space (DIFS). To avoid the possibility of multiple stations transmitting simultaneously, DIFS stations need to wait for a multiple amount of EDSlots (so-called Back-off Time). For every EDSlot duration the wireless medium remains idle after a DIFS period has elapsed, a station decreases its back-off-counter by one.

The Distributed Coordination Function (DCF) is a basic WLAN coordination function. It implements the Carrier Sense Multiple Access (CSMA) and Collision Avoidance (CA) mechanisms. For every frame transmission that fails, DCF doubles the Contention Window (CW) size and draws the new Back-off Time as a random number between 0 and CW. Afterwards, the station may try to send the frame again. After each successful frame transmission, the CW is reset to some (PHY dependent) minimum value. To detect transmission failures, the receiver acknowledges every successfully received unicast frame. The absence of the acknowledgment frame (ACK) indicates a failure. Multicast/broadcast frames are not acknowledged.

The advantage of DCF is that it does not require a central controller to administer the channel access, nevertheless achieving an acceptable level of collisions. So it can be easily used in *ad-hoc* networks. A centrally controlled access mechanism, however, provides better efficiency, since the overhead of large waiting time due to the back-off procedure can be avoided. The WLAN standard defines a number of coordination functions. The so-called Point Coordination Function (PCF) was part of the original IEEE 802.11, but did not gain much popularity. Later, the IEEE802.11e amendment introduced the Enhanced Distributed Channel Access (EDCA) and the Hybrid Coordination Function (HCF). They may guarantee Quality-of-Service (QoS) for the data traffic. The basic principle is that a QoS-aware access point (QAP) assigns transmit opportunities (TXOPs) to the clients in the network, based on their individual QoS requests. This allows reducing of the interframe spacing and other dead times of the data traffic in the network.

In WLAN, interference may occur at the receiver side when solely P-CS is used at the transmitting side. Depending on the network topology stations may become mutually hidden. Consider the C station being in reception range of the B station but outside the P-CS range of the A station. The latter may transmit a frame to B. However, C cannot detect the transmission of A. Thus, it may initiate a frame transmission. Depending on the distance to B, the transmission of C may significantly reduce the Signal to Noise Ratio (SNR) of the transmission from A to B. This effect is known as a hidden station problem. To avoid it, WLAN introduces an optional handshake mechanism as follows:

- Based on a manually set threshold and depending on the actual frame size, the A station sends a Request To Send (RTS) frame. Its duration field sets the NAV in stations nearby A to the duration of the intended frame exchange sequence.
- If the B station receives the frame, it replies by a Clear To Send (CTS) frame. The latter sets the NAV of stations in the surroundings of B. Thus, any station that received the RTS and/or CTS frame holds back from medium access.

The exposed station problem is the counterpart to the hidden station problem. A station that is in range of the transmitter but far from the receiver of a frame exchange, senses the wireless medium as idle or may have its NAV being set. As the station is far from the receiver, it cannot interfere with its frame reception. However, due to a busy wireless medium, the station cannot access the channel for a concurrent transmission. The IEEE802.11 MAC does not provide a solution for the exposed station problem. It relies on the frame acknowledgment. As the frames have arbitrary length, an ACK may follow any time. Both stations that exchange frames constantly change roles of transmitter and receiver. In EDCA, the owner of a transmission opportunity may provide the remaining time to the receiver. Thus, data frames may be sent in both directions.

1.3.2 *WLAN physical layer*

The original IEEE 802.11 specification from 1999 included two physical layers operating in the unlicensed 2.4 GHz RF band: a Frequency-Hopping Spread-Spectrum (FHSS) layer and a Direct-Sequence Spread-Spectrum (DSSS) layer. Both are now completely obsolete. Later revisions (802.11a and 802.11b) added a physical layer for the 5 GHz band based on Orthogonal Frequency Division

Multiplexing (OFDM) and a High-Rate Direct-Sequence Spread-Spectrum (HR/DSSS) layer at 2.4 GHz. The 802.11g amendment offers higher speed by the use of OFDM at 2.4 GHz and backwards compatibility with the 802.11b. All these modes offer data rate up to a few 10 Mbps. The 802.11n amendment adds a Multiple-Input Multiple-Output (MIMO) technology, supporting data rates up to a few 100 Mbps. The most recent amendments 802.11ac and 802.11ad specify physical layers reaching data rates of several Gbps and make the 60 GHz band available for high-speed WLAN. A generic block diagram of a WLAN physical layer is shown in Figure 1.5. In this book, however, we focus on the OFDM PHY for the unlicensed 5 GHz band.

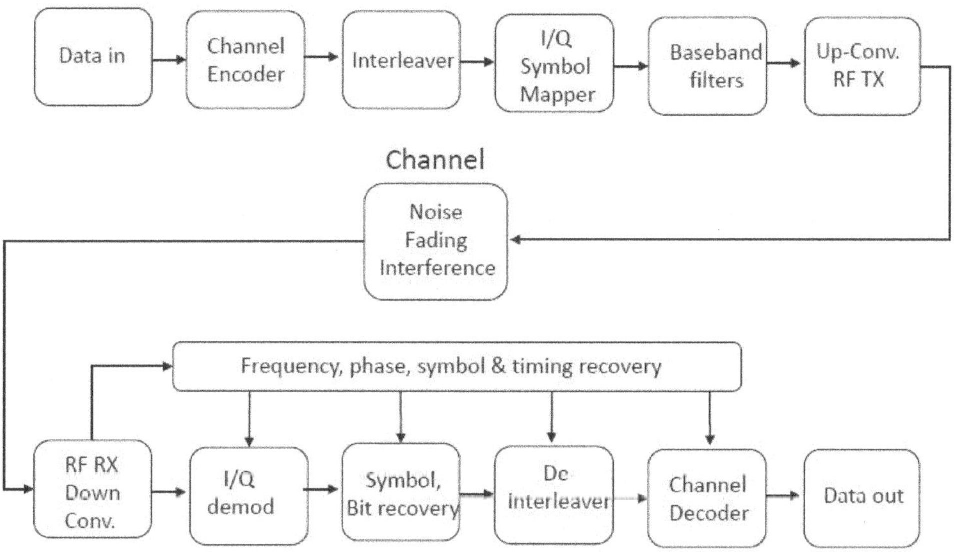

Fig. 1.5. Basic model of a WLAN physical layer

Digital baseband

The digital baseband encodes the original digital data into a form suitable for transmission. The digital stream must be converted into the constellation of amplitudes and phases of the in-phase (I) and quadrature (Q) analog signal components that represents the digital signal. To rise the capacity of a radio channel, multiple baseband signals can be multiplexed over the same channel using different techniques: frequency domain multiplexing, time domain multiplexing, code division multiplexing, or a combination of those.

A PHY frame normally consists of a frame preamble, a PHY header (PLCP header, physical layer convergence procedure) and the PHY payload (PSDU, PLCP service data unit). The latter transfers the data of higher layers, namely the

MAC header, the MAC payload and usually a check sum (cyclic redundancy check) for error detection. The purpose of the preamble is to synchronize incoming frames at the receiver, i.e. to compensate differences in the RF carrier frequency (and also the baseband frequency) at both stations due to quartz crystal tolerances, and to determine the right point in time to start demodulation. Moreover, the receiver's automatic gain control (AGC) must be given some time to settle at a proper signal amplification. The PHY header usually transfers information on the modulation and coding scheme of the frame, i.e. the data rate, Forward Error Correction (FEC), etc. For variable-length frames like in WLAN, it also specifies the frame length (packet size).

Details of the PHY frame structure strongly depend on the respective PHY version. Figure 1.6 depicts the frame format in the IEEE Standard 802.11a (OFDM, 5 GHz band). The signal field as the equivalent of the PHY header specifies the data rate (and modulation scheme) and frame length (theoretically up to 4095 bytes, practically due to restrictions in the MAC protocol up to 2300 bytes). The header is protected by a convolutional error correcting code that requires 6 tail bits and a parity bit allowing error detection. Also the frame payload is protected by the convolutional code, resulting in the need of a Viterbi decoder at the receiver side.

The different versions of IEEE 802.11 WLAN PHY specify quite different modulation schemes [4]. They range from the comparatively simple complementary code keying (CCK, IEEE Standard 802.11b) to very complex MIMO schemes that use many antennas and data streams and require regular training sequences, usually as a part of the frame preamble.

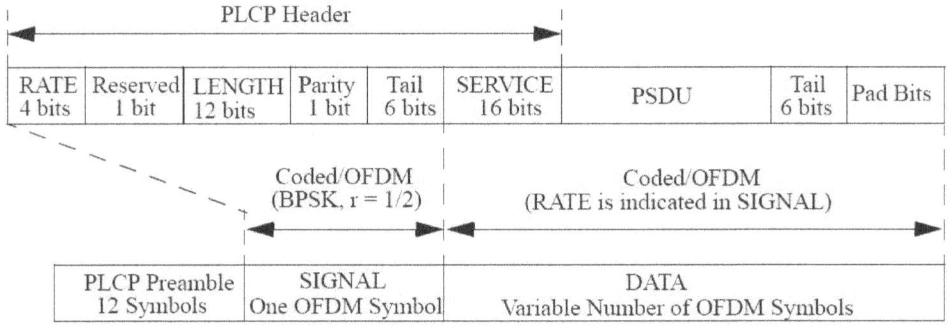

Fig. 1.6. PSDU frame format of the IEEE Standard 802.11a [4]

The OFDM modulation scheme [7] has been specified in many variants in the IEEE Standards 802.11a (5 GHz), 802.11g (2.4 GHz), 802.11n and 802.11ac (both 5 GHz MIMO) and as an option in 802.11ad (60 GHz). It is a very flexible and adaptable modulation scheme that has turned out to be very robust in multi-path RF channels.

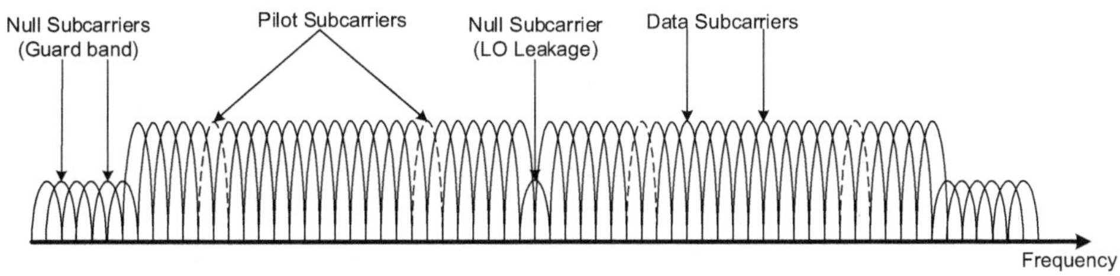

Fig. 1.7. Principle of OFDM modulation in the frequency domain

Basically, the available RF band is split into a number of equally spaced sub-bands. In each sub-band, data are transferred on a RF sub-carrier (Figure 1.7). The IEEE Standard 802.11a, for instance, divides the band into 64 sub-bands. The IEEE Standards 802.11n and 802.11ac define different modes that use up to 468 sub-carriers, resulting in a wide range of supported PHY data rates. Each sub-carrier may be individually modulated in BPSK, QPSK or some QAM, depending on the Signal to Noise Ratio (SNR) in that sub-band. All sub-carriers are modulated and demodulated jointly using the Fast Fourier Transform (FFT). This guarantees orthogonality of the sub-bands, i.e. they do not disturb each other. Some subcarriers are not used for data transmission. *Pilot Subcarriers* (marked as dashed semi-ellipses in Figure 1.7) are spaced throughout the channel in addition to many *Data Subcarriers*. The *Pilot Subcarriers* are used for synchronization and channel estimation. Note that there are a few areas in the channel where subcarriers are forbidden. Typically, these areas are referred to as *Null Subcarriers* (small size semi-ellipses in Figure 1.7). *Null Subcarriers* are placed on the edges of the channel to guard against inter-channel interference. Most notable is the use of a *Null Subcarrier* at the center of the channel. This eliminates issues related to the center frequency contaminating the OFDM modulation in that area, namely *LO leakage* from the mixer.

Analog front-end

The analog front-end performs translation from the digital world to RF signals that can be transmitted by air using antennas. The transmission and reception functions are combined in a so-called transceiver, but the truth is that they can still be clearly identified in the receiver and transmitter sections.

From an architectural viewpoint, the transmitter and receiver are quite symmetrical. The transmitter first converts the digital signals to analog with a digital-to-analog convertor (DAC) and then one or several mixing stages shift the center frequency of the signal towards the desired RF band. This mixing operation also properly combines the in-phase and quadrature channels to get rid of the unwanted signal images. Finally, a power amplifier adapts the signal to the antenna impedance and provides the adequate power. On the other hand, the receiver first amplifies the RF signal from the antenna (which may be weak due to attenuation from the transmission medium) with a low-noise amplifier. Then, the RF signal is down-converted through one or more mixing stages that process two paths, in-phase and quadrature, to properly recover the desired signal. At the end of the signal chain, an analog-to-digital convertor (ADC) finally converts a relatively low-frequency analog signal to digital.

Despite that symmetry, the transmitter and receiver bear an important conceptual difference: while the transmitter only has to deal with a controlled, deterministic signal, the receiver must correctly handle any signal captured by the antenna, attenuating the unwanted sections of the spectrum and amplifying the desired signal up to an adequate signal-to-noise ratio. Chapter 3 will give further details on different architectures of both transmitter and receiver and introduce some important concepts that can be used to specify the function and performance of their building blocks.

References

1. ISO/IEC 7498-4:1989 - Information technology - Open Systems Interconnection - Basic Reference Model: Naming and Addressing, International Organization for Standardization
2. www.etsi.org/standards, European Telecommunications Standards Institute
3. IEEE Standard for Local and Metropolitan Area Networks - Part 16: Air Interface for Fixed Broadband Wireless Access Systems
4. IEEE Standard for Local and Metropolitan Area Networks - Part 11: Wireless LAN Medium Access Control (MAC) and Physical Layer (PHY)
5. https://www.ofcom.org.uk/__data/assets/pdf_file/0022/103297/fat-ism-frequencies.pdf
6. IEEE Standard for Local and Metropolitan Area Networks - Part 15.1a: Wireless Medium Access Control (MAC) and Physical Layer (PHY) Specifications for Wireless Personal Area Networks (WPAN)
7. Goldsmith, A. (2005) *Wireless Communications*, (Cambridge University Press, New York, USA).

https://doi.org/10.1142/9789811210723_0002

Chapter 2

Digital Baseband

Zoran Stamenković
IHP - Leibniz-Institut für innovative Mikroelektronik
Frankfurt (Oder), Germany

Mile Stojčev and **Bojan Dimitrijević**
University of Niš
Niš, Serbia

2.1 Introduction

As we have already mentioned, OFDM is a modulation method that uses simultaneous transmission of data over a large number of adjacent sub-carriers [1-5]. The symbols to be transmitted on each of the sub-carriers are collected in the frequency domain and are all simultaneously transformed to a time domain signal using an Inverse Fast Fourier Transform (IFFT). The idea is that each sub-carrier signal has a very small bandwidth (compared to the coherence bandwidth of the channel) and hence only be subject to flat fading. The channel equalization consists of multiplying each sub-carrier with a complex constant in order to scale and rotate the constellation point to its correct position. This simple channel equalization at the receiver side is possible because of the insertion of a cyclic prefix (CP) to each symbol after the IFFT at the transmitter side. It converts a linear convolution into a circular convolution (the product of FFTs). The cyclic prefix also inhibits the inter-symbol interference (ISI) between adjacent OFDM symbols acting as a guard interval. As long as the delay spread is shorter than the CP, the ISI is not a problem. The architecture of an OFDM digital baseband consists of two main blocks: transmitter and receiver. According to this division, baseband processing happens in two almost independent dataflow directions: transmit and receive [4, 5].

The transmitter consists of an input buffer, scrambler, signal field generator, encoder, interleaver, mapper, IFFT, pilot insertion, CP insertion, and preamble insertion. The receiver consists of a synchronizer, FFT, channel estimator, demapper, deinterleaver, decoder, and descrambler. These baseband components perform calculations based on complex-valued data and bit-based operations:

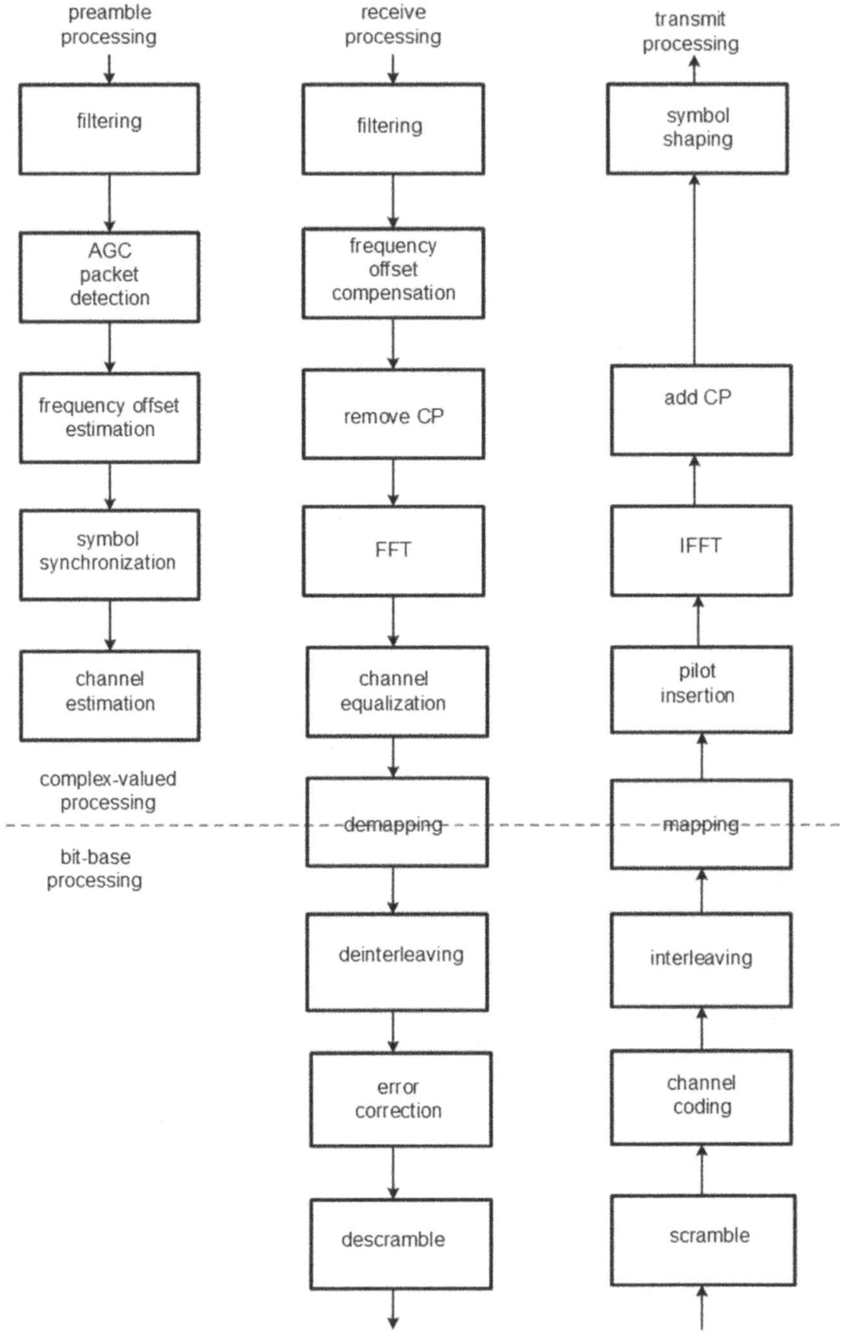

Fig. 2.1. OFDM digital baseband

a) *Calculations based on complex-valued data* – The data between DAC/ADC and mapper/demapper consists of I/Q pairs, which are treated as complex values. The functions to be carried out are modulation/demodulation, synchronization, channel estimation/equalization, IFFT/FFT, cyclic prefix insertion/removal, frequency offset estimation/compensation, etc. (Figure 2.1).
b) *Bit-based operations* – Many processing steps between the interface to MAC-layer and the mapping/demapping (Figure 2.1) are operations based on manipulation of individual data bits (scrambling/descrambling, channel coding, error correction, and interleaving/deinterleaving). This tends to make them rather inefficient for software implementation. On the other hand, many of these operations can be implemented in very small hardware blocks. For example, scrambling and convolutional coding can be implemented as linear feedback shift registers with XOR-gates.

The above mentioned digital baseband components are described in the following sections.

2.2 Scrambling/Descrambling

In many communication systems, the input data is scrambled with the aim to provide frequent transitions and to avoid situations where periodic data and/or repetitive input produce very little change in the transmitted signal. In essence, frequent transitions are inevitably required for timing recovery and for proper operation of automatic gain control circuits. Ideally, a modulated signal should have many transitions and all the points in the constellation should be visited at regular intervals. Scrambling ensures this condition for any input data. In addition, a scrambling function also makes the power spectrum independent of the transmitted data, spreading the power smoothly over the available bandwidth. This reduces radio frequency interference and channel crosstalk, because the power is not concentrated in a narrow frequency band.

When scrambling is implemented, the input data is passed through the scrambler before modulation. After that the modulated signal is passed through the communications channel and is demodulated. Finally, the demodulated data is passed through the descrambler in order to recover the original data.

The main constituents of a scrambler/descrambler system are programmable length shift registers and modulo-2 adders (implemented by using exclusive OR gates). In other words, a scrambler/descrambler is based on linear feedback shift registers (LFSRs) having good statistical properties and efficient hardware implementations.

For correct reconstruction of the original bit-stream, the LFSR in descrambler should be synchronized with the LFSR of scrambler. Depending on the synchronization method, the following two scrambling techniques are used [6]:
a) *Synchronous-scrambling* – This approach uses a sync-word. The sync-word is a pattern that is placed in the data stream through equal intervals in each frame. A receiver searches for a few sync-words in adjacent frames and hence determines the place when its LFSR must be reloaded with a pre-defined initial state. The synchronous-descrambler is just the same device as the synchronous-scrambler.
b) *Self-synchronous scrambling* – The states of scrambler and descrambler LFSRs are automatically synchronized without any additional synchronization process. WLAN transceivers use synchronous-scrambling circuits.

LFSR consists of a group of registers with the output of one register connected to the input of the next so that data is shifted one position down the line at each clock cycle. Figure 2.2 sketches the structure of a scrambler. The feedback is produced by connections (taps) from specific registers. The outputs of these taps are combined using the exclusive-or (XOR) operation (⊕) and the resulting feedback sequence is XOR-ed with the input data stream.

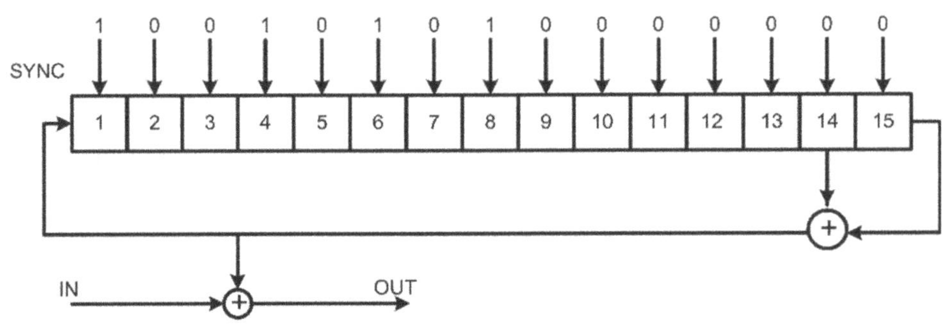

Fig. 2.2. Structure of synchronous-scrambler/descrambler

The specific structures of the synchronous-scrambler and descrambler circuits must match each other in order for the scrambling/descrambling function to be effective. The internal connections of the scrambler/descrambler circuits are defined as a polynomial, and also called characteristic polynomial of the circuit. The general form of the characteristic polynomial is as follows:

$$P(x) = A_n x^n + \ldots + A_2 x^2 + A_1 x^1 + A_0 x^0$$

where n corresponds to the number of registers. With n registers, there are $n+1$ locations where a connection (tap) can be made. The value of each coefficient A_k of the polynomial (where k ranges from 0 to n) is either 0 or 1, depending on whether the tap is absent or present at location *k*. In Figure 2.2, taps are present at locations 0, 14, and 15. There are no taps at locations starting from 1 up to 13. Therefore, the polynomial $P(x)$ equals to $x^{15} + x^{14} + 1$.

A frame synchronous scrambler implemented in IEEE 802.11 WLAN transceivers shown in Figure 2.3 uses the $P(x) = x^7 + x^4 + 1$ generator polynomial.

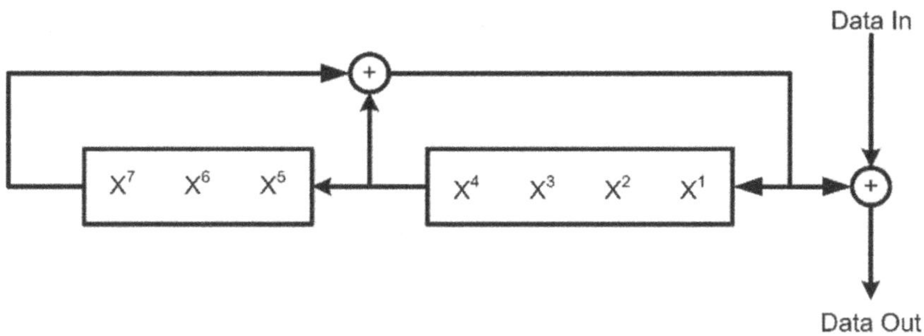

Fig. 2.3. Scrambler and descrambler used in IEEE 802.11 WLAN transceivers

The scrambler and descrambler circuit uses a shifting operation with the aim of introducing delay in the data in order to obtain the required cryptogram. *X* represents the delay operator (x^N indicates the sequence is delayed by *N* units). The symbol '+' in the above equation defines a modulo-2 addition. Data is delayed by seven delay elements (flip-flops). Outputs of 7-th and 4-th delay elements are modulo-2 added with 1 giving the scrambled data.

2.3 Channel Coding and Error Correction

Wireless communications come up with high error rates as a consequence of multipath fading which typifies radio channels. However, WLAN applications need to combat the channel imperfections and significantly reduce the error rates. The Forward Error Correction (FEC) approach [7-10] uses error-correcting codes to deal with the bit errors by adding redundancy to information packets before transmission. This redundancy is used by the receiver, first, to detect and, second, to correct errors. Figure 2.4 depicts the FEC operation where the encoder maps the *k*-bit transmitted data into an *n*-bit codeword and the decoder detects and corrects errors to obtain the original data.

Whenever data is transmitted over a wireless communication channel, there is a chance that one or more bits change value. Such incorrect values are called errors. Error correcting codes insert (parity) check bits into the message sequence in a proper way depending on the type of code being used. These check (parity) bits allow the receiver to correct the error in a message if introduced during data transmission through the channel by noise or interference. The typical metric for quantifying the reliability is the bit error ratio (BER). In principle, error correction can be avoided in cases when retransmission is feasible and effective. But error detection is obligatory in all cases. The main characteristics of the most important error detection schemes are given in the following text.

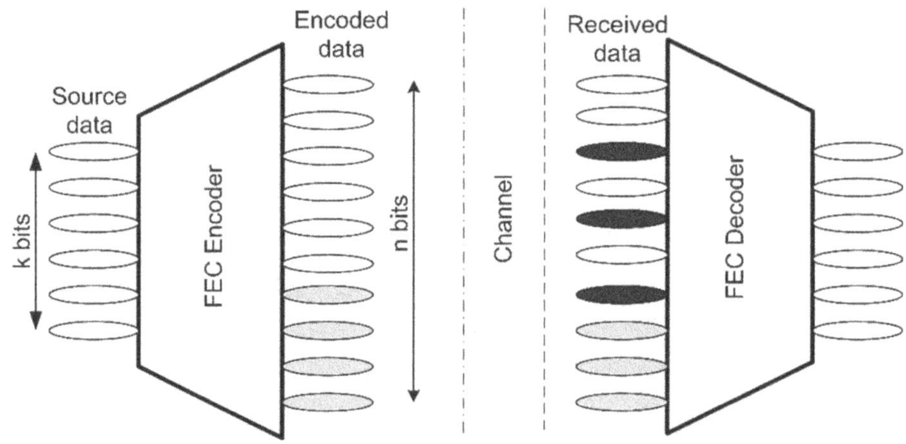

Fig. 2.4. Operation of FEC based on block coding

2.3.1 *Error detection schemes*

The following error detecting schemes are popular in wireless LAN data transfer:
a) *Parity scheme* – Parity bits are allocated (parity can be even or odd) to the transmitted data. The parity of the received data is checked; if it does not satisfy the assigned parity, an error is detected. This scheme is effective only for odd numbers of errors. It cannot detect even numbers of errors since an even number of errors leaves the parity unchanged.
b) *Checksum scheme* – In this approach, a checksum is calculated at the transmitter side and sent with the actual data. The receiver calculates the checksum and compares it with the received checksum. A mismatch indicates the error. If both data and checksum are received with an error, detection is not possible. A more robust checksum is used on the Internet (RFC 1071) where it is the 1's complement of the 1's complement sum of sixteen bit words.

c) *Cyclic Redundancy Check scheme* – An error-detection technique that is widely used in networks is based on Cyclic Redundancy Check (CRC) codes. CRC codes are implemented using linear feedback shift registers, and are used for error detection by link layer protocols where the data transport unit is called a frame. CRC is used to check a block of data for errors. If the calculation at the receiving end differs from the calculation from the sender, the receiver requests the data again. CRCs are popular because they are simple to implement in binary hardware, easy to analyze mathematically, and particularly good at detecting common errors caused by noise in transmission channels. Specification of a CRC code requires definition of a so-called generator polynomial. This polynomial becomes the divisor in a polynomial long division, which takes the message as the dividend and in which the quotient is discarded and the remainder becomes the result. CRC are particularly well suited for the detection of burst errors: contiguous sequences of erroneous data symbols in messages. This is important because burst errors are common transmission errors in many wireless communication channels. This scheme is complex but represents an efficient error detection technique.
d) *Hamming distance based check scheme* – This approach basically uses a parity based scheme but in this case the parities of different combinations of bits are checked for parity. In general, this scheme enables correction of single errors and detection of double errors.
e) *Polarity scheme* – In this scheme the actual data is sent along with its inversion format. At the receiver side, the received data is checked whether the two sets are mutually inverse. If not, it is an indication of error. This scheme is not so popular since the code occupies double the bandwidth for the actual message. Moreover, if corresponding bits in the non-inverted and inverted data are in error then it is not possible to detect the error.

2.3.2 *Error correction codes*

There are numerous ways of classifying forward error correction codes. In the text that follows, we explain the most commonly used classification criteria:
a) *Linear vs non-linear* – Linear codes are those in which the sum of any two valid codewords is also a valid codeword. In the case of non-linear codes, the aforementioned assertion is not always true.
b) *Cyclic vs non-cyclic* – Cyclic codewords are those in which shifting any valid codeword generates another valid codeword. In the case of non-cyclic codewords, the above statement is not always true.
c) *Systematic vs non-systematic* – Systematic codes are those in which the actual information appears unaltered in the encoded data and redundant bits are added for

detection and correction of error. In non-systematic code the actual message does not appear in its original form in the code rather there exists one mapping method from the data word to codeword and vice versa.
d) *Block vs convolutional* – Block codes are those in which one block of data is transformed into a block of code. In this case no memory is required. In the case of convolutional codes, a sequence of data is converted into a sequence of code. Having in mind that the convolutional encoder requires memory the current code represents a combination of present and past data.
e) *Binary vs non-binary* – Binary codes are those in which error detection and correction is done on binary information i.e. on bits. This implies that after the error is located, the bit found in error is corrected by flipping. In non-binary codes, error detection and correction are done on symbols, nevertheless the symbols may be binary. For these codes, both the error location and magnitude are required in order to correct the symbol in error.

2.3.3 *Common metrics of FEC codes*

The following four properties represent a basis according to which a particular Forward Error Correction code can be selected as a convenient metric for data transfer.
a) *Coding gain* – This metric compares (in *dB*) the signal-to-noise ratio per bit required to achieve a given bit error rate in two cases: with encoding and without encoding.
b) *Coding rate* – It is the ratio of the number of transmitted data bits to the total number of transmitted bits (k/n).
c) *Power penalty* – It is a result of sending a large constellation which includes the extra parity or check bits of the error correcting code.
d) *Coding complexity* – It deals with the complexity involved during encoding and decoding. This metric has a direct influence on increments of the design time as well as latency.

2.3.4 *Block codes*

Block codes segment the binary sequence of information bits into k-bit blocks and calculate the redundancy for each block separately. The k bits are transformed into a codeword of n bits, where $n > k$ and $m = n - k$ is the number of redundant bits. In essence, block coding schemes divide a bit stream into non-overlapping blocks and code each block independently. Block coding schemes are frequently referred to as memoryless since successive information blocks are coded independently. The main features of block codes are:

a) The ratio $R = k/n$ is the code rate of the block code, and it is a fraction of the bits sent through the channel that carries actual information.

b) A block code is linear if the sum of two codewords is another codeword.

c) The Hamming distance, d_{min}, is an important metric for describing the resiliency of a code. The Hamming distance is the number of errors that must occur for any codeword to be transformed into any other codeword, i.e. the Hamming distance between two words (of equal length) is the number of bit positions in which they differ. For error correction, a t-error-correcting code must have a Hamming distance of $2t-1 \leq d_{min} \leq 2t + 2$. A code may be defined by a $k \times n$ generator matrix, G, or a $m \times n$ parity check matrix, H, where $m = n-k$. If the k bit vector, u, represents the message sequence, $v=u \cdot G$ gives the corresponding n-bit codeword vector. On the decoding side, the n-bit vector, $r=v+e$ is received, where e is an error vector that contains 1 at error position. Afterwards, $s=r \cdot H^T$ is computed, where s represents the vector syndrome. The exclusive-OR of the new and old check bits is called the syndrome. The syndrome tells us what we need to know about the error. H is the null space of G, so the syndrome will be zero only if r is a codeword. For r to be a codeword, e must be a codeword. The all-zero error vector, which indicates no error, is itself a codeword. Therefore, we must assume that an all-zero syndrome indicates no error. Using this approach, an error will go undetected only if e is a non-zero codeword, and all errors for which e is not a codeword can be detected.

d) A linear block code is said to be a systematic code if the first k bits of the code-words are the data bits and the following n - k bits are the parity check bits.

e) Linear block codes (very often used in wireless technology) are cyclic. It means that shifting any codeword produces another codeword. Cyclic codes can be described by a generator polynomial, $g(X) = 1 + a_1 x + a_2 x^2 + ... + a_m x^m$, where a_i's take on binary values. All codewords in the codes are divisible by the generator polynomial, so the syndrome can be obtained by merely dividing the received vector by $g(X)$. Cyclic codes are important pr actical error correcting codes due to their property and structure. One of the many reasons is that the general class of cyclic codes can be efficiently and easily implemented using linear feedback shift register based encoders and decoders. The cyclical nature of these codes provides them with more structure that helps ease the encoding process. This structure allows us to associate a code with polynomials.

Binary Bose–Chaudhuri–Hocquenghem (BCH) codes and non-binary Reed–Solomon (RS) codes are two types of widely used linear cyclic block codes.

BCH codes

BCH codes [8] form a class of cyclic error-correcting codes that are constructed using finite fields. BCH codes are a generalization of Hamming codes, and can be designed to correct any error pattern of size t or lower. In this sense, the generalization of Hamming codes extends the design of codes for $t = 1$ (Hamming codes) to codes for any desired higher value of t (BCH codes). The design method is based on taking a Least Common Multiple (LCM) of appropriate minimal polynomials. For any positive integer $m \geq 3$ and $t < 2^{m-1}$, there exists a binary BCH code $C_{BCH}(n, k)$ with the following properties: a) Code length: $n = 2^m - 1$; b) Number of parity bits: $n - k \leq mt$; c) Minimum Hamming distance: $d_{min} \geq 2t + 1$; and d) Error-correction capability: t errors. BCH codes are not frequently used in WLANs.

Reed-Solomon codes

Reed Solomon codes are non-binary cyclic error correcting codes that are used for detection and correction of multiple errors [8]. These codes belong to block codes and can be specified as $RS(n, k)$. In a block code we have k individual information bits, r individual parity bits and a total of $n = k + r$ bits. Reed Solomon codes are organized in groups of bits. These groups of bits are referred to as symbols. So we can say this code has n symbols. A symbol comprises m bits, where $n=2^m-1$. These n symbols form a codeword. Out of n symbols, k are information symbols. The rest $(n - k)$ are parity symbols (Figure 2.5).

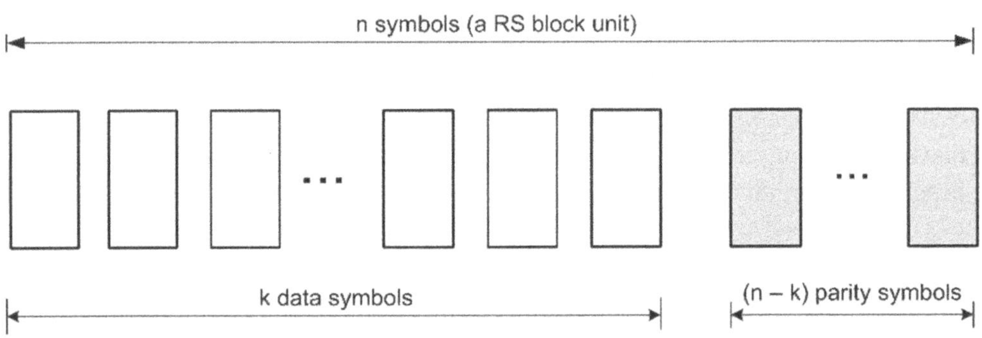

Fig. 2.5. A Reed Solomon block code

The error correcting capability of the code is defined as t, where $t = [(n\text{-}k)/2]$ is the greater integer function. This implies that Reed Solomon code can correct only half of the number of parity symbols because for every error one parity symbol is used to locate the error and the other is used to correct it. The crucial advantage behind Reed Solomon code is that it can correct burst errors. A (n, k, t) RS code with q-ary symbols has the following parameters: 1) block length: $n = q - 1$; 2) number of parity-check bits: $n - k = 2t$; and 3) minimum distance: $d_{min} = 2t + 1$. An (n, k, t) RS code is capable of correcting any combination of t or fewer symbol errors. In practical applications, RS codes with code symbols from $q = 2^m$ are chosen. BCH and RS block coding schemes have a well-defined algebraic structure, which has facilitated the development of efficient coding and decoding schemes. In addition, RS codes have optimal "distance properties", i.e., provide an optimal error correction given a fixed number of parity bits, and excellent "burst error suppression" capabilities.

2.3.5 *Convolutional codes*

The difference between block codes and convolutional codes is the encoding principle [9]. Convolutional codes involve the transmission of parity bits that are computed from message bits. Contrary to block coders, a convolutional encoder does not send the message bits interspersed with the parity bits; it sends just the coded bits. That means that the convolutional codes map information to code bits not block wise, but sequentially convolve the sequence of information bits according to some rule. The code is defined by the circuit consisting of a different number of shift registers allowing building codes of different complexity. Block coding schemes are frequently referred to as memoryless since successive information blocks are coded independently. Convolutional coders are a popular class of coders with memory, i.e., the coding of an information block is a function of the previous blocks. An (n, k, m) convolutional code generates n encoded bits for every k information bit, where m refers to the memory of the encoder. The n encoder outputs at any given time instant, depending not only on the k inputs but also on m previous input blocks (where $m + 1$ is sometimes referred to as the code constraint length). The longer the constraint length, the larger the number of parity bits that are influenced by any given message bit. Because the parity bits are the only bits sent over the channel, a larger constraint length generally implies a greater resilience to bit errors. Usually, the values of n and k are small and k is less than n, but a number of shift registers takes larger values. Figure 2.6 shows a ½-rate binary convolutional encoder with $m = 2$. For each input bit, there are 2 output bits which depend on the previous 2 input bits. The encoder consists of an m-stage shift register together with n modulo-2 adders and a multiplexer for serializing the encoder outputs (not presented in Figure 2.6). If the input sequence is $X = (:::; x[n-2], x[n-1], x[n], :::)$, the two encoder output sequences $p_0[n] = x[n] \oplus x[n-1] \oplus x[n-2]$ and $p_1[n] = x[n] \oplus x[n-1]$ are equal to the convolution of the input sequence X with the two code generator sequences $g_0 = 1,1,1$ and $g_1 = 1,1,0$, i.e., given by the encoding equations $p_0[n] = X \bullet g_0$ and $p_1[n] = X \bullet g_1$. In other words, each parity equation is produced by composing the message bits X, and a generator polynomial g (in the given example, the generator polynomial coefficients are (1,1,1) and (1,1,0)).

Table 2.1. Convolutional encoder: Input, register contents, and outputs

Input	Register contents	p_0p_1
1	00	11
0	10	10
1	01	00
0	10	10
0	01	11

We consider the ½-rate encoder in Figure 2.6 assuming the initial contents of the registers *x[n-1]* and *x[n-2]* is zero and the bit sequence to encode is 101. Before starting the encoder, both registers should be preset to initial values. The output is generated taking into account the register values and the input sequence. Table 2.1 shows the encoding process that produces the output bits in third column.

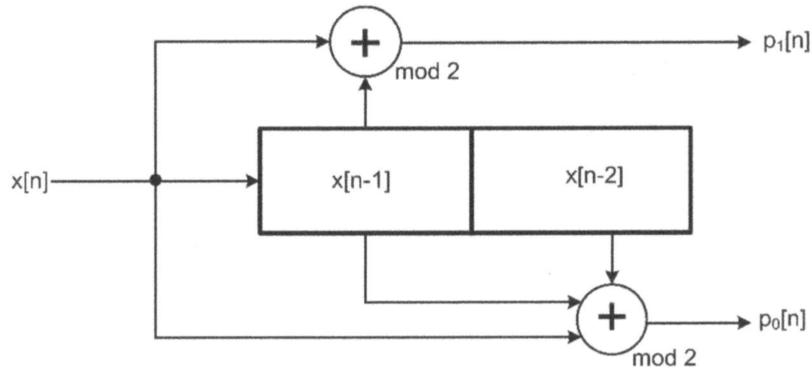

Fig. 2.6. Block diagram of the convolutional coding with a shift register

The rate of a convolutional code is commonly modified via symbol puncturing. Puncturing is a technique used to obtain an $(n_m - p, k_m)$ code from an (n_m, k_m) mother code for any $p < n_m - k_m$, thus increasing the code rate (i.e., removing some of the parity bits after encoding) 10. As a result, p code bits are not transmitted. In other words, puncturing has the same effect as encoding with an error-correction code with a higher rate, or less redundancy. The decoder handles punctured bits as erased, and applies the decoder of the mother code. A brief illustration of the puncturing procedure is presented in Figure 2.7.

Each P consecutive code blocks are combined into a group. Of the nP encoded bits in this group, l encoded bits are deleted (punctured) and not transmitted. Therefore, P information bits are mapped to $nP - l$ code bits. So, the code rate is increased from $R = 1/n$ to

$$R_{punc} = P/(nP - l) \tag{2.1}$$

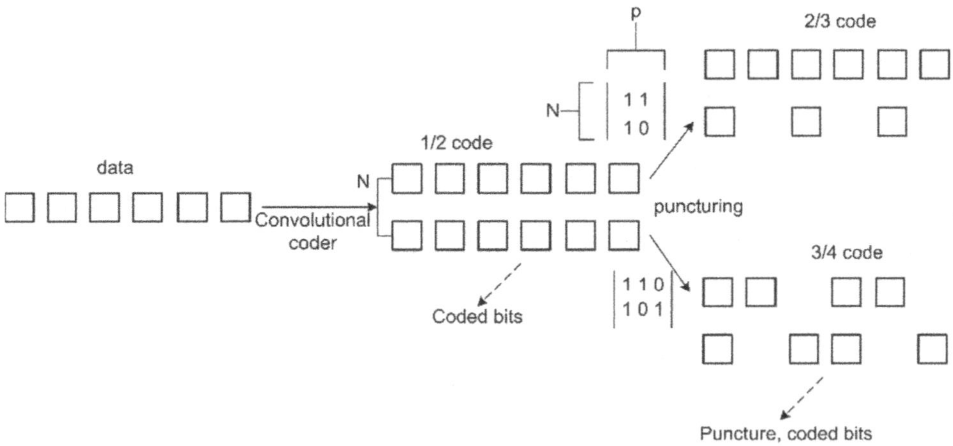

Fig. 2.7. Puncturing procedure

Of course, $l < (n - 1)P$ must be valid to guarantee a code rate smaller than 1 to be able to reconstruct the information bits from the encoded bits. The value P is called a puncturing period and the rate-1/n code is called a mother code. The positions of encoded bits to be punctured are fixed in a puncturing pattern. With increasing P nearly all code rates between $1/n$ and 1 can be achieved. The puncturing process may be described as periodically deleting selected bits from the output of the encoder, thus, creating a periodically time varying trellis code.

1) We begin with a rate $1/n$ parent code and define a puncturing period P, corresponding to P input information bits to the encoder.
2) In one period, the encoder outputs nP coded bits.
3) A puncturing matrix $\boldsymbol{P}$ is associated with the nP encoded bits in the form:

$$P = \begin{bmatrix} p_{11} & \cdots & p_{1p} \\ \vdots & \ddots & \vdots \\ p_{n1} & \cdots & p_{np} \end{bmatrix} \tag{2.2}$$

4) Each column of $\boldsymbol{P}$ corresponds to the n possible output bits from the encoder for each input bit and each element of $\boldsymbol{P}$ is either 0 or 1.
5) When $p_{ij} = 1$, the corresponding output bit from the encoder is transmitted.
6) When $p_{ij} = 0$, the corresponding output bit from the encoder is deleted.
7) The code rate is determined by the period P and the number of bits deleted.
8) If we delete N bits out of nP, the code rate is $P/(nP\text{-}N)$.

2.3.6 *LDPC codes*

Low Density Parity Check (LDPC) code is one of the most promising near-optimal solutions (near-Shannon limit) traditionally used in WLAN systems [11, 12]. An LDPC code consists of several parity-check equations that are applied to the information to be transmitted. Two types of LDPC codes, LDPC block codes and LDPC convolutional codes, have been described in literature. However, most of these studies are focused on LDPC block codes due to many practical considerations such as power and area constraints in real hardware implement-tation. Generally, there are two methods to represent LDPC codes. Like all linear block codes, they can be described via matrices. The second method is a graphical representation based on a Tanner graph. In matrix representation, the LDPC code can be specified by its parity-check matrix H (containing mostly 0‟s and only a small number of 1‟s). Such a structure gives both: a lower decoding complexity and good distance properties. The parity-check matrix can be described as a sparse $M \times N$ matrix, in which each of the M rows represents a parity-check equation over $GF(2^q)$ (*Galois field* of 2^q elements) and the N columns in H correspond to the received symbols. For the binary case with $q = 1$, the symbols are equal to bits, so N also represents the codeword length in terms of bits. The codeword, denoted by $\boldsymbol{c}$, can be computed by means of a generator matrix G that satisfies the equation $G \times H^T = 0$. The codeword is equal to $\boldsymbol{c} = G \times b$, where b is the k-bit information vector. An example of a parity check matrix is given in Figure 2.8.

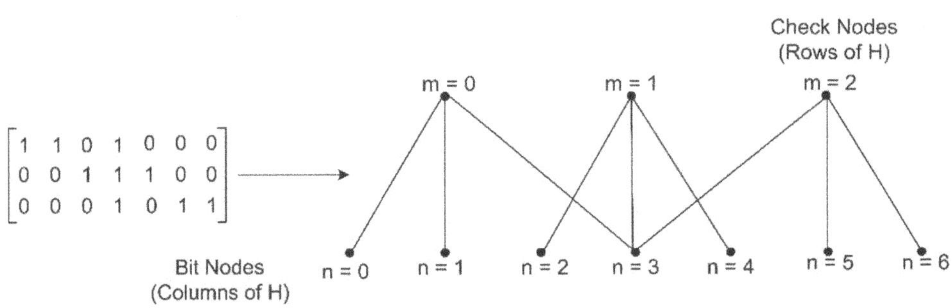

Fig. 2.8. Representation of an LDPC code: parity check matrix H (left) and equivalent Tanner graph (right)

We can now define two numbers describing a parity-check matrix: w_r is the maximum number of 1‟s in each row and w_c is the maximum number of 1‟s in each column. Two conditions ($w_c \ll N$ and $w_r \ll M$) need to be satisfied for a matrix to be called a low-density matrix. A parity check matrix is usually very large, so the example matrix given above is not really of low-density.

LDPC codes can also be represented by a Tanner graph, which is a bipartite graph with two different node types: check nodes and variable nodes. Each check node m_x corresponds to one of the parity-check equations in H. The variable nodes n_x, also called bit nodes, correspond to the columns of H. An edge connects a check node and a variable node if and only if the corresponding row and column share a non-zero element in H. The number of non-zero elements in a row (i.e., the number of variable nodes connected to a check node) is the degree of the check node (d_c), and the number of non-zero elements in a column (i.e., the number of check nodes connected to a variable node) is the degree of the variable node (d_v). In Figure 2.8, the two representations of the same code (parity-check matrix and Tanner graph) can be compared.

LDPC code is decoded iteratively to achieve the optimal decoding performance with moderate complexity. On the other hand, as the number of decoding iterations increases, the decoding throughput decreases and the decoding latency increases. To achieve a high decoding data rate, a large number of computation units is required, which leads to large chip area and high power consumption. Therefore, low power design techniques have become crucial for hardware (and software) implementations of high-speed LDPC decoders.

2.4 Interleaving/Deinterleaving

As we have previously mentioned, error correcting codes (ECCs) can correct errors efficiently as long as there are not too many errors in a single codeword. Essentially, ECC refers to the ability to correct random bit errors within one code-word. While error positions can be random, the number of error bits within one codeword that can be corrected is critical and depends on the selected ECC technique. However, wireless communication channels are subject to burst errors. A burst error is a sequence of unreliably interpreted bits caused by a distortion of the sent signal during a longer period. Namely, errors tend to be bursty (many errors concentrate locally), which typical error correction schemes cannot handle successfully. To combat burst errors in a satisfactory way, interleaving techniques are commonly used [5]. Interleaving, as a standard signal processing technique used in communication systems, addresses this type of errors by reordering the bits at the sender end according to a particular scheme and restoring the original order at the receiver end. As a result, the erroneous bits are spread over the interleaving block or window and therefore the local error density is lowered, allowing more bits to be corrected by the decoder. In the information theory literature, an interleaver is usually understood to be a single input single output finite-state hardware device that takes symbols from a fixed alphabet as the input and produces the identical symbols at the output in a different temporal order. The classical use for interleaving is to disperse sequences of bits in a bitstream so as to minimize the effect of burst errors introduced in transmission. Every interleaver has a corresponding de-interleaver that acts on the output of the original interleaver and puts the symbols back into their original order (with a possible time delay).

There are two classical kinds of interleavers, commonly referred to as block and convolutional. In a block interleaver, the input data is written along the rows of a memory configured as a matrix, and then read out along the columns. A variation of the block interleaver is a pseudorandom block interleaver, in which data is written in the memory sequentially and read from the memory pseudo randomly. In a convolutional interleaver, the data is multiplexed into and out of a fixed number of shift registers. In practice, block interleavers are standardly realized as periodic interleavers (order data in a repeating sequence of bytes) since their structures can be easily accomplished in hardware. At the transmitter side, the input data-stream is first processed by the encoder. The encoder adds redundant bits to the data sequence, enabling correction of possible transmission errors. After that, in order to increase the probability of error correction, the data sequence is reordered by the interleaver. During the next step, the modulator converts the bit sequence in the corresponding carrier wave modulation sequence. At the receiver side, the inverse of every operation is applied to the signal in reversed order. The interleaver implementation comes from the idea that an interleaver can be viewed as a special matrix where symbols are written into the matrix by rows, starting from the position labeled 'I', and read out by columns starting from the 'O' position, as shown in Figure 2.9(a), while the input and output sequences are extended in Figure 2.9(b) for the example considered.

To demonstrate the effectiveness of one-dimensional interleaving technique aimed at handling one-dimensional bursts of errors, we analyzed a simple example. Figure 2.9 shows a codeword that contains four code symbols. In addition, let the code possess error correcting capability, thanks to which a single random error in a codeword can be corrected. Further, we assumed that there are 16 symbols (Figure 2.9(a)) grouped into four codewords (Figure 2.9(b)). Code symbols starting from 1 to 4 are grouped into the first codeword, symbols from 5 to 8 form the second codeword, etc. (the upper part of Figure 2.9(c)). According to the interleaving procedure, a special matrix of order 4 x 4 is formed (Figure 2.9(b)). If a burst of errors involving four consecutive symbols (shaded rectangles in Figure 2.9(c)) is introduced, deinterleaving (Figure 2.9(d)) can effectively spread the error burst among four codewords, resulting in only one code symbol in error for each of the four codewords.

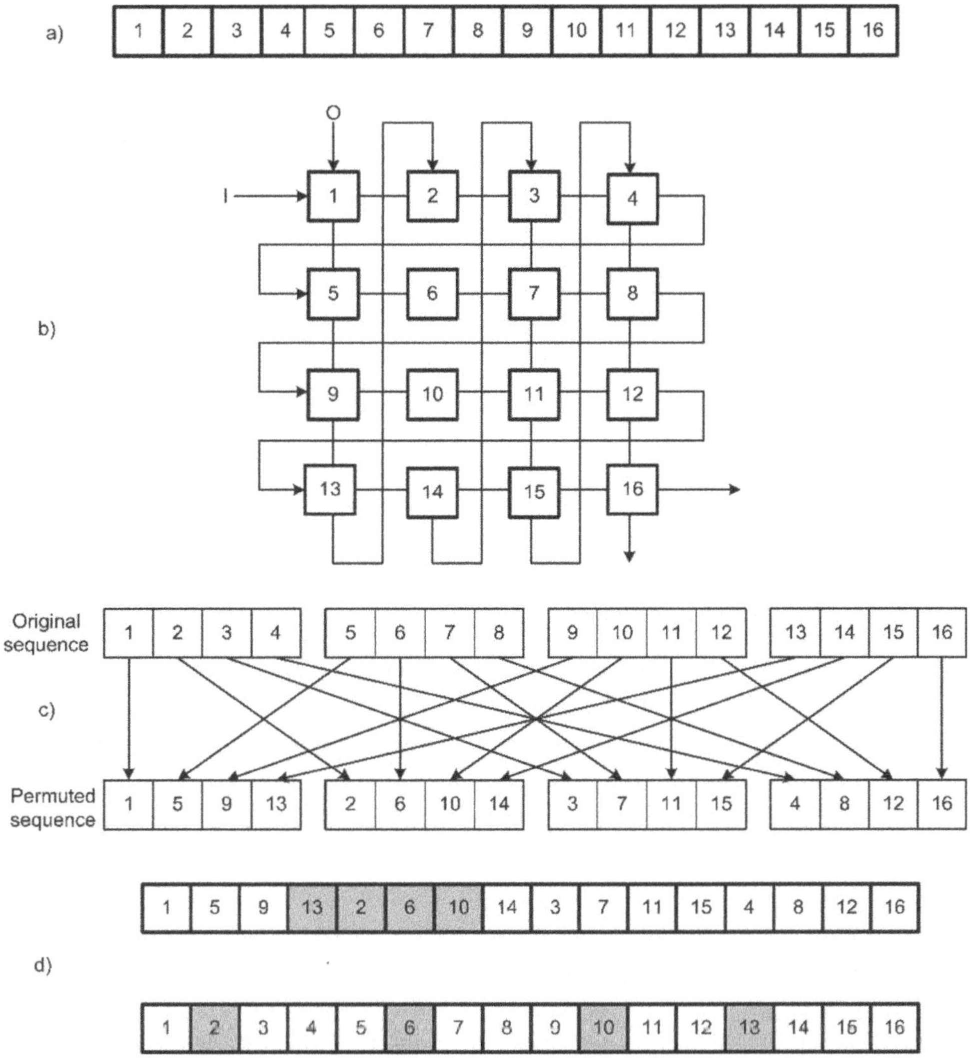

Fig. 2.9. One-dimensional block interleaving/deinterleaving: a) data-stream before interleaving; b) interleaver in the form of a special matrix; c) example of interleaving process (data after interleaving); d) data after deinterleaving

In this way, the combination of interleaving and deinterleaving converts a bursty channel into a random-like channel, and after that random error correction codes can be used. An example of the standard block interleaver typically used in WLAN is shown in Figure 2.10.

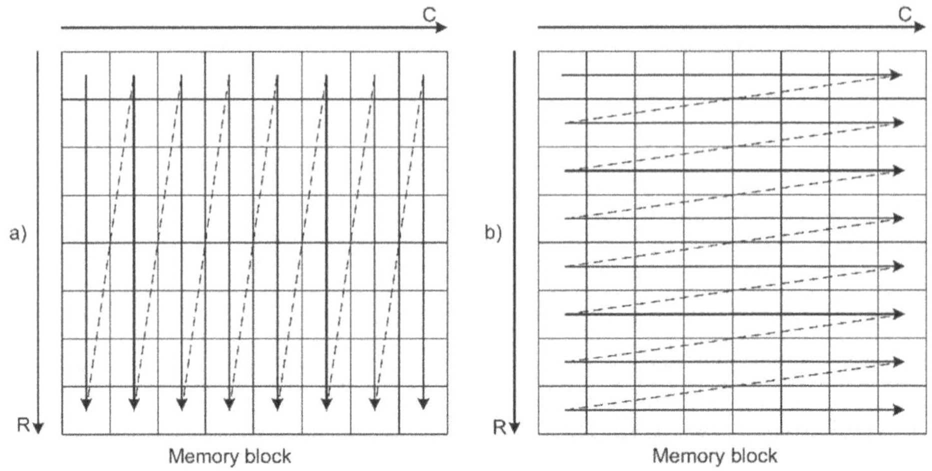

Fig. 2.10. Interleaving in a WLAN transceiver: a) write input data in columns; b) read output data per row

When writing separate bits in a column-wise order and reading the bits in a row-wise order, the distance between two originally adjacent bits is now increased from 0 to $C-1$. For a window (or interleaved block) of C bits and an error tolerance of 1 per window, the maximum burst error length is C. Therefore, to be able to cope with a burst error length of 4 (as shown by the example in Figure 2.9), the number of columns of the standard block interleaving has to be at least 4.

When an OFDM signal is subjected to multipath fading (such as in the case of WLAN), several adjacent carriers can be highly attenuated and the signal can be corrupted by additive noise. In addition, shadow fading deteriorates several consecutive OFDM symbols. Concerning OFDM symbols, two possible solutions can be implemented aiming to cope with a deep fading. In the first one, the separation of closely related bits is realized in the physical dimension of time, thus this separation is called time interleaving. In principle, time interleaving introduces a decoding delay because all closely related bits have to be first received and after that decoded.

In the second one, when multicarrier transmission is considered, there is a way to realize a physical bit separation inside each OFDM symbol in the frequency domain. This is called frequency interleaving. The physical reasons that cause decorrelation in time and frequency domains are independent, so multicarrier transmission with time and frequency interleaving is a very powerful correction technique for radio transmission systems. The frequency interleavers are block interleavers that follow a block structure of the frequency band divided into sub-carriers.

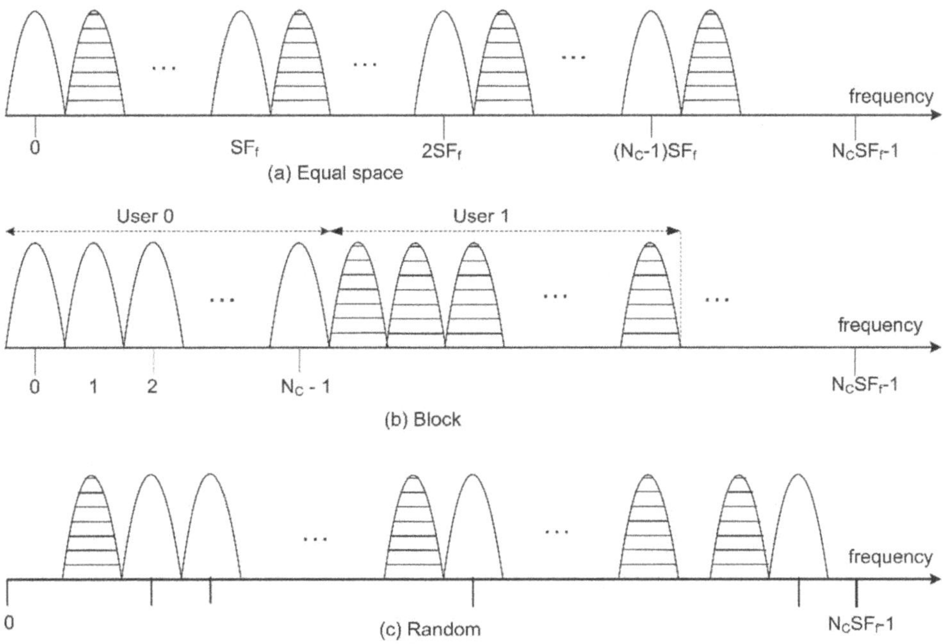

Fig. 2.11. Frequency interleaving patterns

To conclude, we can say that time interleaving is primarily used to exploit the time diversity of channel. After time interleaving, the local time deep fading in some OFDM symbols is averaged over all OFDM symbols. The time interleaving depth should be larger than the maximum burst-error in the time domain. Bearing this in mind, time interleaving is not often applied in WLAN systems because of the slow-fading characteristics of the channel. The following three interleaving patterns (Figure 2.11) are typically used: a) equal space; b) block; and c) random.

Typical implementation of the IEEE 802.11 interleaving requires a standard block interleaver and dedicated hardware. The block size corresponds to the number of bits in a single OFDM symbol. This dedicated component should at least be configurable for 16 rows and a variable number of columns (3, 6, 12 or 18). The number of columns depends on the number of bits per sub-carrier. The data are written in columns and read in rows. The interleaver is defined by a two-step permutation. The first permutation ensures that adjacent coded bits are mapped onto nonadjacent sub-carriers. The second ensures that adjacent coded bits are mapped alternately onto less and more significant bits of the constellation and, thereby, long runs of low reliability bits are avoided.

2.5 Mapping/Demapping

The sub-carrier mapper assigns individual data symbols to individual sub-carriers. It is a building block of most OFDM systems primarily due to the fact that different data symbols may have come from different users, and assigning data symbols to different sub-carriers allows multiple users to share the radio channel. After bit interleaving, the data bits continue serially to a constellation mapper [5]. The constellation mapper accepts a serial bit stream as input and segments the stream into k-bit symbols, which are mapped to coordinates in a signal constellation diagram. The coordinates of each point in a two-dimensional signal constellation diagram represent the baseband in-phase (I) and quadrature (Q) components that modulate the orthogonal sub-carrier signals. The constellation mapper defines the shape and dimension of the signal constellation and, in other words, modulates the signal. It is common practice to encode symbols in such a way that adjacent symbols differ in a single bit (known as the Gray encoding). If the receiver incorrectly detects a given symbol, it is likely that only one of the k bits in the symbol is in error. This encoding scheme in form of the Binary Phase Shift Keying (BPSK), the Quadrature Phase Shift Keying (QPSK), or the Quadrature Amplitude Modulation (QAM) increases the robustness of the transmission scheme in the presence of a noisy channel.

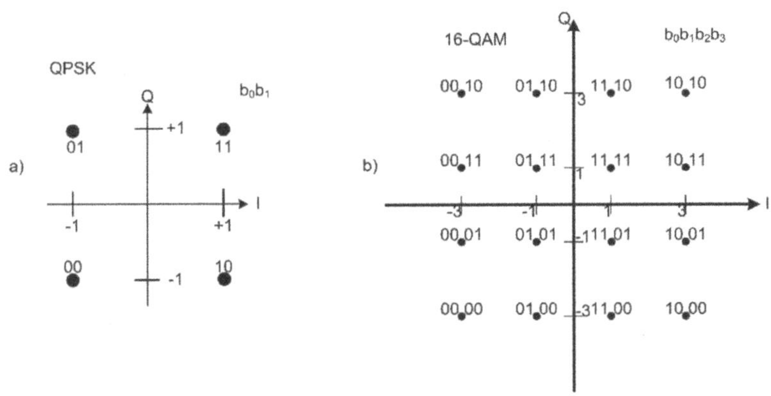

Fig. 2.12. QPSK mapping and corresponding bit pattern (left) and 16-QAM mapping and corresponding bit pattern (right)

For instance, when the data rate is 12 or 18 Mbps, input data bits are formed in 2-bit groups and mapped to complex data as illustrated in the left portion of Figure 2.12. After symbol mapping, the output signal is normalized by the factor $a = \sqrt{2}$. For the data rates of 24, 27, or 36 Mbps, input data bits are formed into 4-bit groups and mapped to complex data as illustrated in the right portion of Figure 2.12. After mapping, the output signal is normalized by the factor $a = \sqrt{10}$.

Constellation demapping is performed in the digital baseband of the receiver and involves translation of the demodulated *I-Q* coordinates back to the k-bit symbols they represent. Demapping can be difficult in case the transmitted symbols are corrupted by a noisy channel. The demodulated *I-Q* coordinates may no longer correspond to the exact location of a signal point in the original constellation diagram. The demapper must detect which symbol was most likely transmitted by finding the smallest distance between the received point and the location of a valid symbol in the signal constellation diagram. Having detected the most likely coordinates, the constellation demapper then performs decoding to output the correct *k*-bit symbol represented by those coordinates.

Once the demapper detects the transmitted *I-Q* coordinates, it can indicate the difference between the actual constellation point and the received coordinates as an error. This information defines the level of confidence in the decision made by the demapper, and can be used by other baseband components as a confidence estimate. It can also be used to compensate for channel impairments such as phase noise, carrier frequency noise, channel dispersion, or suboptimal AD sampling.

2.6 Pilot Insertion

In OFDM systems, dedicated pilot sub-carriers are mixed with data sub-carriers. Figure 2.13 presents some typical pilot patterns, where the gray circles represent pilot sub-carriers and the white circles denote data sub-carriers. Some commonly used pilot patterns in OFDM systems are:

- In a block type, one pilot symbol full of pilot sub-carriers is periodically inserted (in Figure 2.13(a), the period is 7).
- In a comb type, pilot sub-carriers are periodically inserted at some locations for all symbols (every fifth sub-carrier is a pilot, see Figure 2.13(b)).
- Pilot sub-carriers are placed with a fixed spacing in both time and frequency domains (the period is 4 and the frequency period is 4, see Figure 2.13(c)).
- Pilot sub-carriers are placed with a regular shift in the time and frequency domains (the period is 4 and the frequency period is 3, see Figure 2.13(d)).

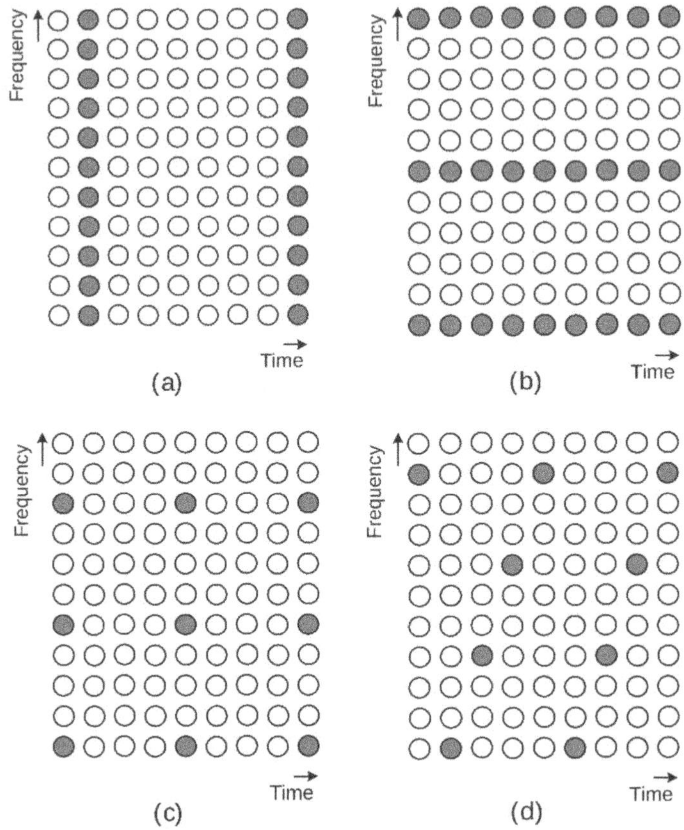

Fig. 2.13. Common pilot patterns: (a) block type, (b) comb type, (c) rectangular grid, (d) parallelogram-shaped grid

In the block-type pilot arrangement, the receiver can obtain channel estimation at all sub-carriers from the pilot symbol and then apply estimated channel gains to equalize the data symbols that follow. The interval between two consecutive pilot symbols must be significantly shorter than the channel coherence time. The block-type pilot pattern is suitable for systems operating under slow-fading channels.

In the comb-type scheme, some sub-carriers are continuously used for pilot signals. Channel estimation can then be performed continuously for every symbol. The spacing of pilot sub-carriers must be less than the coherence bandwidth of the channel. The comb-type pilot arrangement is an effective and accurate channel estimation technique for fast-fading channels.

The remaining pilot patterns use pilot sub-carriers generated by sub-sampling of two-dimensional channel responses. These two-dimensional pilot arrangements reduce the pilot density for a given distance among neighboring pilot signals and thus improve the spectral efficiency.

2.7 IFFT/FFT

OFDM transceivers support high data rates and avoid inter-symbol interference (ISI) by partitioning the broadband into many orthogonal sub-carriers that transmit data in parallel. Thus, the data rate for each sub-carrier is reduced by a factor of N in a system with N sub-carriers. By this method, the channel is divided into many narrowband flat fading sub-channels. This makes an OFDM system more resistant to multi-path frequency selective fading than a single carrier communication system. The sub-carriers are fully independent and orthogonal to each other.

An OFDM signal is generated by implementing the Inverse Fast Fourier Transform (IFFT) [13] at the transmitter side to convert a symbol from the frequency domain to the time domain and the Fast Fourier Transform (FFT) [14] at the receiver side to convert back a time domain symbol to a frequency domain symbol (Figure 2.14).

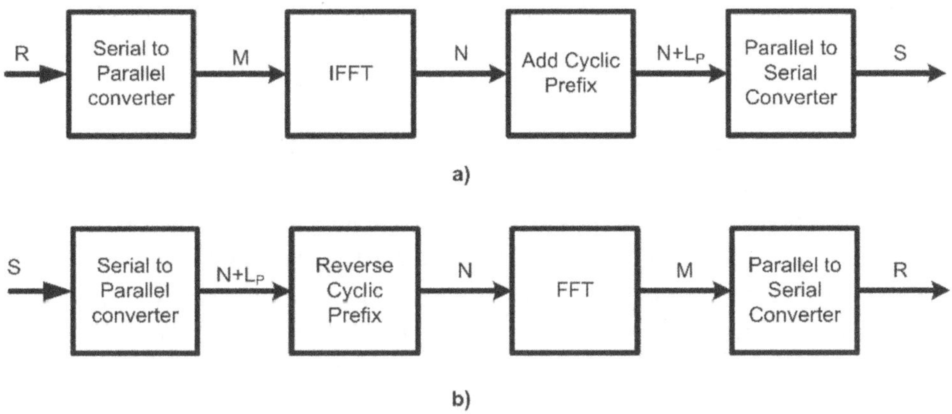

Fig. 2.14. a) OFDM transmitter; b) OFDM receiver

In an OFDM transmitter, the input symbols are fed to the transmitter in series at a rate of R symbols per second. These symbols pass through a serial to parallel converter which outputs the data on N lines in parallel. The symbols are sent to an IFFT block that performs the N-point IFFT operation. This operation transforms the spectrum (the amplitude and phase of the I and Q components) into a time-domain signal. An IFFT converts a number of complex data points (which length is a power of 2) into the same number of points in the time domain and, consequently, generates at the output N time-domain samples. In order to preserve the sub-carrier orthogonality, a cyclic prefix (guard interval) with a length of L_p samples is introduced. If the output of an 8-point IFFT is [1 2 3 4 5 6 7 8], the cyclic prefix will be [7 8]. The cyclically extended symbol would be [7 8 1 2 3 4 5 6 7 8]. Therefore, the length of the transmitted OFDM symbol is $N+L_p$. Prepending the cyclic prefix aids in removing the channel impairments and simplifying the channel equalization at the receiver side. ISI can occur when the multi-path channel causes a delayed version of the original OFDM symbol at the receiver side. If the value of L_p is greater than or equal to the delay spread of the transmission channel, ISI only affects the cyclic prefix.

In an OFDM receiver, the received signal is in the time domain and is distorted due to channel impairments. The received signal goes through a serial to parallel converter and cyclic prefix removal. After cyclic prefix removal, the signal is transformed through an N-point FFT into a frequency-domain symbol. The FFT generates at the output M frequency-domain samples.

As already mentioned, the Fourier transform converts a time-domain function into a frequency-domain function while the inverse Fourier transform converts a frequency-domain function into a time-domain function. In practice, these functions are sampled at equally spaced discrete points. The Discrete Fourier Transform (DFT) and Inverse Discrete Fourier Transform (IDFT) are respectively:

$$X(m) = \sum_{n=0}^{N-1} x(n)\, e^{\frac{2\pi jmn}{N}}, \quad m = 0,1,2, \ldots, N-1 \tag{2.3}$$

$$x(n) = \frac{1}{N}\sum_{m=0}^{N-1} X(m) e^{\frac{-2\pi jmn}{N}}, \quad n = 0,1,2, \ldots, N-1 \tag{2.4}$$

where n represents the sampled points in the time domain, m represents the sampled points in the frequency domain, and N is the number of sampled points. Although DFT is a straightforward mathematical procedure for determining the frequency content of a time-domain sequence, it might be computationally complex to implement. For instance, for computation of an N-point DFT, one requires $(N - 1)^2$ complex multiplications and $N(N - 1)$ complex additions [5]. On the other hand, FFT is an algorithm that decomposes the original N-point sequence into shorter sequence DFTs, thereby reducing the number of computations [5]. For a radix-2 decomposition (where the length of a decomposed sequence is two), the fundamental two-point transformation is known as a butterfly. In this case, $(N/2)\log_2 N$ butterfly operations are needed to complete the FFT algorithm. Other more efficient algorithms also exist: an example is the radix-4, whose input sequence length is a 4^N value.

Generally, FFT analyzes an input signal sequence by using a decimation-in-frequency (DIF) or decimation-in-time (DIT) decomposition to efficiently construct a computational signal-flow graph (SFG). An equivalent decimation-in-frequency FFT structure exists for each decimation-in-time FFT structure. It is important to note that the number of necessary multiplications to implement the decimation-in-frequency FFT is identical to the number of multiplications necessary for the decimation-in-time FFT. There is no difference between the two decimations, so the decision to take one or the other depends only on the designer.

2.8 Cyclic Prefix

An OFDM system is capable of decomposing the high rate data stream (bandwidth = W) into N lower rate data streams and then to transmit them simultaneously over a large number of sub-carriers. The choice of individual sub-carriers is such that they are orthogonal to each other. This allows overlapping of sub-carriers with higher order spectral components of adjacent sub-carriers. The orthogonality ensures optimal separation of sub-carriers and provides minimal spacing between sub-carriers at the same time. In other words, the usage of spectral overlapping provides superior bandwidth utilization.

In a frequency-selective multipath fading channel for a base OFDM system, synchronization mismatches are typically very critical, because every delayed echo component of the signal violates the orthogonality. This leads to severe inter-symbol interference (ISI), and additionally rises the problem of inter-carrier interference (ICI). This property, which is a consequence of the loss of orthogonality due to the overlap of spectral components, would seemingly disqualify narrow-sense ODFM as a useful technique in a multipath channel.

There is, however, a simple technique that modifies the transmitted signal in such a way that the orthogonality is preserved in a certain manner in the presence of multipath signal components. An efficient solution to this problem is to introduce a guard interval (also called a cyclic prefix). ISI is eliminated by introducing a guard interval for each OFDM symbol. The guard interval is chosen to be larger than the expected delay spread so multipath components from one symbol do not interfere with adjacent symbols. This period could be free of signal, but the problem of ICI would still exist. To eliminate ICI, the OFDM symbol is cyclically extended in the guard interval as shown in Figure 2.15. The cyclic prefix (CP) is a copy of the last part of the OFDM frame that is preceded to the beginning of the transmitted frame (Figure 2.15). The length of the cyclic prefix should be longer or equal to the length of the (shortened) channel impulse response. If the cyclic prefix is long enough, the transmitted signal is periodic and orthogonality of the frames is preserved. This ensures that the delayed replicas of OFDM symbols due to the multipath always have an integer number of cycles within the FFT interval, as long as the delay is smaller than the guard period. As a result, multipath signals with delays smaller than the guard period do not cause ICI. In essence, the cyclic extension of the OFDM symbol in the guard interval makes the

OFDM symbol appear to be periodic at the receiver end even though there might be a delay because of the multipath environment.

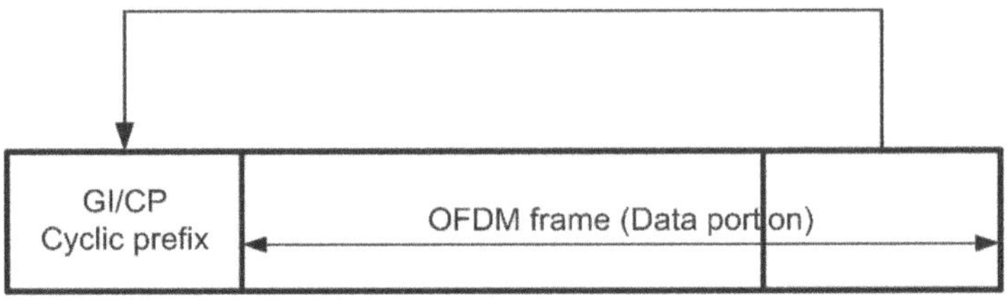

Fig. 2.15. Cyclic prefix (CP) in the guard interval (GI)
Notice: CP is an exact replica of the end segment of the OFDM symbol

To conclude, the usage of CP, with the aim to avoid ISI and ICI, is achieved at the cost of a reduced data rate and increased power consumption.

2.9 Pulse Shaping

The main advantage of OFDM is its ability to decompose a frequency-selective fading channel into several nearly flat fading channels. By doing so, an improved spectral efficiency is achieved along with a greater robustness against the effects of multipath fading and inter-symbol interference, i.e., the OFDM signal is robust against time-dispersion, caused by multipath propagation. ISI represents spreading of one symbol into another as a result of transmission through a band-limited channel. ISI can be eliminated by adding a guard interval (i.e., a cyclic prefix after OFDM modulation) which is longer than the maximum time-dispersion. In essence, the involvement of such a guard interval (cyclic prefix) costs a loss of spectral efficiency and increases power consumption. In general, ISI is relatively low in conventional OFDM signals.

Compared to time-dispersion, the OFDM signal is relatively more sensitive to frequency-dispersion caused by time-variation of the channel. Time-variations corrupt orthogonality of the sub-carriers, thus inter-carrier interference (ICI) occurs. Namely, the sensitivity to carrier frequency offset causes attenuation and rotation of sub-carriers and ultimately leads to ICI between adjacent sub-carriers. In other words, the orthogonality of sub-carriers gets lost due to the carrier frequency offset. Pulse shaping is a well-established technique used to additionally reduce this undesired phenomenon.

In the OFDM spectrum each carrier consists of a main lobe followed by a number of side lobes with reducing amplitude. As long as orthogonality is maintained there is no interference among the carriers because at the peak of every carrier, there exist a spectral null. At that point, the main component of all other carriers is zero. Hence the individual carrier is easily separated.

When there is a frequency offset, the orthogonality is lost because the spectral null does not coincide with individual carrier peaks. So some power of the side lobes exists at the center of individual carriers and it is called ICI power. The ICI power increases as the frequency offset increases. The pulse shaping should increase the width of the main-lobe and/or reduce the amplitude of side-lobes. If we can reduce the side-lobe significantly then the ICI power will also be reduced significantly. Proper pulse shaping techniques enable a digital communication system to transmit data within a limited bandwidth with minimum ISI and ICI. A large stop-band attenuation to diminish ICI and minimization of ISI to attain low bit error rate (BER) are two important properties of the perfect pulse shaping filter. Hence a number of pulse shaping functions are proposed aiming to reduce the side lobe as much as possible. Most conventional pulse shaping filters such as the Raised Cosine (RC) and Square Root Raised Cosine (SRRC) satisfy this condition, although other filters have also been designed to meet more objectives. Pulse shaping filters have been widely used in state of the art communication systems. Various standards such as 802.11a WLAN systems, Wideband Code Division Multiple Access (WCDMA), and 3GPP Long Term Evolution (LTE) propose the square-root raised cosine (SRRC) pulses as transmit and receive filters.

Prior to up conversion, in-phase and quadrature-phase signals are spectrally shaped to conform to the channel bandwidth requirement using a square-root raised cosine filter. The pulse shaping filter at the transmitter performs up-sampling of the signal whereas, the matched filter at the receiver performs down-sampling. The impulse response of the ideal SRRC filter is given by:

$$H(T_x) = H(R_x) = \frac{\sin\left[\frac{\pi t(1-\alpha)}{T}\right]+4\alpha t\left[\frac{\pi t(1+\alpha)}{T}\right]T}{\pi t\left[1-(\frac{4\alpha t}{T})^2\right]/T} \tag{2.5}$$

where *H*(*Tx*) and *H*(*Rx*) are the impulse responses of the transmitter and the receiver respectively, *T* is the symbol interval, and α is the roll-off factor. Typically, WLAN systems employ SRRC pulses with different roll-off factors such as 0.15, 0.25, 0.5, and 0.8. At the end, related to the discussion which deals with justification of pulse shaping implementation we can conclude the following. With rectangular pulse shaping, OFDM has a $\text{sinc}^2(f)$ shaped spectrum with overlapping sub-channels (see Figure 1.7). The consequence of this is that a narrowband interferer will affect not only one sub-carrier but several sub-carriers. The robustness of OFDM to band-limited interference can be increased if the bandwidth of individual sub-channels is strictly limited so that either adjacent sub-channels do not overlap, or each sub-channel spectrum only overlaps with two adjacent sub-channels. As long as the bandwidth of one sub-channel is smaller than $1/T$ the narrowband interferer will only affect one sub-channel. Under such a condition, there is no overlapping between the spectra of adjacent sub-channels, which guarantees the orthogonality between sub-channels. A Nyquist pulse shaping for ISI-free transmission applies on each sub-carrier. This requires oversampling of the received signal and DFT operations at a higher rate than N/T. In other words, the increased robustness to narrowband interference is achieved at the expense of increased complexity.

2.10 Synchronization

The IEEE 802.11a WLAN standard provides data rates up to 54 Mbps in the 5 GHz band using the OFDM modulation and frame structure shown in Figure 2.16.

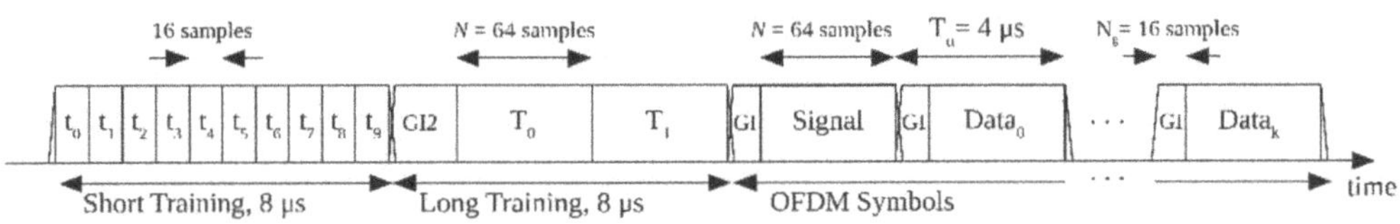

Fig. 2.16. IEEE 802.11a frame structure

The frame structure consists of the following two items: Two training sequences denoted as Short Training (ST) and Long Training (LT), each with a time duration of 8 μs. The ST contains 10 identical parts, denoted as t0 to t9. Each part consists of 16 complex-valued samples. The first seven parts (t0 to t6) are intended for signal detection, automatic gain control and diversity selection (if several antennas are used). The next three parts (t7 to t9) are used for coarse frequency offset estimation and timing synchronization. The LT is formed of two identical parts, marked as T0 and T1 and a guard interval GI2. Each of T0 and T1 consists of 64 complex-values. The last 32 samples of T1 are used for the guard interval GI2 (cyclic prefix). When the frame is susceptible to multipath fading GI2 protects LT from ISI (T0 and T1 are identical), for a delay spread less than the duration of GI2. The LT is used for channel estimation and fine frequency offset estimation. Following the LT is a train of OFDM symbols with a duration of Tu = 4 μs each composed of 80 complex-valued samples. The N=64 information carrying samples (generated by an IFFT at the transmitter) are prefix extended with Ng=16 samples (guard interval denoted as GI) which are identical to the last 16 information carrying samples.

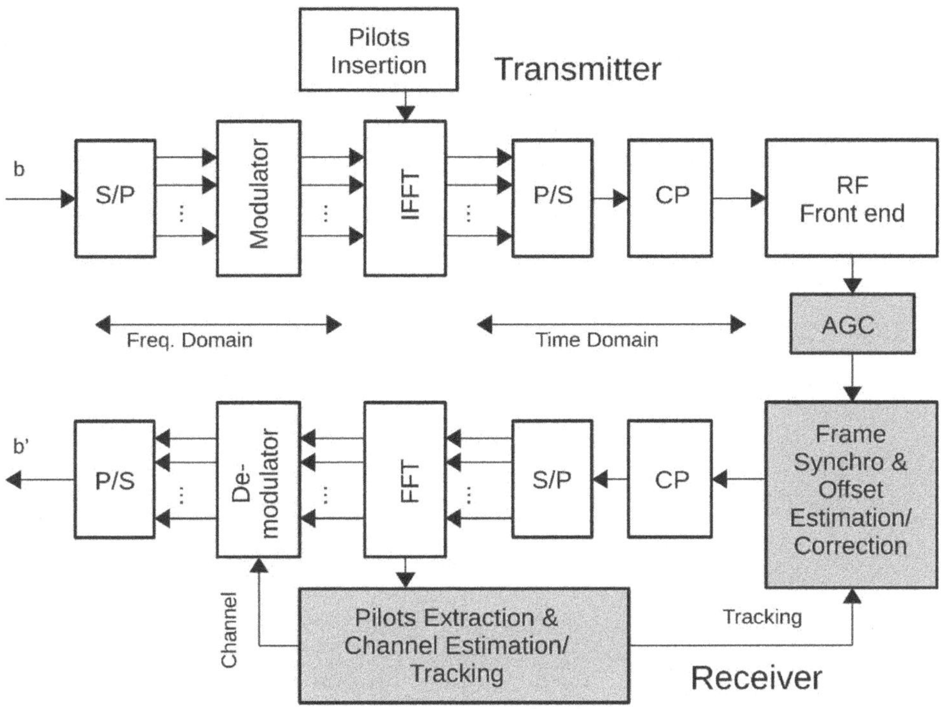

Fig. 2.17. Synchronization at block level in an OFDM system

The first OFDM symbol of fixed format (BPSK) is a special symbol (called Signal) carrying information on the frame length as well as the modulation type (BPSK, QPSK, 16-QAM or 64-QAM) and data rate of the encoded data. The remaining OFDM Symbols ($Data_0$ - $Data_k$) are intended for data. The OFDM symbol (in frequency domain) includes four pilot symbols within 64-point IFFT which generates 64 information carrying samples. Pilot symbols are generated pseudo-randomly (with amplitudes ± 1).

At the transmitter side (Figure 2.17), the already encoded (scrambled, interleaved, and encoded) binary stream *b* is converted from serial to parallel form and modulated (mapping and modulation) before IFFT pilot symbols are inserted. At the IFFT output, time domain signals are obtained which are converted back (by summing) to serial (to a single signal in the time domain). After that the building block CP adds appropriate cyclic prefixes (GI2 and GI given in the definition of the frame structure). The output of CP drives the RF front-end transmitter. The generated RF signal passes through a wireless communication channel and reaches the receiver (Figure 2.17). After RF front-end processing, the baseband signal is normalized by an automatic gain control (AGC) block providing a constant signal amplitude. After that, the signal has to be synchronized in time and frequency (by the Frame Synchro and Channel Estimation/Correction block). The remaining processes have to be carried out inversely (by the CP, S/P, FFT, Demodulator, and P/S blocks). The channel can be tracked using scattered pilots at certain positions and, with this additional information, demodulation can be realized more efficiently reducing offset estimation errors (by the Pilot Estimation and Channel Estimation/Tracking blocks).

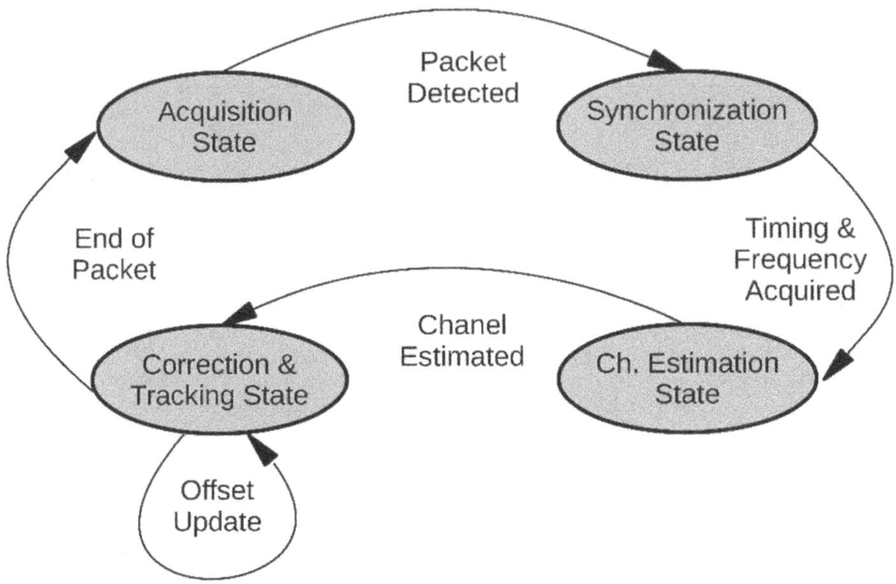

Fig. 2.18. OFDM receiver states

During the phase of packet acceptance and decoding, the synchronization and estimation subsystem goes from one state to another. The complete process can be described by the flowchart shown in Figure 2.18.

Initially, the subsystem is in the *Acquisition State* (AS). In this state, the AGC measures the average power, i.e., the noise power before OFDM packet arrival, and normalizes the signal amplitude after packet arrival (including noise and signal power). If a packet is successfully detected, the subsystem moves to the *Synchronization State* (SS). This transition is marked as the Packet Detected. During SS, the time and frequency are estimated and synchronized. Having this completed (Timing & Frequency Acquired), the subsystem enters into the *Channel Estimation State* (CES). During this activity, state channel coefficients for each sub-carrier of the OFDM symbol are estimated and stored. They are needed for coherent detection of BPSK, QPSK, 16-QAM, and 64-QAM signals. The Channel Estimated transition leads the subsystem into a final state marked as the *Correction & Tracking State* (CTS). In this state, the following tasks are completed: i) OFDM symbols are corrected according to the estimated channel coefficients and, then, demodulated; ii) offsets (for each sub-channel in the OFDM symbol) are tracked and the values of estimated channel coefficients are updated aiming to improve the results. After receiving the last OFDM symbol, the subsystem moves back to the first state (AS) and is ready for receiving a new packet. Typical synchronization errors in OFDM WLANs are the following:

1. *Symbol timing offset* (STO) T_d deals with the error in the estimated symbol boundary at receiver in reference to the actual boundary in the received signal.
2. *Carrier frequency offset* (CFO) is the frequency deviation Δf causing the received complex baseband signal to rotate according to Δf.
3. *Sampling clock offset* (SCO) δ appears as a result of sampling the received continuous-time input signal at an interval of $(1+\delta)T_s$ instead of the ideal time interval T_s.
4. *Carrier phase error* $\Phi(t)$ introduces an additional rotation in the received complex baseband signal.

In the text that follows, we start with definition of the OFDM baseband signal, and after that, give brief explanations for each of the above mentioned errors.

An OFDM baseband symbol is generated from N complex data using the IFFT on N sub-carriers. The sub-carrier spacing is $1/T$, where T is the duration of N time-domain samples. The cyclic prefix of L_p samples is inserted at the beginning of each symbol. As a result, each OFDM symbol has $N+L_p$ samples. The n-th time-domain sample of the i-th transmitted symbol can be written as

$$x(i(N+L_p)T_s + L_pT_s + nT_s) = x_{i,n} = \frac{1}{N}\sum_{k=-N/2+1}^{N/2} X_{i,k}e^{j2\pi nk/N} \quad n = -L_p, \ldots, N-1 \tag{2.6}$$

where $x(t)$ corresponds to the transmitted signal, and $X_{i,k}$ is the k-th complex-value frequency-domain signal of the i-th symbol.

When the received signal is deteriorated by the time-varying multipath fading and the impulse at time t is:

$$h(t,\tau) = \sum_r h_r(t)\delta(\tau - \tau_r(t)) \tag{2.7}$$

where τ is the time delay; $h_r(t)$ is the gain of the r-th path; and $\tau_r(t)$ is the delay of the r-th path. At the receiver end, the signal is defined as:

$$z(\tau) = \sum_r h_r(t)x(\tau - \tau_r(t)) + \nu(\tau) \tag{2.8}$$

where $\upsilon(\tau)$ is the channel noise. If we consider the channel as (close to) stationary ($h_r(t) = h_r$, $\tau_r(t) = \tau_r$), the frequency-domain channel distortion H_k is:

$$H_k = \sum_r h_r e^{-j2\pi\frac{\tau_r k}{T_s N}} \tag{2.9}$$

Symbol timing offset

When the signal arrives earlier for a time interval of T_d, then an additional signal phase shift is introduced. If T_d is less than the maximum excess delay, the received signal will be free of inter-symbol interference (ISI). Under this condition, the received signal can be described by:

$$Z_k = X_k H_k e^{-j2\pi T_d k/(NT_s)} + V_k \tag{2.10}$$

where Z_k, X_k, and V_k are the received frequency-domain signal, the transmitted frequency-domain signal and the noise signal of the *k*-th sub-carrier, respectively. During this we consider only one symbol and therefore we omit the index *i*.

STO estimation

Based on the structure of an IEEE802.11a data packet, the time synchronization algorithm consists of the following steps:
i) *Coarse Time Synchronization* (CTS) identifies the preamble in order to detect a packet arrival, i.e., to provide a reference point for a fine time synchronization.
ii) *Fine Time Synchronization* (FTS) estimates the remaining time offset.
A preamble can be utilized in the time synchronization process by conducting one of the following approaches:
a) Implementing an auto-correlation function that is based on a repetition structure of the preamble (ST symbols, LT symbols, and GIs) with the aim to acquire timing synchronization.
b) Implementing a cross-correlation function that uses a good auto-correlation property of the predefined training sequences to obtain a precise timing.

In addition, other synchronization algorithms based on a maximum-likelihood approach are described in [15-17]. However, the computational complexity of these algorithms is very high and, therefore, they are rarely used.
CTS based on cross-correlation approach uses the ST symbol as a local reference signal at the receiver, and cross-correlates it with the incoming signal. The estimated cross-correlation is defined as:

$$M_{cs}(n) = \frac{\sum_{m=0}^{Nw-1} r(n+m)S^*(m)}{\sum_{m=0}^{Nw-1} |S(m)|^2} \tag{2.11}$$

where *Nw* is the length of the correlation window, and *S*(m)* stands for the conjugate time domain samples of the short preamble signal defined (by the standard IEEE 802.11a) in the frequency domain as:

$$S_{-26,26} = 13/6 * \{0,0,1+j,0,0,0,-1-j,0,0,0,\\ 1+j,0,0,0,-1-j,0,0,0,1+j,0,0,0,0,0,0,0,-1-j,\\ 0,0,0,-1-j,0,0,0,1+j,0,0,0,1+j,0,0,0,1+j,0,0\} \tag{2.12}$$

A more simplified method for implementation of CTS is based on auto-correlation and uses repeated ST symbols. The estimated auto-correlation is defined as:

$$M_{as}(n) = \frac{\sum_{m=0}^{N_w-1} r(n+m)r^*(n+m+N_w)}{\sum_{m=0}^{N_w-1} |r(n+m)|^2} \tag{2.13}$$

For FTS, a cross-correlation approach is used. At the receiver, two LT symbols are used and cross-correlated with the incoming signal. The cross-correlation is estimated by:

$$M_{cl}(n) = \frac{\sum_{m=0}^{N_w-1} r(n+m)L^*(m)}{\sum_{m=0}^{N_w-1} |L(m)|^2} \tag{2.14}$$

where $L^*(m)$ stands for the conjugate time domain samples of the long preamble signal defined (by the standard IEEE 802.11a) in the frequency domain as:

$$L_{-26,26} = \{1,1,-1,-1,1,1,-1,1,-1,1,1,1,1,1,1,-1,\\ -1,1,1,-1,1,-1,1,1,1,1,0,1,-1,-1,1,1,-1,1,-1,\\ 1,-1,-1,-1,-1,-1,1,1,-1,-1,1,-1,1,-1,1,1,1,1\} \tag{2.15}$$

Carrier frequency offset (CFO)

In the presence of a carrier frequency offset Δf, the received input signal has the following form:

$$z_{i,n} = z(t)e^{j2\pi\Delta f t}|_{t=i(N+N_g)T_s+N_gT_s+nT_s} \tag{2.16}$$

In regard to sub-carrier spacing $f_s = 1/(NT_s)$, the carrier frequency offset can be decomposed into the integral component ε_i and fractional component ε_f, and can be expressed as $\Delta f = (\varepsilon_I + \varepsilon_f)f_s$ and $-0.5 \leq \varepsilon_f < 0.5$. The received signal now has the following form:

$$Z_{i,k} = X_{i,k-\varepsilon_I} H_{k-\varepsilon_I} \frac{\sin(\pi\varepsilon_f)}{N\sin(\frac{\pi\varepsilon_f}{N})} e^{j2\pi\frac{i(N+N_g)+N_g}{N}(\varepsilon_I+\varepsilon_f)} e^{j2\pi\frac{N-1}{N}\varepsilon_f}$$

$$+ \sum_{l=-N/2+1, l\neq\varepsilon_I}^{N/2} X_{i,l} H_l \frac{\sin\left(\pi(\varepsilon_I+\varepsilon_f+l-k)\right)}{N\sin\left(\frac{\pi(\varepsilon_I+\varepsilon_f+l-k)}{N}\right)} e^{j2\pi\frac{i(N+N_g)+N_g}{N}(\varepsilon_I+\varepsilon_f)} e^{j2\pi\frac{N-1}{N}(\varepsilon_I+\varepsilon_f+l-k)}$$

$$+V_{i,k} \tag{2.17}$$

The second term in Equation (2.17) points to inter-channel interference (ICI) and represents the influence from other sub-carriers on the desired sub-carrier signal. $V_{i,k}$ is the channel noise component in the k-th sub-carrier of the i-th symbol. The term ε_f causes a magnitude attenuation, phase shift and ICI, while ε_I (if not compensated) causes a channel index shift as well as a phase shift in the received frequency domain signals. The phase shift is identical in every sub-carrier and is also proportional to the symbol index i. According to Equation (2.17), we can conclude that ICI exists as a consequence of the fractional part of CFO, i.e. ε_f.
With the aim of reducing ICI and improving SNR, the residual CFO has to be as small as possible. For instance, for a 64-QAM constellation, the residual CFO should be below $0.01 f_s$ in order to provide DSNR < 0.3 dB for relevant SNR levels. In the case of a QPSK, the residual CFO can be up to $0.03 f_s$.

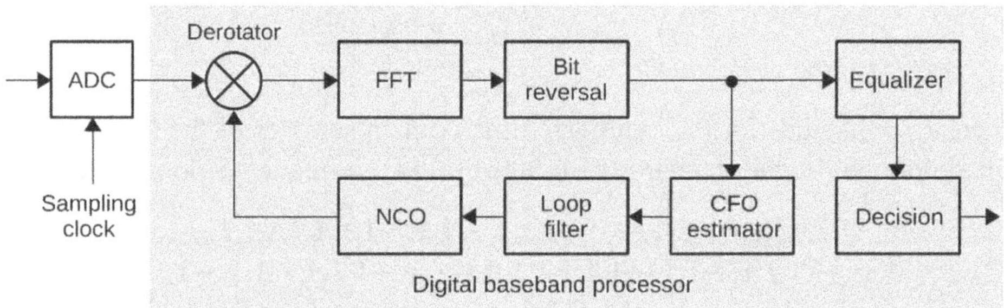

Fig. 2.19. OFDM baseband receiver architecture with a CFO synchronization phase-locked loop that uses the time domain compensation

In a practical receiver realization like shown in Figure 2.19, a time domain de-rotator initially reduces the residual CFO. The de-rotator is implemented as a complex multiplier (rotates the complex valued input signal by a phase). A numerically controlled oscillator (NCO) is used for phase rotation. Since the CFO is not a constant and is influenced by noises and interferences, a phase-locked loop (PLL) is adopted in the receiver for tracking and compensating the estimated CFO. Thanks to the appropriate PLL feedback loop, the residual CFO is maintained within an acceptable level.

In order to avoid the impact of long delay in the CFO estimation and compensation loop, the receiver usually compensates CFO in the frequency domain. In such a solution, a frequency domain interpolator is used instead of a phase rotator. The interpolator width should be appropriately chosen. An interpolator with more taps can eliminate ICI more effectively for a price of higher complexity. Let us note that a trade-off exists between tap count and receiver hardware complexity. In general, an interpolator with an adequate number of taps requires higher complexity than a time domain de-rotator.

CFO estimation

As recommended by the IEEE 802.11a standard, the frequency synchronization is performed in two steps: a *Coarse Frequency Synchronization* (CFS) step is followed by a *Fine Frequency Synchronization* (FFS) step. During the CFS step, the auto-correlation based on repetition in ST symbols is calculated according to the following expression:

$$Z(N_p) = \sum_{m=0}^{N_w-1} r^*(m)r(m+N_p) \tag{2.18}$$

where N_p is the repetition period (or multiple of this period) in ST symbols, and N_w is the window length corresponding to the multiple repetitions in ST symbols (i.e. 2 × 16 samples). The coarse estimation of the CFO is given by

$$\varepsilon_c = \frac{N}{2\pi N_p} arg(Z(N_p)) \tag{2.19}$$

where N denotes the number of FFT points.

After applying the coarse offset compensation (estimated by CFS step) some residual CFO will remain. Next, in the FFS step the auto-correlation based on repetition in LT symbols is calculated, and is given by the following equation:

$$Z(M_p) = \sum_{m=0}^{M_w-1} r^*(m)r(m+M_p) \tag{2.20}$$

where M_p is the period of the LT symbol equal to 64 samples; M_w is the window length corresponding to one repetition in the LT symbol (i.e. 1 × 64 samples).

The fine estimation of CFO is thus given by:

$$\varepsilon_f = \frac{N}{2\pi M_p} arg(Z(M_p)) \tag{2.21}$$

Apart from the above algorithms, algorithms based on (*a posteriori probability*) MAP criterion for improving accuracy of the CFO estimation can also be used (for more details see [15, 16, 19]).

Sampling clock offset (SCO)

When the received signal is sampled at an interval of $(1+\delta)T_s$ instead of the ideal time interval T_s, then the n-th received sample of the i-th symbol can be written as:

$$z_{i,n} = z(t)|_{t=i(N+N_g)(1+\delta)T_s+N_g(1+\delta)T_s+n(1+\delta)T_s} \tag{2.22}$$

If there is no ISI inside the FFT window, then the k-th frequency-domain received signal of the i-th symbol is given by:

$$Z_{i,k} = X_{i,k}H_k \frac{\sin(\pi\delta k)}{N\sin(\frac{\pi\delta k}{N})} e^{j2\pi\frac{i(N+N_g)+N_g}{N}\delta k} e^{j2\pi\frac{N-1}{N}\delta k}$$

$$+\sum_{l=-N/2+1,l\neq\varepsilon_l}^{N/2} X_{i,l}H_l \frac{\sin\left(\pi((1+\delta)l-k)\right)}{N\sin\left(\frac{\pi((1+\delta)l-k)}{N}\right)} e^{j2\pi\frac{i(N+N_g)+N_g}{N}\delta l} e^{j2\pi\frac{N-1}{N}((1+\delta)l-k)}$$

$$+V_{i,k} \tag{2.23}$$

From Equation (2.23), it follows that: i) the sampling clock offset δ causes magnitude attenuation; and ii) the phase shift is inserted into the signal (the phase shift has a constant increment proportional to k and δ as the symbol index i increases). In addition, ICI (the second term in Equation (2.23)) is also introduced. The signal attenuation and magnitude of ICI linearly depend on δ and the sub-carrier index *k*.

SCO compensation can be realized using one of the following three approaches: a) Time-Domain Interpolator; b) All-Pass Fractional-Delay Filter; and c) Frequency-Domain Rotator. A brief explanation of the above mentioned approaches follows.

When the number of sub-carriers *k* is large, a time domain SCO compensation should be adopted. In this case, a possible solution is based on an interpolator which is capable of calculating samples with fractional delay. The ideal interpolator uses *sinc* function to implement fractional delay:

$$sinc(x) = \frac{\sin(\pi x)}{\pi x} \tag{2.24}$$

However, a *sinc* function is infinitely long and non-causal and, therefore, is not suitable for practical realization. To cope with this problem in an effective way, polynomial based interpolators with good frequency characteristics are commonly used. Several interpolators with different polynomial orders can be used (linear, quadratic, cubic, etc.). Having in mind the tradeoff between hardware complexity and magnitude distortion, the second-order (quadratic) piecewise-parabolic interpolator offers a promising solution [3].

A finite-order all-pass digital filter with a near-linear phase response that corres-ponds to a fractional-sample delay can be a very good design solution for SCO compensation. The transfer function of this filter is:

$$H(z) = \frac{z^{-N}A(z^{-1})}{A(z)} = \frac{z^{-N}(a_0+a_1z+...+a_Nz^N)}{a_0+a_1z^{-1}+...+a_Nz^{-N}} \tag{2.25}$$

Where N is the order of the filter, the numerator polynomial is a mirrored version of the denominator *A(z)*, while the coefficients are real values. The phase response of the filter should be as linear as possible. Additionally, the stability of the filter should be provided [3].

This design solution is more convenient when the OFDM system has a small number of sub-carriers or in cases when SCO can be kept within a very small range. In this case, the ICI term can be ignored and only the phase shift of the received signal needs to be corrected. A frequency-domain phase rotator is inserted before the equalizer in order to reduce the loop latency and to avoid magnitude distortion. At the receiver side, the SCO estimate is first filtered and then integrated with the aim of generating fractional delay. Next, fractional delay is used to correct the phase shift of the sub-carrier signal. The FFT window is adjusted when the fractional delay overflows (> 1) or underflows (< 0). The estimation of SCO is based on pilot sub-carriers. The SCO is estimated using the fact of a linear relationship between the phase rotation caused by the offset and the pilot sub-carrier index.

Carrier phase offset

The carrier phase offset contains an initial constant phase offset between the transmitter and the receiver as well as the phase noise from oscillators and other transceiver components. The received signal with a carrier phase offset in the frequency domain is defined as:

$$\begin{aligned} Z_{i,k} &= X_{i,k} H_k e^{j2\pi} e^{j\phi_0} \frac{1}{N} \sum_{n=0}^{N-1} e^{j\phi_{i,n}} \\ &+ \sum_{l=-N/2+1, l\neq \varepsilon_I}^{N/2} X_{i,l} H_l e^{j\phi_0} \frac{1}{N} \sum_{n=0}^{N-1} e^{j\phi_{i,n}} e^{j2\pi n(k-l)/N} \\ &+ V_{i,k} \end{aligned} \tag{2.26}$$

where $\Phi_{i,n}$ is the phase noise sampled at the n-th sample in the i-th symbol and Φ_0 is the constant phase offset at the beginning of the i-th symbol.

By analyzing Equation (2.26), we can draw the following conclusions: a) the first term, due to the inserted constant phase error Φ_0, causes incorrect *I-Q* signal detection; b) the second term represents the influence of ICI; and c) the third term corresponds to noise. If the phase noise is band-limited (relatively constant within an OFDM symbol) then there is very little ICI.

2.11 Channel Estimation

The goal of channel estimation is to provide information related to distortion of the signal when it propagates through the channel. The obtained information is then used by the equalizers in order to remove the fading effect and ICI. There are two types of channel estimators: non-data-aided and data-aided.

Non-data-aided estimators (also called blind channel estimators) use only received signal statistical properties in order to realize channel estimation. No specialized training signals (no additional redundancy) are needed. This approach retains the transmission efficiency. However, it requires large data collection in order to generate accurate channel estimation.

On the other hand, data-aided estimators use known training signals and can provide rapid and accurate channel estimation. In real situations, a sufficient number of training signals must be used (the number of training signals depends on channel variation i.e. coherence time and coherence bandwidth of the channel). This type of channel estimation is employed in an IEEE 802.11a OFDM based communication system.

In the standard IEEE 802.11a, a comb-type pilot-aided channel estimation scheme is implemented. To this end, the commonly used methods are: a) time-domain windowing and b) frequency-domain interpolation. Let us note that both methods have different implementations, but generate equivalent results.

2.11.1 *Time-domain windowing*

In an OFDM receiver, incorrect timing operation often causes poor demodulation performance. Bearing this in mind, channel estimation needs to take into account the timing of the channel impulse response (CIR). In practice, due to the signal propagation, both pre- and post-cursors exist in the reconstructed CIR. A selected time-domain window must, first, retain the major portion of the reconstructed CIR, and second, reject the aliased and noisy components. It is evident from Figure 2.20 that the center of the time domain window must be shifted to the right in respect to the origin (denoted with 0 on the time axis).

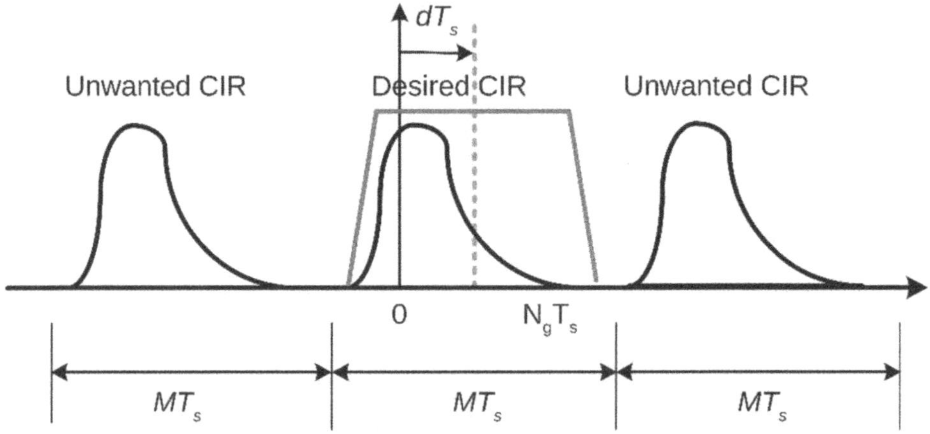

Fig. 2.20. Periodic CIR in the time domain

The time domain windowing algorithm consists of the following steps:

- For generating, the time-domain channel impulse response (CIR), IFFT of the frequency-domain channel response over pilot sub-carriers only is performed (the rest of data sub-carriers are equal to zero).
- To reduce the noise level and aliasing effect, the rectangular windowing and minimum mean-squared error (MMSE) windowing are applied. Additionally, the zero padding is implemented outside the windowing function.
- The 'cleaned-up' time domain CIR is transformed back to the frequency domain by FFT (and channel coefficients are estimated).

Rectangular windowing (RW)

In the RW technique, the time-domain CIR samples that are below a certain threshold are rejected. This approach can be regarded as applying a rectangular window to the reconstructed time-domain CIR. In low-SNR scenarios, involving sample selection according to amplitude level causes reduction of the estimated error level.

MMSE windowing (MMSEW)

In the MMSEW technique, in addition to window size selection, an appropriate weighting vector is applied to the time domain reconstructed CIR. The weighting vector is obtained using the MMSE algorithm. MMSE window is characterized with larger weights for the strong channel taps and smaller weights for the weak channel taps. In respect to SNR, very good estimation performance can be obtained by implementing this algorithm.

2.11.2 *Frequency-domain interpolation*

The frequency-domain interpolation algorithms sample the channel frequency responses at pilot sub-carriers and after that apply various interpolation techniques in order to estimate the channel responses at data sub-carriers. The channel responses at the pilot sub-carriers (for IEEE 802.11a) can be given by:

$$\widetilde{H}_j = \frac{z_j}{x_j} \quad j = -26, -7, 7, 26 \tag{2.27}$$

The channel responses at non-pilot sub-carriers are then interpolated from the responses at pilot sub-carriers. Two commonly used interpolation techniques called polynomial-based and raised-cosine-function-based are further discussed.

Polynomial interpolator (PI)

PI is often used in frequency-domain interpolation algorithms due to its low implementation complexity. Linear PI is usually used in order to estimate the frequency-domain channel responses at data sub-carriers. In general, the estimation quality can be improved by using higher-order polynomials, at the cost of higher hardware complexity and power dissipation.

Raised-cosine-function interpolator (RCI)

This interpolator utilizes the raised-cosine function [22] and defines the frequency-domain interpolation coefficients by:

$$W_l = \frac{\sin(\pi Ml/N)}{\pi Ml/N} \frac{\cos(\pi \beta Ml/N)}{1-4\beta^2(Ml/N)^2} e^{-j2\pi dl/N} \tag{2.28}$$

where M is the number of pilot sub-carriers, β is the roll-off factor, d is the time shift and it is decided by the worst-case channel delay spread. In a channel with short delay spread, the estimation error is insensitive to d since the major portion of the CIR will be covered by the window with a wide range of d. For more details, see Figure 2.20. The raised-cosine interpolation coefficients have good attenuation on noise and aliasing and preserve most CIR information. They also provide a very good estimation accuracy. In principle, the complexity of RCI is lower than complexity of time domain channel estimators and, therefore, it finds wide applications in OFDM communication systems.

2.11.3 *Selection of a comb-type estimator*

In comparison to the above mentioned time domain windowing algorithms, the MMSEW achieves the best estimation performance. Its main drawback deals with the fact that the receiver must have the channel statistics available. Contrary to MMSEW, the RW algorithm discards all weak channel taps and, consequently, cannot obtain precise channel estimates. Generally speaking, polynomial-based interpolation methods are of a significantly lower complexity at the cost of lower estimation precision. The RCI has a window with a flat top and no side lobe, and is free of waveform distortion and aliasing.

The RCI is also capable of shifting the window position to allow full coverage of the CIR. The RCI channel estimator is attractive as a design solution especially when the system is operating with a limited number of pilot sub-carriers. When the number of RCI estimator coefficients is between 6 and 8, RCI performance is very close to MMSEW performance.

2.11.4 *Decision-directed channel estimators*

In order to secure high spectral efficiency, OFDM systems commonly use a small number of pilot sub-carriers (for example, IEEE 802.11a uses 4 pilot sub-carriers). However, it can limit effectiveness and precision of the channel estimation. A common solution is a decision-directed procedure which uses demodulated and decoded information in order to generate virtual pilots. In [23], the authors initially proposed the coded decision-directed demodulation (CD3) scheme presented in Figure 2.21. According to the CD3 concept, processing is done as follows.

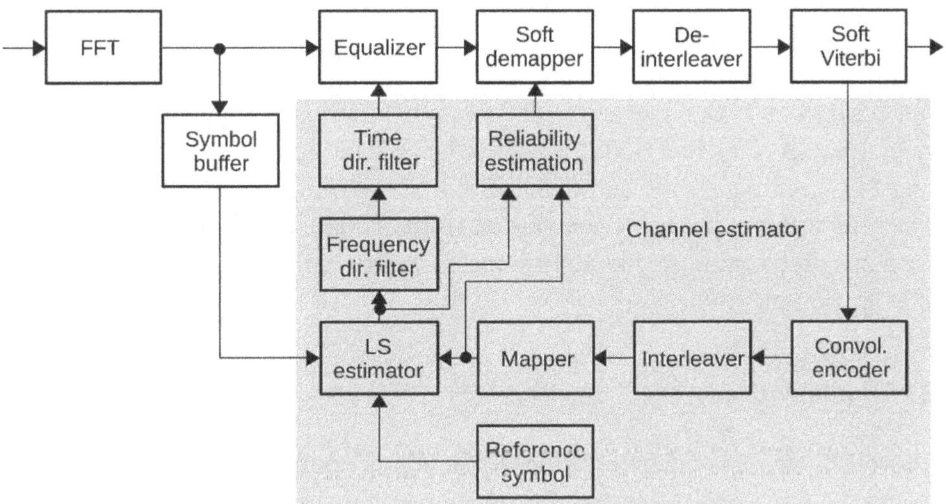

Fig. 2.21. Block diagram of the CD3 scheme

The procedure begins with the usage of a reference symbol, which is transmitted in the preamble of OFDM frame. The content of these two symbols (LT sequence, as shown in Figure 2.16) is generally a BPSK modulated PN sequence. Having in mind that this sequence is known by the receiver, a reference estimation for the CTF can be easily obtained. Next, filtering is applied in order to reduce noise from the estimated channel response. Obtained results are stored for further use in a next processed symbol.

The stored reference is applied to the next incoming symbol (first data symbol) by the equalizer in order to correct the phase and amplitude of the adequate sub-carrier signals. Afterwards, the generated signal is demodulated, deinterleaved, and decoded. Simultaneously, the uncorrected symbol is stored in the symbol buffer.

The output bits from the decoder are fed back into the channel estimator. The decoded signal is inversely processed by encoding, interleaving, and mapping. The generated signal is now used as a pseudo-pilot aiming to obtain a new estimation of CTF. In the absence of noise, the pseudo-pilots should correspond to the equalized input signals.

The new CTF is generated using NLMS or RLS adaptation algorithms. The estimation noise present in a new CTF is dependent on parameters of the applied adaptation algorithm. Optimization of the adaptation algorithm for low noise increases tracking latency.

The final CTF achieved after filtering in time and frequency is taken as the new CTF to be used for correction of the next data symbol. The procedure is repeated from step 2 onwards until the end of the frame is reached.

The presented scheme in Figure 2.21 is of the feedback type and is characterized by high noise sensitivity and potential instability (the system can be inoperative for very low SNR). Noise reduction in the CD3 scheme is achieved using FEC techniques and additional filtering in the time and frequency domains as well as using appropriate integration in the equalizer's adaptation algorithm. The main drawbacks of this algorithm are high level of signal processing and hardware complexity.

2.12 Equalization

The well-known advantage of OFDM systems in comparison with single carrier systems is the simple one-tap frequency-domain equalizer (FDE) which can equalize OFDM signals in the presence of frequency-selective fading. The impulse response of the channel should remain constant within one OFDM symbol period in order to use one-tap FDE. The received signal at each sub-carrier is given as:

$$Z_{i,k} = H_{i,k} X_{i,k} + v_{i,k} \tag{2.29}$$

One-tap FDE restores the transmitted signal as:

$$\hat{X}_{i,k} = G_{i,k} Z_{i,k} \tag{2.30}$$

where $G_{I,k}$ is the equalizer coefficient at the k-th sub-carrier during the i-th symbol. The coefficient $G_{I,k}$ is usually obtained using the NLMS (normalized least mean squares) or RLS (recursive least squares) algorithms. The reference signal for adaptation of NLMS/RLS can be obtained using a training sequence or already detected signals (decision directed approach).

References

1. Rabiner, L. R. and Gold, B. (1975) *Theory and Application of Digital Signal Processing*, (Prentice-Hall, Englewood Cliffs, USA).
2. Oppenheim, A. V. and Schafer, R. W. (1989) *Discrete-Time Signal Processing*, (Prentice Hall, Englewood Cliffs, USA).
3. Chiueh, T.-D. and Tsai, P.-Y. (2007) *OFDM Baseband Receiver Design for Wireless Communications*, (John Wiley & Sons, Singapore).
4. Lyons, R. G. (2011) *Understanding Digital Signal Processing*, (Prentice Hall, Englewood Cliffs, USA).
5. Kienle, F. (2014) *Architectures for Baseband Signal Processing*, (Springer, New York, USA).
6. Lee, B. G. and Kim, S. C. (2012) *Scrambling Techniques for Digital Transmission*, (Springer-Verlag, London, United Kingdom).
7. Liu, H., Ma, H., El Zarki, M. and Gupta S. (1997). Error control schemes for networks: An overview, Mobile Networks and Applications, 2, pp. 167–182.
8. Lin, S. and Costello, D. J. (2004) *Error Control Coding*, 2nd Ed. (Prentice Hall, Englewood Cliffs, USA).
9. Moon, T. K. (2005) *Error Correction Coding Mathematical Methods and Algorithms*, (John Wiley & Sons, Hoboken, USA).
10. Chen, X. (2012) *Coding in 802.11 WLANs*, PhD Thesis (National University of Ireland, Maynooth, Republic of Ireland).
11. Gunnam, K., Marc, J., Pérez, C. and Garcia-Herrero, F. (2016). Algorithms and VLSI architectures for low-density parity-check codes, *IEEE Solid-State Circuits Magazine*, 8, pp. 57–63.
12. Wang, Z., Cui, Z. and Sha, J. (2011). VLSI design for low-density parity-check code decoding, *IEEE Circuits and Systems Magazine*, 11, pp. 52–69.
13. Bhavani, R. D. and Sudhakar, D. (2013). Design and implementation of inverse Fast Fourier Transform for OFDM, *International Journal of Science and Engineering Applications*, 2, pp. 155–158.
14. Kalavani, R., Jose, D. and Kumar, P. N. (2016). An area efficient FFT implementation for OFDM, *International Journal of Advanced Research in Biology Engineering Science and Technology*, 2, pp. 1–5.
15. Canet, M. J., Almenar, V., Marin-Roig, J. and Valls, J. (2007). Time synchronization for the IEEE 802.11a/g WLAN standard, *Proc. 18th IEEE International Symposium on Personal, Indoor and Mobile Radio Communications*, pp. 1–5, doi: 10.1109/PIMRC.2007.4394066.

16. Manusani, S. K., Kshetrimayum, R. S. and Bhattacharjee. R. (2006). Robust time and frequency synchronization in OFDM based 802.11a WLAN systems, *Proc. Annual IEEE India Conference*, pp. 1–4, doi: 10.1109/INDICON. 2006.302775.
17. Speth, M., Classen, F. and Meyr. H. (1997). Frame synchronization of OFDM systems in frequency selective fading channels, *Proc. 47th IEEE Vehicular Technology Conference*, pp. 1807–1811, doi: 10.1109/VETEC.1997.605870.
18. Van de Beek, J. J., Sandell, M. and Borjesson, P. O. (1997). ML estimation of time and frequency offset in OFDM systems, *IEEE Transactions on Signal Processing*, 45, pp. 1800–1805.
19. Jian, L., Guoqing, L. and Giannakis, G. B. (2001). Carrier frequency offset estimation for OFDM-based WLANs, *IEEE Signal Processing Letters*, 8, pp. 80–82, doi: 10.1109/97.905946.
20. Pollet, T., van Bladel, M. and Moeneclaey, M. (1995). BER sensitivity of OFDM systems to carrier frequency offset and Wiener phase noise, *IEEE Transactions on Communications*, 43, pp. 191–193.
21. Pollet, T., Spruyt, P. and Moeneclaey, M. (1994). The BER performance of OFDM systems using non-synchronized sampling, *Proc. IEEE Global Telecommunications Conference*, pp. 253–257.
22. Ai, B., Ge, J., Wang, Y., Yang, S., Liu, P. and Liu, G. (2004). Frequency offset estimation for OFDM in wireless communications, *IEEE Transactions on Consumer Electronics*, 50, pp. 73–77.
23. Mignone, V. and Morello, A. (1996). CD3-OFDM: A novel demodulation scheme for fixed and mobile receivers, *IEEE Transactions on Communications*, 44, pp. 1144–1151.

https://doi.org/10.1142/9789811210723_0003

Chapter 3

Analog Front-End

Gildas Leger and **Antonio Gines**
Consejo Superior de Investigaciones Científicas,
Universidad de Sevilla
Instituto de Microelectrónica de Sevilla
Seville, Spain

3.1 Introduction

The purpose of the analog front-end is to interface the realm of signal processing and information theory to that of electromagnetism. Indeed, most of the communication protocol is handled in the digital baseband, as defined in the previous chapter. But for WLAN, the communication channel from device to device is the air and the information carriers are electromagnetic waves transmitted and received by antennas. These transmission and reception operations actually lend their names to the most important device of the analog front-end: the transceiver.

The transceiver must thus appropriately transform the digital signals to and from RF signals. This is usually (but not necessarily) done in two steps: first from digital to analog and then from analog to RF as we will further see.

There are obviously many similarities between the transmitter and the receiver channels. The applied concepts and architectures are almost symmetrical. However, there are also substantial differences that justify a separated treatment. The main cause of these differences is that few assumptions on the quality of the transmission medium can be made on the transmitter side, while the receiver has to cope with its non-idealities (attenuation, congestion of the spectrum, noise, etc.).

The tasks of the RF front-end and mixed-signal section of the transceiver are: 1) to adapt the base-band data information coming from the digital section for transmission though the available band in the communication channel (around 2.4

GHz and 5 GHz in the IEEE 802.11 standard [1]), and 2) to assure a correct reconstruction for a given transmission rate and signal power in the receiver path.

3.1.1 *Basic concepts in communication*

The process of modulation can be conceptually understood as a frequency shift of the signal band of interest toward the channel frequency f_c [2, 3]. In an ideal mixer, the frequency up-conversion is achieved, as shown in Figure 3.1 by a direct multiplication of the based-band message signal *m*(t) by the carrier using a amplitude modulation (AM) scheme:

$$x(t) = m(t) \cdot \cos \omega_c t \tag{3.1}$$

The resulting modulated signal, $x(t)$, known in the bibliography as double-sideband suppressed-carrier (DSB-SC), shows a spectrum which is symmetrically spaced above and below the carrier frequency, as a consequence that the signals in baseband are real.

In this scheme, the demodulation task (frequency down-conversion) can be achieved as shown in Figure 3.1 by simple multiplying the signal by the carrier using a low-pass filter (LPF) to remove the high-frequency components, as it can be derived from:

$$x(t) \cdot \cos \omega_c t = m(t) \cdot \cos^2 \omega_c t = \underbrace{\frac{m(t)}{2}}_{LP} + \underbrace{\frac{m(t)}{2} \cdot \cos 2\omega_c t}_{High-frequency\ component} \tag{3.2}$$

This process requires some information on the carrier phase for efficient coherent demodulation, since any phase difference between the modulator (up-conversion mixer) and demodulator (down-conversion) affect negatively to the receiver gain, eventually leading to zero when $\theta = \pi/2$:

$$\begin{aligned} x(t) \cdot \cos(\omega_c t + \theta) &= m(t) \cdot \cos(\omega_c t) \cdot \cos(\omega_c t + \theta) \\ &= \underbrace{\frac{m(t)}{2} \cdot \cos \theta}_{LP} + \underbrace{\frac{m(t)}{2} \cdot \cos(2\omega_c t + \theta)}_{High-frequency\ component} \end{aligned} \tag{3.3}$$

Phase estimation can be practically done providing a reference signal to the receiver by including the carrier in the transmitted spectrum, $A(1 + m(t)) \cdot \cos \omega_c t$, as in classical AM radio modulation, or transmitting a pilot signal in an adjacent channel to simplify the demodulation process. Alternatively, phase can be estimated from the receiver information using a relatively complex algorithm in the receiver digital section (DSP).

Although conceptually simple, the previous DSB-SC approach suffers from several practical drawbacks that limit the transmission throughput due to the inefficient use of the bandwidth — the DSB frequency redundancy due to the symmetric property leads a factor 2 excess in radio band occupation.

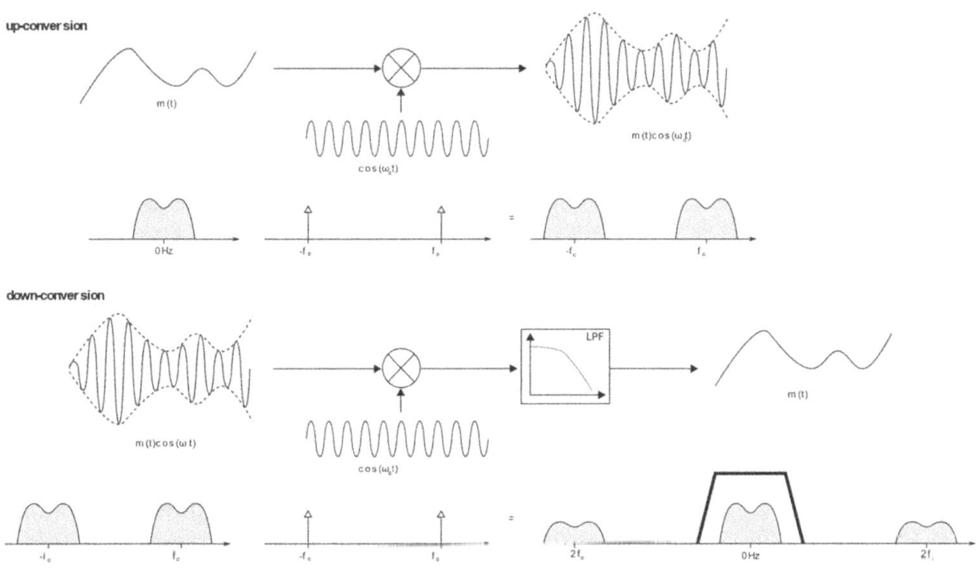

Fig. 3.1. Conceptual scheme of the frequency conversion process in a transceiver using and ideal multiplier-based mixer

The need of optimizing the spectral efficiency has motivated the development of alternative modulation schemes. Early solutions, such the *Single-Side Band* (SSB) modulation relied on analog filtering to suppress the redundant spectrum information, at the expense of *signal-to-noise ratio* (SNR) degradation, limitation on the achievable bandwidth and constraining the achievable up-conversion frequency range. To overcome this drawback, modern communication systems (Bluetooth, GSM, WLAN, etc.) rely on *In-phase and Quadrature* (IQ) modulation, also known in the bibliography as complex (or just quadrature modulation), to allow two base-band messages, $I(t)$ and $Q(t)$ to be transmitted simultaneously through a channel sharing the same radio band without interfering with each other, hence improving the channel capacity in a factor 2.

Quadrature Modulation: in this method the transmitted signal $x(t)$ is generated (see Figure 3.2a) as additive contribution of two in-phase and quadrature components,

$$x(t) = I(t)\cdot\cos(\omega_c t) + Q(t)\cdot\underbrace{\cos(\omega_c t + \pi/2)}_{-\sin(\omega_c t)} \tag{3.4}$$

Given the orthogonal characteristic of the sinusoidal carriers, the message signals, $I(t)$ and $Q(t)$, can be recovered multiplying by the modulation carrier considering a low-pass filter for removing the high-frequency components, as follows,

$$\begin{aligned} x(t)\cdot\cos\omega_c t &= \underbrace{\frac{I(t)}{2}}_{LP} + \underbrace{\frac{I(t)}{2}\cdot\cos 2\omega_c t - \frac{Q(t)}{2}\cdot\sin 2\omega_c t}_{High-frequency\ component} \\ x(t)\cdot\sin\omega_c t &= \underbrace{-\frac{Q(t)}{2}}_{LP} + \underbrace{\frac{Q(t)}{2}\cdot\cos 2\omega_c t + \frac{I(t)}{2}\cdot\sin 2\omega_c t}_{High-frequency\ component} \end{aligned} \tag{3.5}$$

The modulation signal can carry either analog and or digital information. As an example of digital communication, the 16-QAM transmission codifies the data stream in groups of 4-bits, determining a specific symbol in IQ constellation diagram of Figure 3.2b, which acting on the amplitude of two carriers with phase shift of 90º,

$$x(t) = A\cdot\left[\alpha_I\cdot\cos(\omega_c t) - \alpha_Q\cdot\sin(\omega_c t)\right] \quad \text{with} \quad \alpha_{I,Q}\in\{\pm 1, \pm 3\} \tag{3.6}$$

Each of the 16 combinations is codified in the amplitude and phase of the resulting modulated signal. Figure 3.2c shows the resulting output for the 16-QAM signal. For simplicity, in the example the symbol rate ($f_{\text{symb}} = 1/T_{\text{symb}}$) is equal to the carrier frequency, that is, each symbol time results modulated with a single carrier period. In a general situation, multiples carry cycles can be transmitted per symbol time to simplify the demodulation process improving immunity against noise.

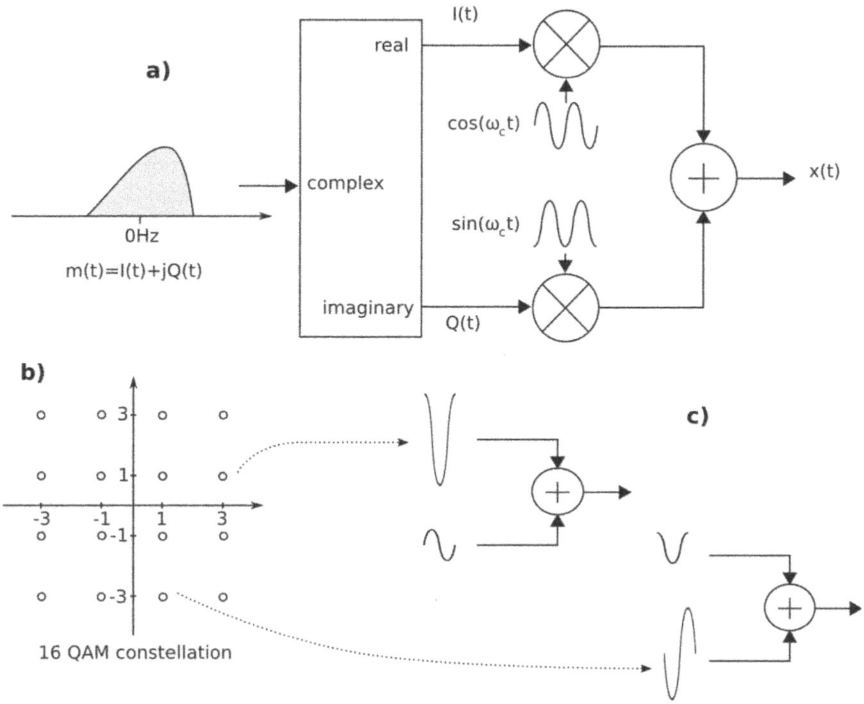

Fig. 3.2. a) Simplified diagram block of a quadrature modulator; b) example of the symbol constellation for 16-QAM case study, and c) resulting waveform of the modulated signal

3.1.2 *Modulation schemes in IEEE 802.11bag/n/ac*

As mentioned before, IEEE 802.11 is a quite flexible standard in term of modulation schemes and maximum data rate, depending on the channel characteristics (noise, fading, etc.), the available power, or the operating band (ISM radio bands around 2.4 GHz and 5 GHz.).
For maximum throughput, as shown in Table 3.1, the standard relies on OFDM with high-order modulation schemes (up to 256-QAM in 802.11ac[1]), channel bandwidths in the range of 20/40/80/160MHz and operation in the less congested 5GHz radio band, since more crowded 2.4GHz band is shared with other standards such as Bluetooth ZigBee, etc.

A good understanding of the standard implications of these factors [4,5,6] is crucial during the physical design and test of both the transmitter and the receiver

---

[1] Higher modulation schemes (1024QAM, 2048QAM) are expected to be standardized in a future release of the standard IEEE 802.11ax (currently in the last draft iterations).

channels, especially when dealing with maximum transmission rate based on OFDM in which the design margins are more constrained.

Table 3.1. Versions of the IEEE 802.11 Standard

Standard	Radio Band (GHz)	Channel Bandwidth (BW, MHz)	Modulation Scheme	Maximum Order	Spatial Stream	Max. Throughput (Mb/s)
802.11 b	2.4	20	DSSS	DQPSK	SISO	11
802.11 a/g	2.4 / 5	20	OFDM	64-QAM	SISO	54
802.11 n	2.4 / 5	20/40	HT-OFDM	64-QAM	4x4 MIMO	600
802.11 ac	5	20/40/80/160	VHT-OFDM	256-QAM	8x8 MIMO	6933

IEEE 802.11 OFDM scheme: From the point of view of the transceiver RF and mixed-signal front-end, the OFDM technique can be understood as an analog modulation scheme in which the messages, $I(t)$ and $Q(t)$, are built at the same time as an IQ (complex) base-band signal which considers multiples sub-carriers with a fixed symbol rate equal to T_{sym} = 3.2us. Guard Intervals (GI), ranging between 800ns and 400ns depending on the wanted transmission rate, are considered between symbols to optimize transmission.

To assure that the modulation messages between sub-carriers do not interfere each other, the frequency separation between sub-carriers is equal to the inverse of the symbol time, $\Delta f_{\text{sub-carry}} = 1/3.2\mu s = 312.5kHz$, with a maximum frequency error tolerance lower than 20-25ppm depending on the operation band and modulation scheme. In such a way, orthogonal discrimination between different messages, partially sharing the same radio band, can be achieved, based on the following properties when integrating in the symbol time for any sub-carrier,

$$\begin{aligned} &\int_0^{T_{sym}} \cos(\omega_n \cdot t)\cdot\cos(\omega_m \cdot t)\, dt = \delta(n-m) \\ &\int_0^{T_{sym}} \cos(\omega_n \cdot t)\cdot sen(\omega_m \cdot t)\, dt = 0 \end{aligned} \tag{3.7}$$

where δ $(n\text{-}m)$ is the Kronecker'delta which take the unit value when the integer indexes n and m are equal, that is when the received signal is correlated with its own carry reference, and zero in the other cases.

Table 3.2. Sub-carrier configurations and Guard Intervals (GI) in the IEEE 802.11bag/n/ac

Bandwidth (BW) (MHz)	FFT Points (N_{FFT})	Data Subcarriers (*DataCarriers*)	Pilot Subcarriers	Guard Interval (GI) (ns)
20 MHz	64	48 / 52	4	400 / 800
40 MHz	128	108	6	400 / 800
80 MHz	256	234	8	400 / 800
160 MHz	512	468	16	400 / 800

With a fixed spacing between sub-carriers, its number depends on the selected channel bandwidth, as shown in Table 3.2. IEEE 802.11n defines two bandwidth modes (20 MHz and 40 MHz) providing retro-compatibility with previous IEEE 802.11b and IEEE 802.11ag which considers 20MHz within the 2.4GHz and 5GHz ISM bands. To increase throughput, 802.11ac introduces two new channel widths (see Figure 3.3). In stage 1 mode, support for 80MHz channels width is provided doubling the size of the spectral channel with respect to 802.11n. In stage 2, 160MHz channel option is further defined for higher transmission rates in both contiguous and non-contiguous modes to deal with radio band congestions.

As an example, Figure 3.4 shows channel configuration for the 20MHz bandwidth case (BW = 20MHz). According to the number of FFT points (N_{FFT}) in Table 3.2, the total bandwidth is distributed in a set of equally spaced frequency bins (BW = $N_{FFT} \times \Delta f_{sub\text{-}carry}$), being the center and extreme points of the band (null carriers) reserved to reduce offset and inter-channel interference issues, respectively. Some sub-carriers are reserved as pilot for simplifying the reception (synchronization, phase-estimation, and channel characterization).

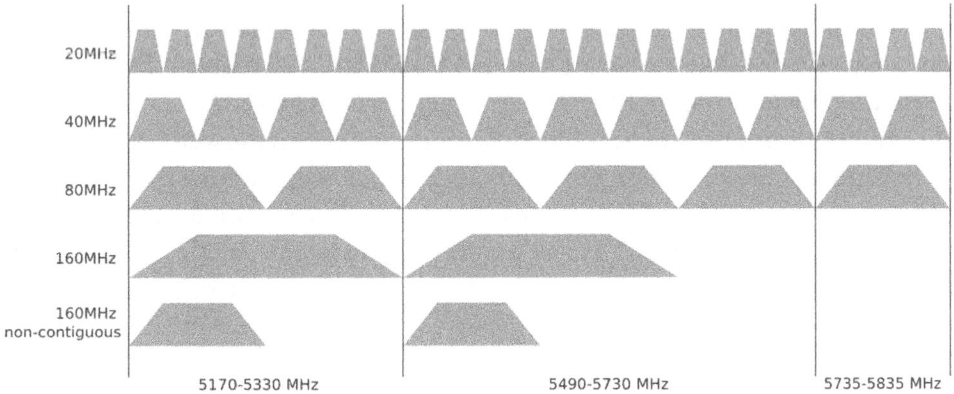

Fig. 3.3. Channel configurations in the IEEE 802.11ac standard

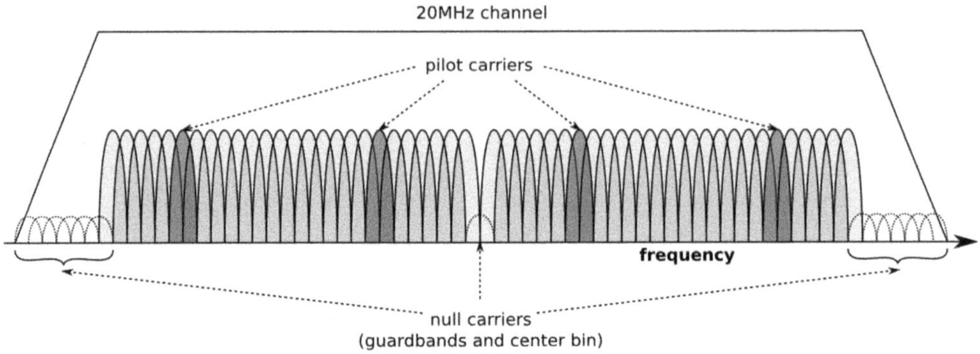

Fig. 3.4. Channel configuration for the 20MHz bandwidth case (BW = 20MHz)

Maximum theoretical throughout rate for a given configuration can be evaluated according to the following expression,

$$DataRate_{\max} = \frac{DataCarriers \cdot SymbolBits \cdot CodeRate \cdot SpatialStreams}{T_{sym,eff}} \quad (3.8)$$

where *DataCarriers* defines the number of data subcarriers used in the transmission (for instance: 48 in the 802.11a/g), *SymbolBits* shows the number of bits per symbol depending on the modulation order (f.e., 6 bits in 64-QAM), *CodeRate* indicates the codification redundancy ratio (3/4 at higher order modulation in 802.11a/g), *SpatialStreams* is the number of simultaneous streams (1 for SISO versions), and the variable $T_{\mathrm{sym,eff}}$ is the effective symbol duration including transient Guard Intervals (GI), that is, 3.6μs or 4.0μs in total depending on the considered extra time (400ns or 800ns, respectively).

Particularizing this expression for the 20MHz-bandwith transmission in the IEEE 802.1ag standard (constant GI equal to 800ns) with the higher order 64-QAM modulation, it leads to a maximum rate of 54Msps. For a given 20MHz transmission bandwidth, improvements in data rate can be achieved by: a) incrementing the number of data sub-carriers (up to 52) at expenses of narrower guard bands, b) incrementing the code ratio reducing the redundancy margin (up to 5/6), and additionally, c) reducing the time Guard Intervals down to 400ns. Following this approach data rates up to 72 Mb/s are theoretically achievable in IEEE 802.11n. Further improvements in transmission speed require, as previously introduced, incrementing the bandwidth and/or using higher modulation order (up to 256-QAM in IEEE 802.1ac). Unluckily, both strategies have specific implications for the RF analog and mixed-signal section of the transceivers, since this parameter has direct correlation at the circuit level imposing extra constrains in terms of circuit bandwidth, tolerance to noise (SNR) and distortion, already highlighted in the standard itself, as summarized in the following sub-sections for the main constraints of both transmitter and the receiver paths.

3.2 Specifying Transceiver

3.2.1 *IEEE 802.11 transmitter constraints*

Error Vector Magnitude (EVM): it is the primary metric of modulation quality for an 802.11 transmitter for higher speed configurations based on OFDM. It is defined, as shown in Figure 3.5, as the ratio of the amplitude of the error vector to the ideal symbol reference in the IQ constellation plane,

$$EVM_{dB} = 20\log_{10} e_i \tag{3.9}$$

where the effective error e_i is evaluated as the root mean square (RMS) value over a large number of symbols. Notice that EVM is thus a statistical metric. Indeed, it accounts for noise but also for distortion contributions and its value can thus vary for different data packets (if symbols are more concentrated to the origin, the denominator is lower and the EVM higher).

Table 3.3 shows the IEEE 802.11 maximum EVM specification for the different modulation schemes. As expected, the higher order of modulation, the smaller margins between symbols in the constellation, and the lower the admissible EVM.

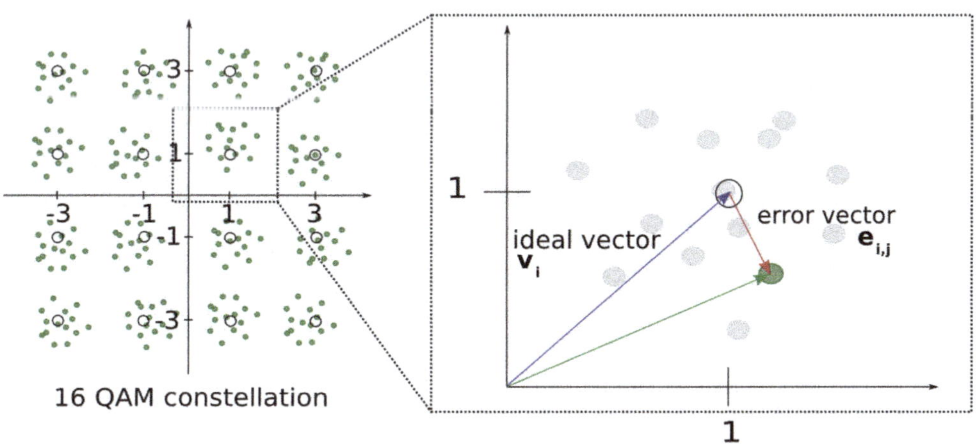

Fig. 3.5. Error vector example for a 16-QAM constellation

Table 3.3. EVM specifications for the different modulation schemes in IEEE 802.11

Modulation Scheme*	Code Rate	EVM Specification
BPSK	1/2	-5 dB
BPSK	3/4	-8 dB
QPSK	1/2	-10 dB
QPSK	3/4	-13 dB
16-QAM	1/2	-16 dB
16-QAM	3/4	-19 dB
64-QAM	2/3	-22 dB
64-QAM	3/4	-25 dB
64-QAM	5/6	-27 dB
256-QAM	3/4	-30 dB
256-QAM	5/6	-32 dB

*The maximum order of modulation is limited for each specific version of the standard as shown in Table 3.1. A similar constraint is found for the maximum code rate.

Delivered Power and Spectral Emission Mask (SEM): the IEEE 802.11 standard does not specify the maximum transmission power available at the antenna, since this parameter depends on local regulations. Typical values in most of the countries ranges from 60mW to 1W (17.78dBm to 30dBm at 50Ω) with a clear distinction between in-door and out-door applications.

In spite of the flexibility on the maximum delivered power level, the standard imposes a hard constraint in the relative amount of power than can be transmitted to the adjacent channel to minimize the inter-channel interferences. Figure 3.6 shows the specified spectral mask depending on the selected channel bandwidth for OFDM transmission. The filtering profile is specified as a function of attenuation levels at -20dBr, -28dBr and -40dBr with respect to the maximum power spectral density of the signal. This relative attenuation can be relaxed to an absolute power spectral density at any frequency offset, whose value depends on the bandwidth (-53dBm/√Hz for a 20MHz bandwidth, -56dBm/√Hz for a 40MHz bandwidth and -59dBm/√Hz for 80 and 160MHz). Similar constraints can be found in 802.11b for the DSSM scheme.

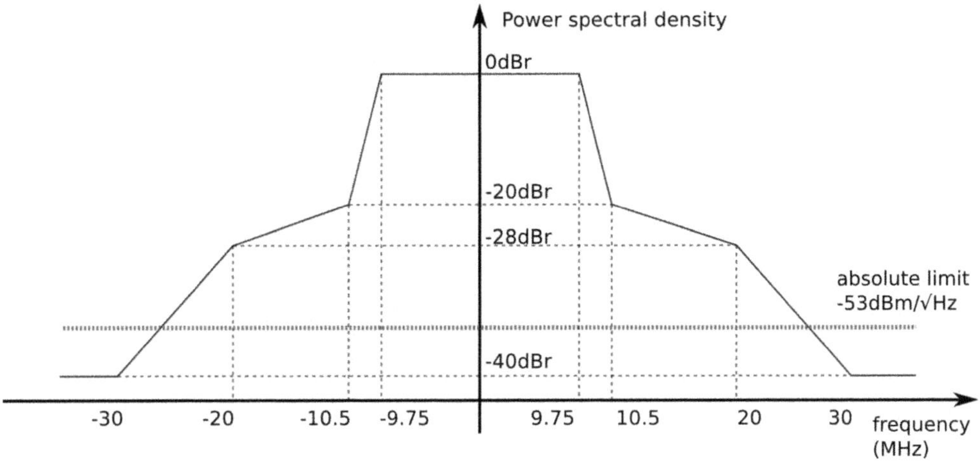

Fig. 3.6. Spectral mask definition for a 20MHz bandwidth OFDM transmission

Additional Constrains: aside from the previous key specifications in terms of EVM tolerance and power constraints, several additional aspects should be considered during the RF analog and mixed signal front-end design phase. Among them, it is worth highlighting the following considerations completed in the IEEE 802.11 standard:

- Stability of the carrier frequency: In order to properly recover the signal, the carrier frequency which defines the center of the signal band must be precisely defined. A tolerance of ±20 ppm is considered in the 5GHz band and of ±25 ppm in the 2.4GHz band.
- Spectral flatness: OFDM modulation scheme considers independent sub-carriers across the band, which makes the transmission more resilient to frequency errors. In spite of this, a constraint is set in the standard for a minimum spectral flatness. This constraint can negatively impact the filtering trade-offs in the transmitter, but fortunately systematic deviations can be handled in the digital domain by proper equalization. The requirements are illustrated in Figure 3.7, where the reference power is taken as the average power over the center half of the bandwidth.
- Power wake-up time: In low-power applications, it may occur that the transmitter is powered-down when no transmission is required. This occurs, for instance, in Time Division Duplex. In such cases, a maximum wake-up time must be specified for the transmitter to avoid undue latencies.

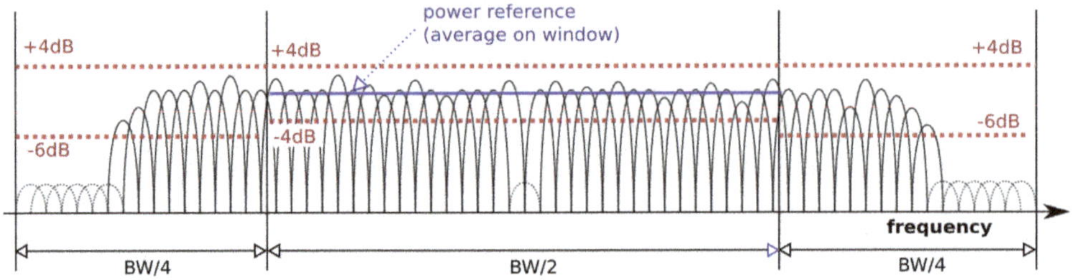

Fig. 3.7. Spectral flatness requirements

3.2.2 *IEEE 802.11 receiver constraints*

Sensitivity Requirements: The sensitivity of a receiver is defined as the minimum input power level, measured at the antenna, which is compliant with the target transmission quality.

In the IEEE 802.11 this specification is derived from the maximum package error rate (PER) when transmitting a data unit of 1024 octets. For 802.11b based on DSSS modulation, the specified maximum PER is 8% leading to sensitivity levels of −80dBm and −76dBm at 2Mb/s and 11Mb/s, respectively. In OFDM scheme, the sensitivity levels compliant with the standard requirements (BER $\leq$ 10%) result in a function of the modulation scheme, code rate and channel bandwidth.

Under the assumption of a typical NF = 10dB and considering 5dB of extra margin for dealing with channel and transceiver non-idealities, the standard itself evaluates in Table 3.4 the minimum sensitivity values for 10% BER compliance. As expected for a given bandwidth configuration, the minimum power level increases with transmission throughput rate (higher modulation scheme with lower redundancy). From column to column, a factor 2 is found in the processing bandwidth (doubling the integration noise), which is translated to a 3dBm difference in the sensitivity, under the assumption of constant SNR in the receiver.

Table 3.4. Standard Receiver Sensitivity based on OFDM modulation

Modulation Scheme	Code Rate	Minimum Sensitivity (BW=20MHz)	Minimum Sensitivity (BW=40MHz)	Minimum Sensitivity (BW=80MHz)	Minimum Sensitivity (BW=160MHz)
BPSK	1/2	-82 dBm	-79 dBm	-76 dBm	-73 dBm
BPSK*	3/4	-81 dBm	-	-	-
QPSK	1/2	-79 dBm	-76 dBm	-73 dBm	-70 dBm
QPSK	3/4	-77 dBm	-74 dBm	-71 dBm	-68 dBm
16-QAM	1/2	-74 dBm	-71 dBm	-68 dBm	-65 dBm
16-QAM	3/4	-70 dBm	-67 dBm	-64 dBm	-61 dBm
64-QAM	2/3	-66 dBm	-63 dBm	-60 dBm	-57 dBm
64-QAM	3/4	-65 dBm	-62 dBm	-59 dBm	-56 dBm
64-QAM	5/6	-64 dBm	-61 dBm	-58 dBm	-55 dBm
256-QAM	3/4	-59 dBm	-56 dBm	-53 dBm	-50 dBm
256-QAM	5/6	-57 dBm	-54 dBm	-51 dBm	-48 dBm

*No valid for some bandwidth configurations

Maximum Input Level: the maximum input power available at the antenna is a key specification to minimize receiver de-sensitization in high strength conditions. IEEE 802.11 standard contemplates this scenario with the definition of a set of power thresholds, shown in Table 3.5, depending on the operation band and modulation scheme. For input power levels below these values (and above the sensitivity value), the receiver should guarantee the BER compliance for certification. As a general trend, maximum power level at the input is relaxed for higher transmission throughput rate (higher modulation scheme with lower redundancy).

Table 3.5. IEEE 802.11Thresholds on the Maximum Input Power Level

Modulation scheme*	PER Target (%)	Threshold Power (dBm) (2.4 GHz Radio Band)	Threshold Power (dBm) (5 GHz Radio Band)
DSSS (2Mb/s)	8	-4	-
DSSS (11Mb/s)	8	-10	-
OFDM	10	-30	-30
HT-OFDM	10	-20	-30
VHT-OFDM	10	-	-30

* Some modulations schemes are restricted to a specific radio band

Adjacent/non-adjacent Channel Rejection: the channels of the communication in the 2.4GHz and 5GHz ISM band can be shared between different users and devices. IEEE 802.11 receivers should be able to achieve the target PER in the presence of strong interferences, that could exceed, as illustrated in Figure 3.8, in several tens of decibels the power associated to the signal of interest.

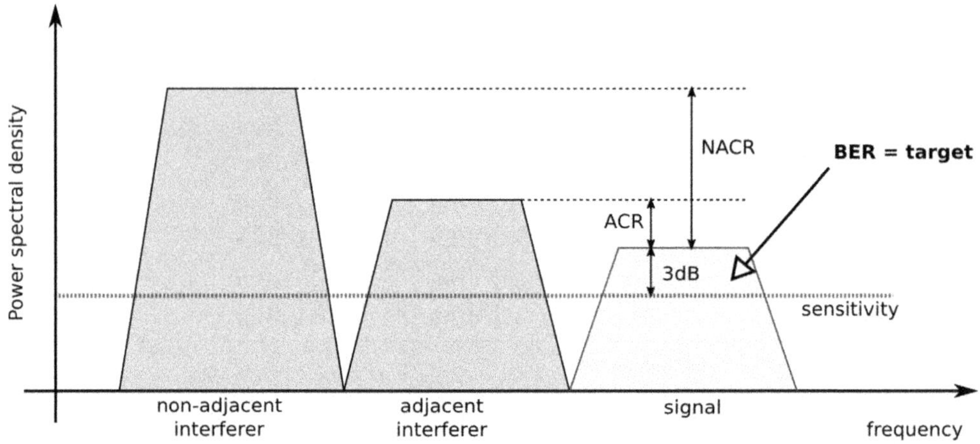

Fig. 3.8. Definition of the Adjacent Channel Rejection (ACR)

The ability of a receiver of filtering any interference in the adjacent channel is characterized by the *Adjacent Channel Rejection* (ACR) metric. This metric is defined as the minimum amount of power that should be affordable in the adjacent channel such that a correct transmission can be guaranteed (i.e. BER < target), when the input power is slightly above the sensitivity. In OFDM modulation, as shown in Table 3.6, the standard reference power is just +3dB above sensitivity. In DSSS modulation (802.11b), a +6dBm margin is considered for a target ACR threshold of 35dB.

For the non-adjacent aggressor, the Non-Adjacent Channel Rejection metric (N-ACR) can be equivalently defined, as shown in Table 3.7.

Table 3.6. IEEE 802.11 specifications of Adjacent Channel Rejection (ACR) based on OFDM transmission

Modulation Scheme	Code Rate	Reference Input Power	Adjacent Channel Rejection (ACR)		
			802.11a/g BW = 20 MHz	802.11n/ac BW = 20/40/80/160 MHz	802.11ac BW = 80+80 MHz
BPSK	1/2	Sensitivity + 3dB	28 dB	16 dB	13 dB
BPSK*	3/4		27 dB	-	-
QPSK	1/2		25 dB	13 dB	10 dB
QPSK	3/4		23 dB	11 dB	7 dB
16-QAM	1/2		20 dB	8 dB	5 dB
16-QAM	3/4		16 dB	4 dB	1 dB
64-QAM	2/3		12 dB	0 dB	-3 dB
64-QAM	3/4		11 dB	-1 dB	-4 dB
64-QAM	5/6		-	-2 dB	-5 dB
256-QAM	3/4		-	-7 dB	-10 dB
256-QAM	5/6		-	-9 dB	-12 dB

Table 3.7. IEEE 802.11 specifications of Non-Adjacent Channel Rejection (N-ACR) based on OFDM transmission

Modulation Scheme	Code Rate	Reference Input Power	Adjacent Channel Rejection (ACR)		
			802.11a/g BW = 20 MHz	802.11n/ac BW = 20/40/80/160 MHz	802.11ac BW = 80+80 MHz
BPSK	1/2	Sensitivity + 3dB	42 dB	32 dB	29 dB
BPSK*	3/4		41 dB	-	-
QPSK	1/2		39 dB	29 dB	26 dB
QPSK	3/4		37 dB	27 dB	24 dB
16-QAM	1/2		34 dB	24 dB	21 dB
16-QAM	3/4		30 dB	20 dB	17 dB
64-QAM	2/3		26 dB	16 dB	13 dB
64-QAM	3/4		25 dB	15 dB	12 dB
64-QAM	5/6		-	14 dB	11 dB
256-QAM	3/4		-	9 dB	6 dB
256-QAM	5/6		-	7 dB	4 dB

3.3 Transmitter Architecture

The basic functions realized by a transmitter are three: The conversion of a complex modulated signal into a real signal, its frequency shifting to the carrier frequency and its amplification and adaptation to the radiating antenna. Any of these functions can bear several practical implementations and we can thus imagine as many transmitter architectures as possible combinations.

Let us describe, at a sufficiently high description level, the most relevant architectures.

3.3.1 *Direct conversion transmitter*

The direct conversion architecture shown in Figure 3.9 (also called ZIF: zero intermediate frequency) is possibly the most natural one. The complex modulated signal is transformed into a real signal by a direct IQ modulation to the carrier frequency. That IQ modulation also performs the frequency shifting function. The signal is then amplified and adapted to the antenna.

From the perspective of the building blocks, the real part and the imaginary part of the complex modulated signal are converted from digital to analog by the corresponding data converters. The analog signal must then be low-pass filtered to remove the images generated at multiples of the sampling frequency. These two signals are then shifted to the carrier frequency by mixers. This is where the IQ modulation takes place, since the real part is multiplied by an in-phase version of the local oscillator while the imaginary part is multiplied by a quadrature version

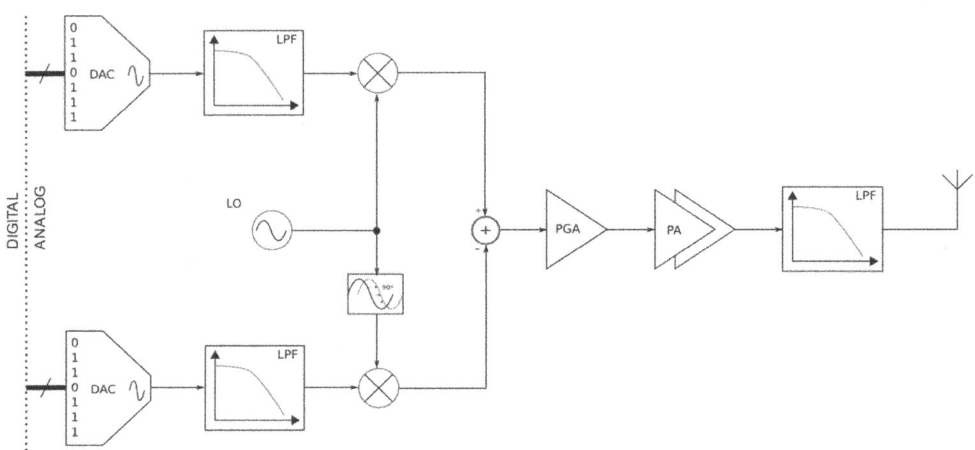

Fig. 3.9. Basic diagram of a direct conversion transmitter

(i.e. phase shifted by $\pi/4$) of the local oscillator. These two bandpass signals are then summed up and amplified by a programmable gain amplifier and sent to the power amplifier in charge of the antenna adaptation. Another low-pass filter can be used to clean up the output spectrum of the PA from high-frequency components induced by distortion or spurious coupling of high frequency tones.

Strength of the architecture: The main strength of the direct conversion transmitter is its compactness, which lends itself perfectly to low-cost, area-efficient embedded implementations.

Weakness of the architecture: Such an architecture is particularly sensitive to RF coupling. The mixers have to work at high frequency since they are driven by the RF carrier signal. In RF circuits, it is very cumbersome to completely isolate the output from the input. It is thus likely that the carrier frequency leak into the output spectrum. Such a leakage cannot be removed by further filtering since it is in the bandwidth of the useful signal. An offset in any of the DACs will produce the same effect in the output spectrum: an unwanted tone at the carrier frequency. Modulation schemes like OFDM can afford some tolerance to that carrier leakage since the sub-carrier associated to DC in the baseband is not used for data.

In addition, the direct conversion architecture is prone to injection pull-in. The presence of high-power RF components in the vicinity (consider 20% as a rule of thumb) of the Local Oscillator frequency can indeed disturb its spectrum. A possible solution to mitigate this issue consists in using an LO at twice the carrier frequency and divide its output by two. This has the additional benefit of the direct generation of almost perfect quadrature mixing signals. However, due to non-linearity of the RF blocks, harmonic distortion occurs generating some RF power around multiples of the carrier frequency. This is particularly true for chopper-based mixers, for which the power of odd harmonics is significant. As a result, there may still be significant RF power around LO frequency. The design of the LO and mixers is thus critical for direct conversion architectures.

3.3.2 *Heterodyne transmitter*

3.3.2.1 *Conventional architecture*

The well-known heterodyne transmitter, depicted in Figure 3.10, takes an additional step with respect to the direct conversion. The real and imaginary part of the signal are also combined through an IQ modulation, but that modulation is performed at an intermediate frequency, different from the carrier frequency. It is the combined signal that is then shifted to the carrier frequency.

Multiplying by two sinewaves of different frequencies, ω_1 and ω_2, images are generated at $\omega_1+\omega_2$, $\omega_1-\omega_2$, $\omega_2-\omega_1$ and $-\omega_2-\omega_1$. In principle, any of these combinations could be chosen to transmit the signal of interest. However, in practice, the first combination is preferred such that both ω_1 and ω_2 can be set at a fraction of the carrier frequency, since practical implementations of building blocks at higher frequencies is more demanding.

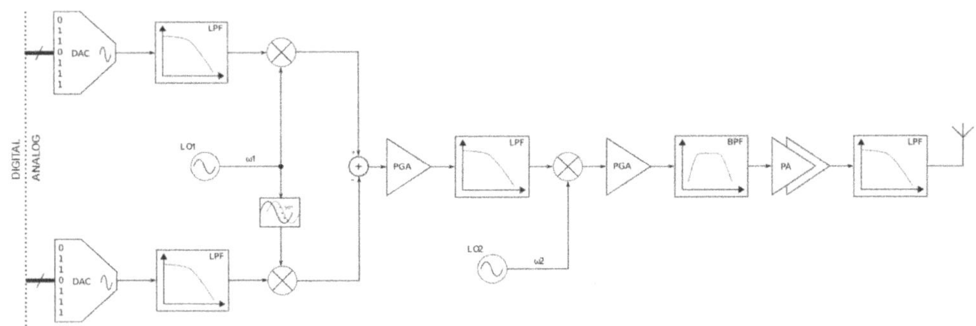

Fig. 3.10. Basic diagram of a heterodyne transmitter

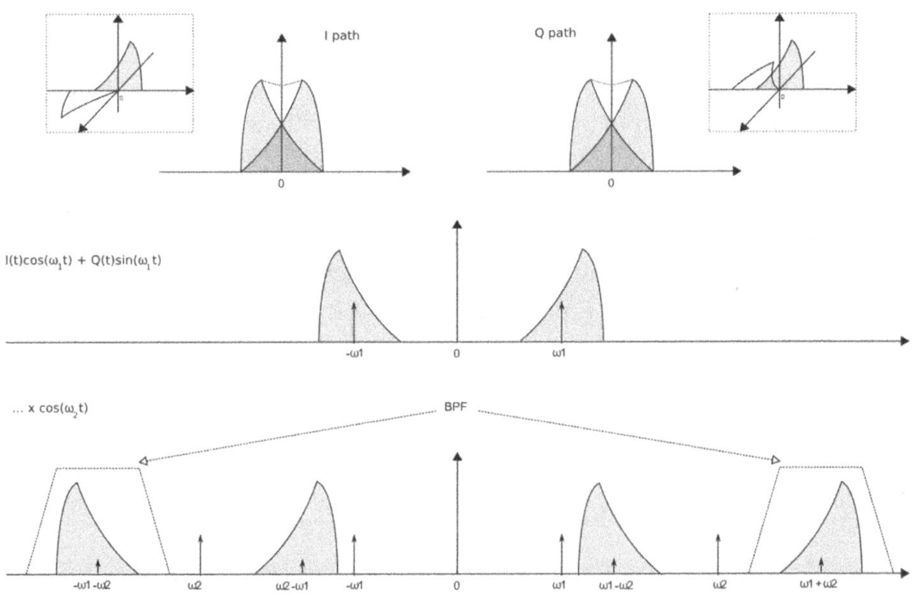

Fig. 3.11. Conceptual representation of signal and image combination in a heterodyne receiver

Obviously, the frequency planning must take into account the bandwidth of the signal to be transmitted: for the information to be recovered, the signal must remain a bandpass signal. This means that no spectral overlap can be admitted between the different images. Figure 3.11 shows an example of the power spectral density after modulation. Practical filter implementations limit the selectivity, and additional margin should thus be taken into account for the width of the stopband. This is not the only frequency restriction. Indeed, the mixing stage will also generate images around the harmonics of the mixing frequencies. In the transmitter output spectrum these images can possibly appear around any frequency ($l.\omega_1+k.\omega_2$), where l and k are signed integers. Careful frequency planning can help selecting the frequencies such that these images can best be filtered out. In addition, a low-pass filter can (should) also be added after the first mixing stage to minimizes the power of the images around multiples of ω_1.

Strength of the architecture: The heterodyne transmitter has been thought to overcome some of the issues related to the direct conversion. The first mixing process is left unchanged, so a spurious tone at the intermediate frequency can still leak into the output of the mixers, that will be impossible to filter out. However, it is much easier to avoid such a leakage during the mixer design for moderate frequencies. From that point of view, it is thus desirable that ω_1 be chosen as small as possible. For the second mixer, the modulation carrier at ω_2 (and its harmonics) will not necessarily fall into the RF signal bandwidth and can thus be filtered out.

Weaknesses of the architecture: The obvious counterpart of these benefits is an increased complexity. One more mixer is necessary, as well as a bandpass filter. And overall, the filtering requirements are increased.

In addition, two separate mixing frequencies must be accurately generated. Here again, careful frequency planning can minimize the risk of injection pull-in, but this comes at the cost of two independent RF synthesizers.

3.3.2.2 *RF synthesizer sharing: Variable-IF Transmitter*

In order to reduce the cost of the mixing frequencies generation, these two frequencies can be generated by integer division of a single high-speed RF oscillator, as shown in Figure 3.12.

The two integer M and N must be properly chosen to avoid the spectral contamination. In particular, the carrier frequency (supposing the case of a supradyne configuration) will be,

$$\omega_{TX} = \omega_1 + \omega_2 = \omega_{osc} \frac{M+N}{MN} \tag{3.10}$$

It appears that the transmission center frequency is not necessarily a simple fraction of the oscillator frequency. In that way, injection pull-in from harmonics of the transmit frequency can be avoided. That being said, the main drawback of this approach is that the oscillator frequency has to be adjusted if the transmit frequency must address several transmission bands. This implies that the intermediate frequency ω1 will also vary. In a conventional heterodyne receiver, the intermediate frequency is usually fixed, so that the first mixers and the following bandpass filter can be fine-tuned and optimized. Adopting that variable IF strategy would thus reduce the requirements on the local oscillators but increase those on the first mixers and filters (or alternatively, increase their complexity by making then programmable).

Fig. 3.12. Implementation of the mixing frequencies from a single oscillator

3.3.2.3 *Digital IQ implementation: Real-IF Transmitter*

As technology evolves toward smaller nodes, analog design is becoming more and more cumbersome. Low-voltage techniques have to be implemented and the ever-decreasing supply voltage limits the number of transistor stacks. One way to fully realize the potential of the smallest nodes thus consists in shifting the maximum number of operations to the digital domain. The remaining analog blocks usually implement some form of programmability and calibration to mitigate the impact of variability, aging and environmental conditions.

In that line, it appears interesting to perform the first modulation stage in the digital domain. A high-speed Digital-to-Analog Converter is used to directly generate a bandpass signal centered on the intermediate frequency, as depicted in Figure 3.13. Just like in the conventional heterodyne architecture, this real signal is then shifted to the transmission frequency by mixer.

The main benefit of such an approach is that the impairments related to the non-ideal implementation of the IQ modulation are removed. For instance, there will be no mismatch between the in-phase and quadrature channels. In addition, the first mixing operation can be considered ideal and there will thus be no images around the multiples of the intermediate frequency. However, this architecture does put a lot of pressure onto the design of the DAC. Indeed, the sampling frequency must be sufficiently higher than twice the signal maximum frequency to let some room for the reconstruction low-pass filter to remove the images associated with the sampling operation. Notice that the maximum signal frequency, here, is not the bandwidth of the signal to be transmitted but that signal modulated at the intermediate frequency (BW+ω_1). An alternative would consist in using the images of the signal at multiple of the sampling frequency, but in that case the lowpass filter must be substituted by a bandpass filter to eliminate the unwanted images in the first unused Nyquist zones.

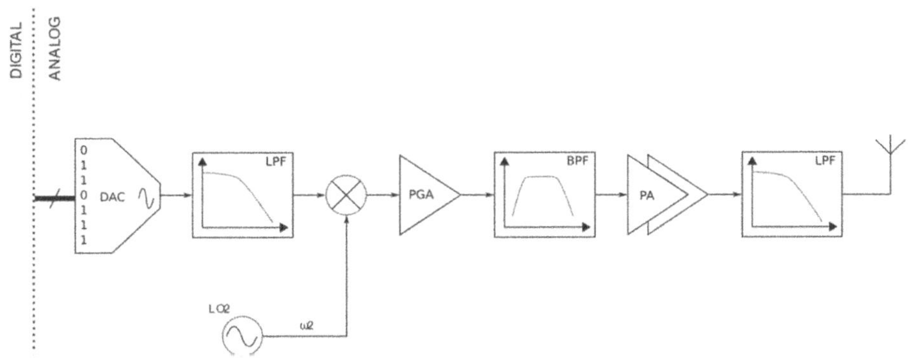

Fig. 3.13. Simplified diagram of a Real-IF transmitter

Data converters face a trade-off between resolution and speed. It is thus expected that a high-speed DAC will exhibit a quite reduced resolution. That may be adequate for some modulation schemes with reduced amplitude variations. Unfortunately, many modern high data-rate standards rely on modulation schemes like 64QAM where a medium resolution is mandatory.

3.3.3 *Polar transmitter*

Up to this point we have discussed some architectures that have been deduced form the Cartesian representation of the complex signal to be transmitted: one channel represents the real part of the signal (modulated by the in-phase signal) and the other the imaginary part (modulated by the quadrature signal). In both channels, the components of the signal are amplitude modulated.

However, it is also possible to resort to the polar representation of the complex signal. In that case, the signal must be amplitude- and phase-modulated. An implementation of a polar transmitter is sketched in Figure 3.14. In the digital domain, the complex signal is converted from Cartesian to polar coordinates. The phase is then differentiated in order to convert it into a frequency which can then be summed to the transmit frequency. An all-digital PLL is used to convert this time-dependent frequency into a frequency-modulated carrier that drives a mixer. On the other hand, the amplitude signal is generated by a DAC and sent to the other input of the mixer. In first order, the RF signal can thus be written:

$$\begin{aligned} s(t) &= \rho(t)\left(\cos\left(\omega_{TX} + \dot{\phi}(t)\right)t\right) \\ s(t) &= Re\{\rho(t)e^{-i\omega_{RX}t+\phi(t)}\} \end{aligned} \tag{3.11}$$

After the mixer, the amplified signal must be low-pass filtered to get rid of the distortion components generated at high frequency. In that case, we cannot properly speak of images around the carrier frequency harmonics, since there is not such carrier, but a frequency modulated signal. However, the intermodulation components are conceptually similar.

Strength of the architecture: The main benefit of the polar transmitter is that it can use an RF oscillator directly at the transmit frequency, since there is no need for generating quadrature phases.

In addition, the mixing operation can in some cases be implemented in a very compact way. If the dynamic range of the amplitude signal (*ρ(t)*) is not excessive, the mixing operation can be combined with the Power Amplifier itself. The PA takes as an input the frequency-modulated signal out of the PLL and its amplitude is modulated through the attenuation of its power supply. The linearity of such an

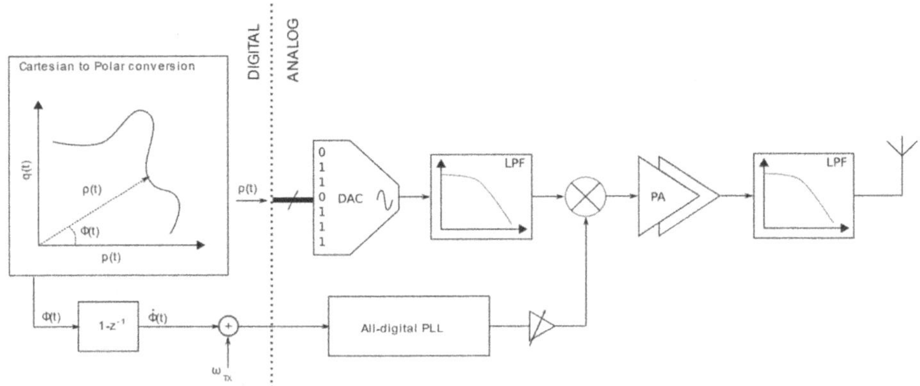

Fig. 3.14. Simplified diagram of a polar transmitter

approach is obviously limited, which restricts its use to the standards where the symbol constellation in the complex plane is not close to the origin.

Weakness of the architecture: The main drawback associated to polar transmitters is that, while the bandwidth of the amplitude may be of the same order as that of the complex signal itself, the bandwidth of the phase signal can be significantly larger. As an illustration to support that claim, consider the cases when the complex signal crosses 0. A phase change occurs abruptly and the differentiation (or discrete derivative) of such a step is a Dirac function, whose Fourier Transform is a complex exponential of infinite bandwidth. Obviously, the complex signal is not always close to zero and the phase is not always abruptly changing. However, this simple example shows that the phase signal does contain high frequency components. Such a wide bandwidth must thus be taken into account in the design of the PLL which must implement a wide tuning range.

Another significant drawback of polar transmitters is the time alignment of the two channels. Indeed, the amplitude and phase time-domain signals must be correctly aligned since a time offset would change the modulation. Actually, the inverse of the time offset should be significantly lower than the effective bandwidth of the phase signal. The timing alignment is also relevant for the IQ modulation but, since the signal paths of the two channels are similar, it is much easier to limit mismatching.

3.4 Receiver Architectures

3.4.1 *Direct conversion*

The *Direct Conversion* architecture, also known as "*Homodyne*" or "*Zero-IF*" receiver, translates the RF information to base-band using a single frequency down-conversion step, as shown in Figure 3.15. The frequency of the local oscillator (LO) is equal to center frequency of the channel of interest.

This approach produces that the non-symmetric spectra of the IQ modulation overlaps in baseband. However, as demonstrated in Eq. (3.5), the In-phase and Quadrature components can be ideally recovered without loss of information considering two mixing paths with 90° degrees phase shift between them[2].

[2] The distinction between IQ paths is often implicitly assumed in some of the transceivers in the bibliography [7,9,10]. For sake of simplicity, some simplified block diagrams show a single mixer.

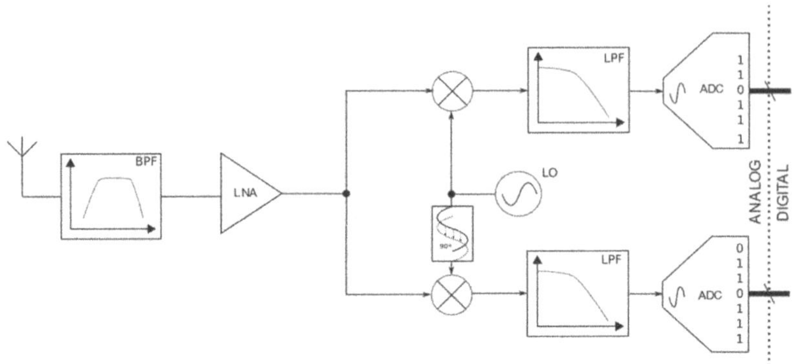

Fig. 3.15. Simplified diagram of a Homodyne receiver

3.4.1.1 *Pros and Cons*

Strength of the architecture: the main advantages of the architecture stem from the topology simplicity. The reduction of the number of processing stages in the receiver has notable impact on power consumption, area and performance in terms of sensitivity, noise figure and achievable linearity.

The direct conversion architecture has the additional benefit of bandwidth reduction for the base-band circuitry, since half the effective bandwidth at RF is needed after translation. This relaxes the circuit requirements of critical building blocks at baseband, such as ADC, Programmable Gain Amplifiers (PGAs) and Filters, which can be optimized toward low power consumption improving the linearity and noise features.

Weakness of the architecture: in a general wireless communication system, several drawbacks have been traditionally identified in the direct conversion architecture [2, 3], such as: LO leakage, greater sensitivity to IQ mismatching or degradation due to offset and flicker noise. Based on these limitations, early receivers compliant with the first releases of the IEEE 802.11 standard did not often consider this architecture, mainly due to former technological limitations [11, 16], or a priori prejudices.

At the present time, however, the direct conversion has been found to be the optimum selection in modern WLAN transceivers operating in the 2.4GHz/5GHz radio bands [7, 8, 9, 10, 12, 13, 14, 15]. Understanding the justification to this architecture change requires a good knowledge of some countermeasures at the design level [2, 3], in addition to some particularities of the IEEE 802.11standard itself that make the direct conversion particularly suitable.

3.4.1.2 *Non-ideal Effects and Countermeasures in the Architecture.*

LO Leakage: the coupling of the local oscillator (LO) with the antenna produces that part of the LO power is translated, as shown in Figure 3.16, to the receiver input as an additive disturbance on the primary input. This coupling is mainly produced by: 1) limited reverse isolation in the RF front-end (LNA and mixers), 2) substrate coupling, and 3) layout imbalance. Also, magnetic coupling between antenna and on-chip inductors in the VCO (Voltage Controlled Oscillator) and the LNA could contribute negatively [10].

The effect of the LO leakage is especially relevant when the VCO works at the same frequency as the mixers. This causes, according to Eq. (3.2), that the coupling is down-converted, and a DC component proportional to the amplitude of the disturbance is produced at the input of the base-band section contributing to the total offset. If this offset is not handled properly, it would be amplified by the back-end PGA eventually leading to saturation or significant receiver desensitization and distortion. DC offset compensation techniques based on AC coupling structures or feedback loops (either analog or mixed-signal) are mandatory in this scenario [2].

An alternative approach to reduce the impact of LO leakage considers a VCO working at twice the frequency of interest, from which the 90° out of phases for the mixers are derived using a frequency divider-by-2. This solution was formerly discarded in receivers operating in the 5GHz radio-band due to the difficulty of designing a VCO working at 10GHz with the required phase noise and tuning range trade-offs [11,16]. However, at the moment, this not a technological problem anymore [21], and this approach has become and standard the facto in modern WLAN receivers.

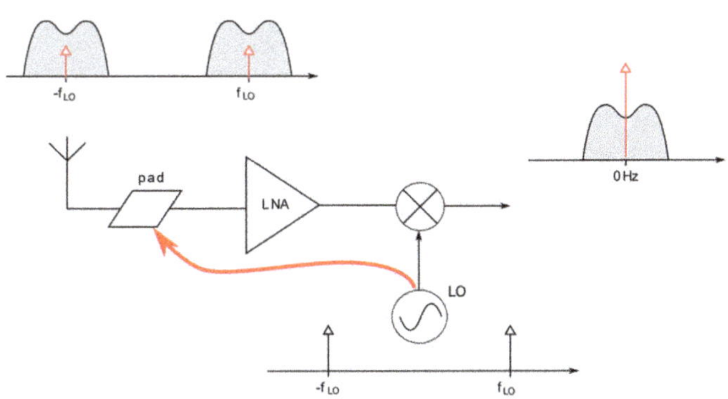

Fig. 3.16. Effect of LO leakage on the receiver

Offset and Flicker Noise of Building Blocks: In the same way as for the offset induced by the LO leakage, the Direct Conversion architecture is in a general situation quite sensitive to the block offset and flicker noise of the base-band section, since their power contributions could interfere with the signal of interest. This effect prevents the use of this architecture in communication systems where this contribution is not bearable, such as GSM. Nevertheless, in the conception of the IEEE 802.11 standard itself, these aspects were considered locating a null subcarrier at center of the channel (see example of Figure 3.4) to relax the offset and flicker requirements, making them easier to handle by design.

IQ Mismatching: the ideal expressions for the IQ demodulation in Eq. (3.2) assume a perfect matching between paths in the receiver. In practice, some non-ideal effects, such as the phase and gain imbalances (see example in Figure 3.17 for a QPSK case study), produce a deviation of the effective constellation with respect to the ideal positions, affecting to the achievable EVM. From the spectra point of view, this error can be understood as an uncontrolled interference in the information that folds in the baseband, according to the example in Figure 3.15. Phenomenologically, the quality of a receiver against the self-image mixing is characterized by the *Image Rejection Ratio* (IMR), defined as the power ratio between the desired signal and the spurs produced by the imbalance (typically, expressed in decibels).

This problematic increases with the order of modulation as the distance between constellation points is reduced. Actually, maximum throughputs in the IEEE 802.11ac standard impose stringent imbalance requirements which are above the technological limits of pure analog design. The situation is even more complex when increasing bandwidth, since frequency-dependent could be not negligible [7]. Luckily, the drawback can be overcome in the digital domain using calibration [7, 8, 9, 10]. Following this approach, IMR values below -50dB has been recently reported [7], safe enough even for 1024-QAM modulation in the expected 802.11ax standard release.

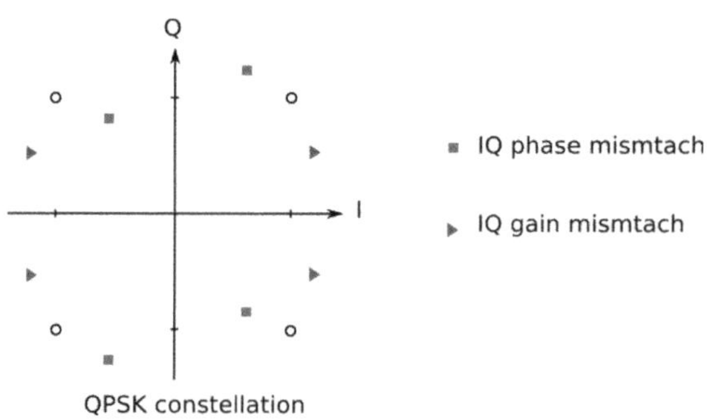

Fig. 3.17. Effect of IQ mismatching in a QPSK example

3.4.2 *Heterodyne receiver*

The heterodyne receiver performs the frequency translation of the RF signal in more than one step using several mixing stages. It is specially indicated in those applications with stringent requirements in terms of spectral purity around the center channel frequency, such as GSM, or in those technologies when the VCO dynamic response is not suitable for a direct conversion architecture [11,16]. Although obsolete for modern WLAN receivers operating in the 2.4GHz/5GHz radio bands[3], a good comprehension of this architecture is recommended in any RF design.

The key idea in the classical Heterodyne receiver overcomes the need of a high speed LO, using a mixing frequency which is lower f_{LO} than the center frequency of the channel of interest f_c. This selection translates the RF spectrum to a lower frequency, denoted by *Intermediate Frequency:* $f_{IF} = |f_c - f_{LO}|$, generating an up-converted version at $(f_{LO}+f_c)$ that needs to be later filtered out, as shown in Figure 3.18. The frequency shift can be evaluated considering the trigonometric identity, $2\cos(x)\cos(y) = \cos(x+y)+\cos(x-y)$, in the form,

$$m(t)\cdot\cos\omega_c t\cdot\cos\omega_{LO}t = \underbrace{\frac{m(t)}{2}\cdot\cos\left(\omega_c-\omega_{LO}\right)t}_{Down-conversion}+\underbrace{\frac{m(t)}{2}\cdot\cos\left(\omega_c+\omega_{LO}\right)t}_{Up-conversion} \quad (3.12)$$

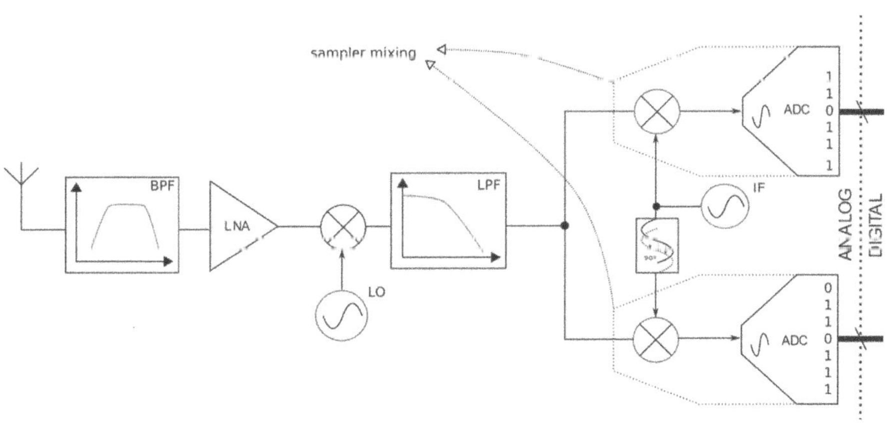

Fig. 3.18. Simplified diagram block of a Heterodyne receiver

[3] Even in the IEEE 802.11ad/ay extension working in the 60GHz ISM band, the Direct Conversion architecture [21] is nowadays the optimum choice to reduce power consumption and area. In [21], an insertion-locked VCO generated the 60GHz mixing signals in a standard 65nm CMOS process.

To arrive at the final base-band signal, the process can be iteratively performed. However, due to practical limitations, the number of frequency translations in the analog domain is effectively limited to one or two mixing stages. In the case of a single translation, the final base band signal will be estimated in the digital domain, after A/D conversion at *Intermediate Frequency*, using the sampling process in the ADC as an additional mixing stage.

From the mathematical point of view, taking into account the parity property of the cosine function in Eq. (3.12), there is not any restriction regarding the relative position between f_c and f_{LO} to achieve the wanted frequency displacement. Actually, whenever the distance $|f_c - f_{LO}|$ remains constant, it is possible to consider a frequency in the LO above the carrier, $f_{LO} > f_c$. Obviously, this approach, called in the bibliography "*High-side Injection*", does not relax block requirements, in contrast to the "*Low-side*" counterpart $f_{LO} < f_c$, the VCO dynamic requirements, but in some situation, as described below, it can result positive.

Input Image Band Issue: one of the main problems in the classical heterodyne receiver comes from the spectral content in the input image band. Notice, as shown in Figure 3.19, that any signal with a spectrum symmetrically located around the LO frequency would be translated by the mixer to the same *Intermediate Frequency* corrupting the desired information.

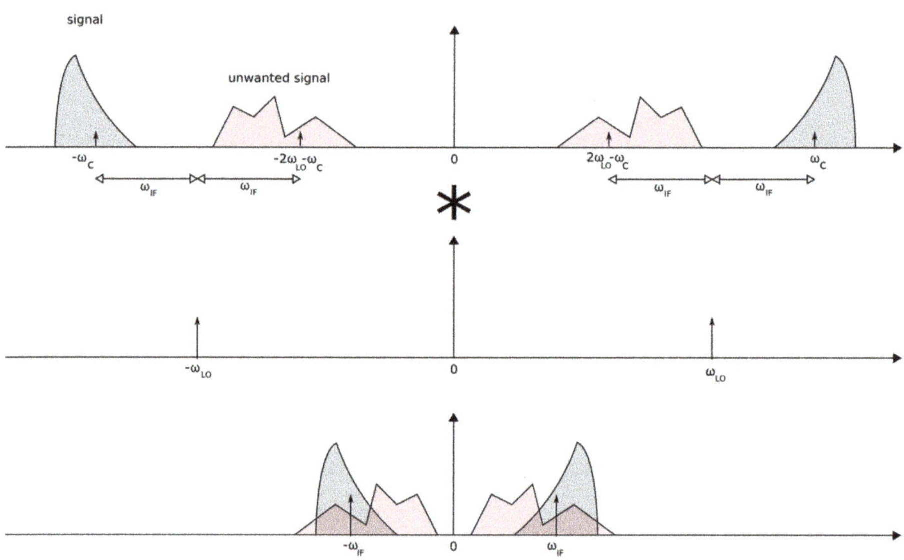

Fig. 3.19. Problems of image band in Heterodyne receivers

To demonstrate this behavior, let us assume that the effective signal at the input of the receiver, $x_{RF}(t)$, has two different contributions: 1) one related to the wanted message $m(t)$, and 2) an additive term modelling the interference at the image band $s(t)$,

$$x_{RF}(t) = m(t) \cdot \cos \omega_c t + s(t) \cdot \cos(\omega_{LO} \pm \omega_{IF}) t \tag{3.13}$$

Under these conditions, we can demonstrate similarly to Eq. (3.12) that the output of the mixer will have a disrupting term associated to the image at the intermediate frequency,

$$x_{RF}(t) \cdot \cos \omega_{LO} t = \underbrace{\frac{1}{2}\big(m(t) + s(t)\big) \cdot \cos \omega_{IF} t}_{Down-conversion} + \underbrace{f_{UP}(m(t), s(t))}_{Up-conversion} \tag{3.14}$$

Traditionally, this problem was overcome, as shown in Figure 3.20, considering an image rejection filter at the input of the mixer. Bulky off-chip components were needed to achieve the required selectivity at high-frequency, with the consequent drawbacks in terms of integrability [17]. To make monolithic approach possible, modern heterodyne receivers eliminate the image rejection at the input of the mixer following different strategies [2,3]. One possibility consists in selecting a LO frequency [11], either "*Low-side*" or "*High-side*" Injections, which assures that the image band falls in a region of the output spectrum without much activity — for instance using the reserved frequency slots for low-noise satellite communications. Unluckily, this option has a limited applicability in WLAN due to the crowded spectrum allocations, which are periodically subjected to assignment revisions depending on the country regulations [18]. An alternative way to deal with the image problem takes advantage of the phase difference between the desired IQ and the image band signals to cancel out the undesired component. This approach, known in the bibliography as image reject receivers [2,3,19], is used in different approaches, such as Hartley, Weaver and complex filter-based architectures.

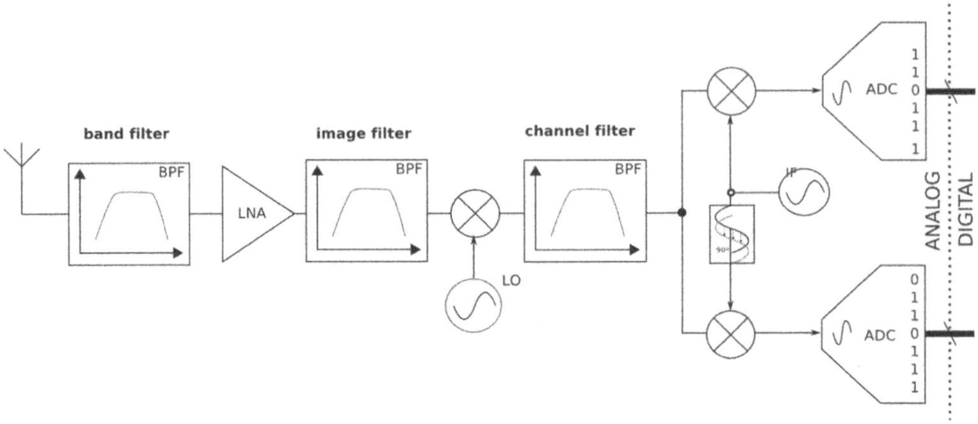

Fig. 3.20. Simplified block diagram of a Heterodyne receiver using off-chip image rejection filters

Image Reject Receivers: the key concept in all these architectures takes advantage of the phase difference between the wanted signal and the image component, as a result of the original symmetry around the LO frequency. Under these conditions, the quadrature receiver makes it possible to postpone the image signal suppression after the down-conversion process (without a dedicated image filter at high frequency), when both the wanted and the image signals are situated at lower intermediate frequency.

Let us illustrate the concept considering the "Low-side Injection" in the Harley receiver of Figure 3.21 as case study. In this circuit, a 90 degrees phase shift is intentionally introduced after the low pass filters in one of the IQ paths. According to Eq. (3.14), the signal at node A can be expressed as,

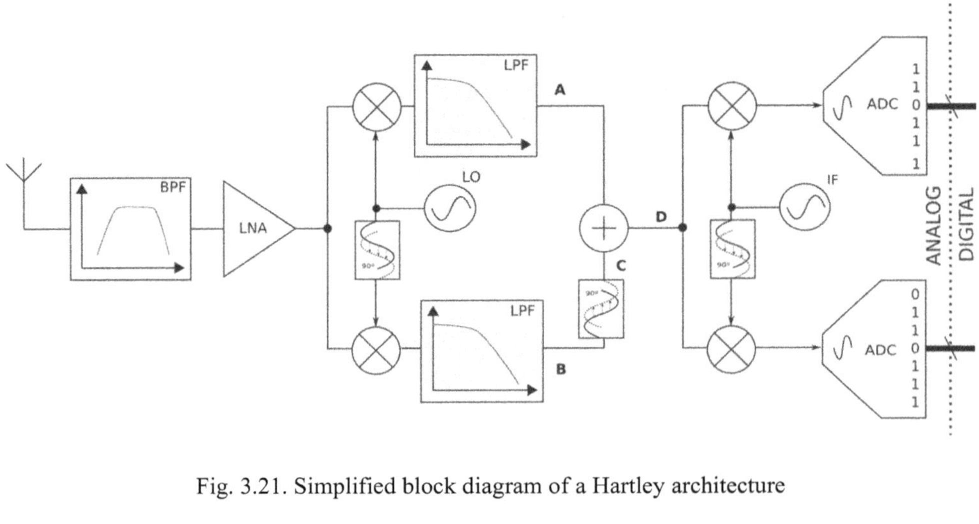

Fig. 3.21. Simplified block diagram of a Hartley architecture

$$x_A(t) \approx \frac{1}{2} \cdot \left(m(t) + s(t)\right) \cdot \cos \omega_{IF} t \tag{3.15}$$

In the same conditions, the resulting signals at node B and C become,

$$\begin{aligned} x_B(t) &\approx \frac{1}{2} \cdot \left(-m(t) + s(t)\right) \cdot \sin \omega_{IF} t \\ x_C(t) &\approx \frac{1}{2} \cdot \left(+m(t) - s(t)\right) \cdot \cos \omega_{IF} t \end{aligned} \tag{3.16}$$

which leads to suppression of the image band interference when combined the information of the two paths at node D,

$$x_C(t) = x_A(t) + x_D(t) \approx m(t) \cdot \cos \omega_{IF} t \tag{3.17}$$

Notice that in practical the 90º difference can be effectively implemented introducing ±45º phase degrees between the two signal paths using an RC and CR ladders, respectively. The concept can be generalized combining the low-pass filter with the delay based on either poly-phase or complex filter topologies [20].

Strength of the architecture: as summary of the heterodyne receiver, the architecture is specially indicated for applications in which the offset and flicker noise requirements in the center of the channel make not feasible a direct conversion scheme. The architecture also offers advantages in applications where the RF band of interest is outside the technological limit for LO frequency synthesis, being relatively immune to LO leakage with a proper selection of the intermediate frequency.

Weakness of the architecture: the drawback in the architecture stems, as previously highlighted, from the extra complexity of the receiver channel with additional blocks contributing to the total die area, power consumption and error budget. In addition, the architecture is sensitive to the RF spectral contents at the image band[4], requiring some post-processing to minimize its impact.

Among the Image-rejection architectures, the complex filter approach [20] is specially indicated in quadrature receivers with a low-IF selection as optimum trade-off between signal bandwidth, image rejection and immunity against offset and flicker noise. This solution, however, is not competitive in modern WLAN

[4] The optimum selection of the mixing frequencies should consider the contribution of the main and higher order harmonics (called "Mixing Spurs" in the bibliography) in the all the down-conversion steps.

receivers, since it duplicates the signal bandwidth of the baseband section with respect the Direct Conversion approach. Moreover, the linearity constrains for the mixer and input filter increases since the effective dynamic range should be dimensioned to deal with the power associated to the image band. All these constrains, in combination with the more complex frequency generation scheme to generate the two mixing signals, makes at the present time the Heterodyne receiver uncompetitive with respect to the Direct Conversion architecture in WLAN receivers.

References

1. IEEE Std 802.11TM-2016 (2016), IEEE Standard for Information technology—Telecommunications and information exchange between systems—Local and metropolitan area networks—Specific requirements—Part 11: Wireless LAN Medium Access Control
2. Razavi, B. (2011) *RF Microelectronics*, 2nd Ed. (Prentice Hall Press, USA)
3. Lee, T. H. (2003) *The Design of CMOS Radio-Frequency Integrated Circuits*, (Cambridge University Press).
4. National Instruments, *Introduction to Wireless LAN Measurements. From 802.11a to 802.11ac*, Available at: www.ni.com.
5. Keysight Technologies (2017), *IEEE 802.11 Wireless LAN PHY Layer (RF) Operation and Measurement, Application Note.* (Keysight Technologies, USA).
6. Rohde & Schwarz (2004), *WLAN Tests According to Standard 802.11a/b/g. Application Note: 1MA69*, Available at: www.keysight.com
7. Kawai, S. *et al.* (2018). An 802.11ax 4x4 high-efficiency WLAN AP transceiver SoCs supporting 1024-QAM with frequency-dependent IQ calibration and integrated interference analyzer, *IEEE Journal of Solid-State Circuits*, 53, pp. 3688–3699.
8. Yan, S.T. *et al.* (2017). An 802.11a/b/g/n/ac WLAN transceiver for 2×2 MIMO and simultaneous dual-band operation with +29 dBm P_{sat} integrated power amplifiers, *IEEE Journal of Solid-State Circuits*, 52, pp. 1798–1813.
9. Chen, T.M. *et al.* (2017). 7.1 An 802.11ac dual-band reconfigurable transceiver supporting up to four VHT80 spatial streams with 116fsrms-jitter frequency synthesizer and integrated LNA/PA delivering 256QAM 19dBm per stream achieving 1.733Gb/s PHY rate, *Proc. IEEE International Solid-State Circuits Conference, ISSCC*, pp. 126–127.
10. Chung, Y.-H. *et al.* (2015). A dual-band 802.11abgn/ac transceiver with integrated PA and T/R switch in a digital noise controlled SoC, *Proc. IEEE Custom Integrated Circuits Conference, CICC*, pp. 1–8.

11. Han, D.O. *et al.* (2011). Fully integrated dual-band transceiver for IEEE 802.11a/b/g/j/n wireless local area network applications with hybrid up/down conversion architecture, *IET Circuits, Devices & Systems*, 5, p. 433–441.
12. Ingels, M. *et al.* (2010). A 5 mm^2 40 nm LP CMOS transceiver for a software-defined radio platform, *IEEE Journal of Solid-State Circuits*, 45, pp. 2794–2806.
13. Debaillie, B. *et al.* (2009). Calibration of direct-conversion transceivers, *IEEE Journal of Selected Topics in Signal Processing*, 3, pp. 488–498.
14. Degani, O. *et al.* (2008). A 1×2 MIMO multi-band CMOS transceiver with an integrated front-end in 90nm CMOS for 802.11a/g/n WLAN applications, *Proc. IEEE International Solid-State Circuits Conference, ISSCC*, pp. 356–619.
15. Afsahi A. *et al.* (2008). A low-power single-weight-combiner 802.11abg SoC in 0.13 μm CMOS for embedded applications utilizing an area and power efficient cartesian phase shifter and mixer circuit, *IEEE Journal of Solid-State Circuits*, 43, pp. 1101–1118.
16. Leung Lai Kan L. *et al.* (2007). A 1-V 86-mW-RX 53-mW-TX single-chip CMOS transceiver for WLAN IEEE 802.11a, *IEEE Journal of Solid-State Circuits*, 42, pp. 1986–1998.
17. Rudell, J.C. *et al.* (1997). A 1.9-GHz *Wide-Band* IF double conversion CMOS receiver for cordless telephone applications, *IEEE Journal of Solid-State Circuits*, 32, pp. 2071–2088.
18. Federal Communications Commission Office of Engineering and Technology Policy and Rules Division (2019). *FCC ONLINE TABLE OF FREQUENCY ALLOCATIONS*, Available at: www.fcc.gov.
19. Samavati, H. *et al.* (2000). A 5-GHz CMOS wireless LAN receiver front-end, *IEEE Journal of Solid-State Circuits*, 35, pp. 1198-1204.
20. Martin K.W. (2004). Complex signal processing is not complex, *IEEE Transactions on Circuits and Systems I*, 51, pp. 1823-1836.
21. Wu R. *et al.* (2017). 64-QAM 60-GHz CMOS transceivers for IEEE 802.11ad/ay, *IEEE Journal of Solid-State Circuits*, 52, pp. 2871–2891.

Part 2

Modeling

https://doi.org/10.1142/9789811210723_0004

Chapter 4

System Models

Zoran Stamenković
IHP - Leibniz-Institut für innovative Mikroelektronik
Frankfurt (Oder), Germany

Bojan Dimitrijević and **Mile Stojčev**
University of Niš
Niš, Serbia

4.1 Introduction

First, a human language description of the main system functions is necessary. This simple description needs translation to a system model understandable to computers and electronic design automation (EDA) tools. The system model has to capture the intent of the design and define what the design should do.

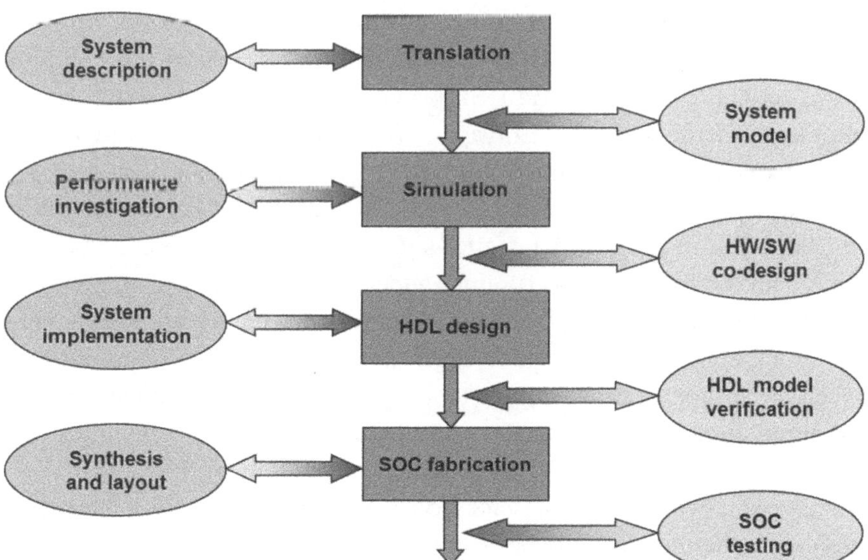

Fig. 4.1. System design and implementation flow

It allows designers to understand the implications of their choices and decisions, describes system behavior and performance, defines system architecture and framework for implementation, as well as verifies all the requirements regarding performance, energy consumption, cost, and maintenance effort. Using such a model, designers can consider different design styles and partition the system into hardware and software components doing hardware/software co-design [1-27]. Figure 4.1 shows a block diagram of the system design and implementation flow.

Hardware/software co-design investigates the concurrent design of hardware and software components of complex electronic systems. It tries to integrate hardware and software design techniques in a consistent system-level methodology allowing higher security (specifications and developments are verifiable), higher efficiency (cost analyzable) and better automation (through system synthesis). The era of hardware/software co-design started in 1985 when Smith et al. [2] described a computer-aided design system developed to support the co-design of hardware and software architectures for high-performance digital-signal processors (DSP) based on a directed graph methodology. About eight years later, Gupta and De Micheli [3] demonstrated the feasibility of synthesizing heterogeneous systems by using timing constraints to delegate tasks between hardware and software so that performance meets the requirements. In 2000, Keutzer et al. [10] introduced architecture-function communication-based co-design platforms applying them to the design of wireless systems. In the next several years, hardware/software co-design became a recognized research field and moved from an emerging discipline to a mainstream technology [13]. Up to date, many authors dealt with this technology pushing its limits to a standard in the field of electronic system design. Hosemann et al. [14] increased the computational power of an application-specific DSP providing multiple data paths operating in parallel in a Viterbi decoder. Jerraya and Wolf [15] defined a long-term roadmap for the successful integration of components in heterogeneous multiprocessors. Wiangtong et al. [16] introduced a systematic approach to hardware/software co-design targeting data-intensive digital-signal processing applications. Sung and Park [17] analyzed timing constraints and processing time of major functional components of subscriber station and determined which components should be implemented in hardware to support high-speed TDMA MAC operations. Monton et al. [18] presented a set of plug-ins to an open-source CPU emulator that enables mixed simulations between platform emulators and hardware modules described in SystemC. Kim et al. [19] proposed a novel hardware/software co-design technique of network protocol stacks on a system-on-chip (SOC) platform that should guarantee the real-time performance of periodic packets and minimize the average response time of aperiodic packets. Bertels et al. [20] developed a heterogeneous multicore platform that selects the best hardware configuration for mapping the application and modifies that application so that different system parts execute on the most appropriate hardware component. Shalf et al. [21] discussed hardware/software co-simulation and code analysis tools that facilitate deeper collaboration between hardware architects and application teams as an essential component of the co-design process. Khan and Kundu [22] proposed an on-line testing scheme for detecting and debugging hard errors in multiprocessors based on a hardware/software virtual machine concept. Teich [23] reviewed the importance of hardware/software co-design techniques used extensively to develop embedded electronic systems for automobiles, industrial automation, avionics, mobile devices, and home appliances. Sredojevic et al. [24] looked across the design stack of a numerical accelerator, from algorithm formulation down to accelerator microarchitecture, and showed a high degree of flexibility and design reuse without significant performance sacrifice. El-Moursy et al. [25] presented a complete framework and methodology (virtual platform) for design, simulation, and debugging of large SOCs. Drozdenko et al. [26, 27] introduced a method for modeling a generic orthogonal frequency division multiplexing (OFDM) wireless transceiver on the Zynq system-on-chip by decomposing the standard specifications into a set of functional blocks used in multiple protocols. Li et al. [28] proposed a methodology that facilitates development and integration of signal processing hardware and software modules involving heterogeneous programming languages and platforms. The proposed design process was exercised to implement a vehicle classification algorithm using dataflow-based deep neural networks (DNN).

Dedicated high-level modeling tools such as Matlab/Simulink [29] and Catapult [30] and system-level design languages such as SystemC [31] and SystemVerilog [32] are able to support hardware/software co-design. These software tools include also automatic generators for hardware description language (HDL) code (for hardware implementations) and high-level programming language (most often C/C++) code (for software implementations). Criteria for partitioning the system into hardware and software components are usually based on the required performance (data rate and latency), functionality (simple bit operations of the data encryption, error detection and correction, and baseband processing need hardware implementations), and flexibility (reprogramming of a hardware implementation is not or hardly possible). This approach can yield substantial reductions in time-to-market for embedded systems. The complete design is specified, partitioned, developed and debugged all without fabricating a real prototype. The generated model is further verified using C/HDL co-simulation. It ensures the correct functionality of the system as a whole and enables a concurrent development of software and hardware system components. An effective environment for hardware/software co-simulation must combine tools familiar to both the software and hardware engineers and capable of design performance and cost evaluation. There are companies that provide such solutions and commercial products [33-36].

The system models provide the ability to evaluate rapidly architectures based on measured results, and to perform hardware/software co-simulation that provides engineers with the accurate data needed for optimal decision-making at the system level against known constraints. The additional ability to execute real system software prior to building any hardware ensures that systems will meet their functional and performance requirements the first time they are built. Software developers can begin testing and debugging their work while the hardware is being designed and built. Each team member has a copy, so they can concurrently evaluate and debug their portions of the system design resulting in a shorter development cycle.

Experimental results have shown that hand written HDL (such as VHDL and Verilog) models consume significantly less hardware resources (digital logic). However, the hardware models generated by system level design tools can be optimized and perform close to their hand written counterparts. If the system is large in size and complex in nature, generating hardware by high-level modeling tools provides an adequate compromise between system development time and performance.

After the design is verified (Part 4, Verification), the synthesized net-list and physical layout can be generated. The next step is system-on-chip fabrication and packaging (Part 3, Implementation and Integration). Finally, the fabricated chip can be electrically tested (Part 4, Testing).

4.2 WLAN MAC Model

In this section, we present a strategy of hardware/software partitioning of IEEE 802.11 MAC functionality and a complete system solution for the MAC sublayer. The IEEE 802.11 MAC sublayer provides functionality to allow reliable data delivery for the upper layers over the wireless medium with data rate up to 54 Mbps. It specifies how a computer on the wireless network gains access to receiving and transmitting data, and how to maintain the already established communication. The basic medium access control method used in the IEEE 802.11 standard is a carrier sense multiple access with collision avoidance (CSMA/CA).

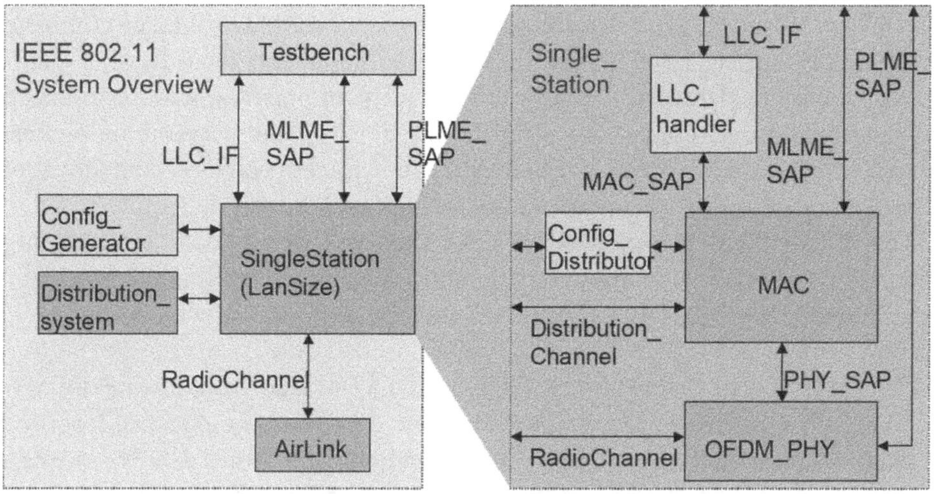

Fig. 4.2. Structure of a system level WLAN model

The first step is development of a detailed simulation model of the IEEE 802.11 MAC protocol using a system description language [37]. This model can be used to verify functional correctness of the proposed solution. On the other hand, it can serve as a basis for performance investigations to identify those parts of the protocol that require either software optimization (hand-optimized C code) or implementation in hardware in order to meet the real-time requirements. Figure 4.2 shows a top-level structure of the model.

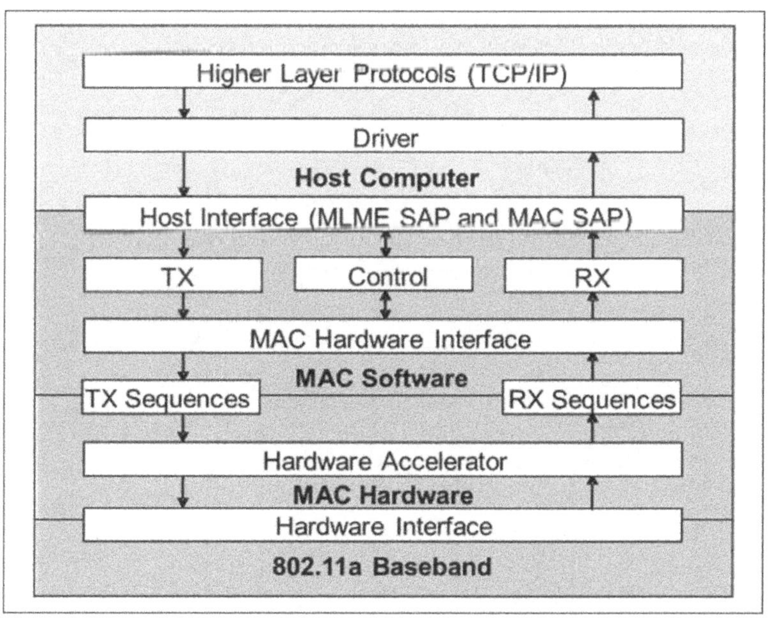

Fig. 4.3. Hardware/software partitioning of MAC sublayer

A sophisticated hardware/software partitioning of the MAC sublayer eliminated performance bottlenecks (Figure 4.3). Software implements the following control blocks having low data rate, complex functionality, or potential for later extension and modification: 1) connection setup between access point and mobile clients with administration of power-save modes; 2) authentication and de-authentication; 3) association and disassociation; and 4) transmit and receive data queues.

The architectural blocks necessary for transport of payload-data with high throughput and relatively simple functionality need hardware implementation: 1) cyclic redundancy check (CRC) and RC4-encryption and 2) acknowledgements and retransmissions.

The hardware accelerator functions (Figure 4.4) include the hardware driver for the MAC-PHY interface, retrieval of Rx frame data from the physical layer byte by byte, address filtering and CRC check, and storage of the data by means of direct memory access. Moreover, the accelerator retrieves Tx frame data from a memory location, calculates and appends the check sum, and pushes the data to the physical layer. It indicates a successful reception or transmission of a frame by an interrupt, handles received and transmitted beacons with respect to timing in the WLAN cell, and extracts information on channel time allocations. The hardware accelerator also manages acknowledgments, RTS, and CTS frames. Finally, the hardware accelerator conducts the back-off procedure in the contention access period, and calculates the actual duration of a frame transmission based on its payload length and data rate.

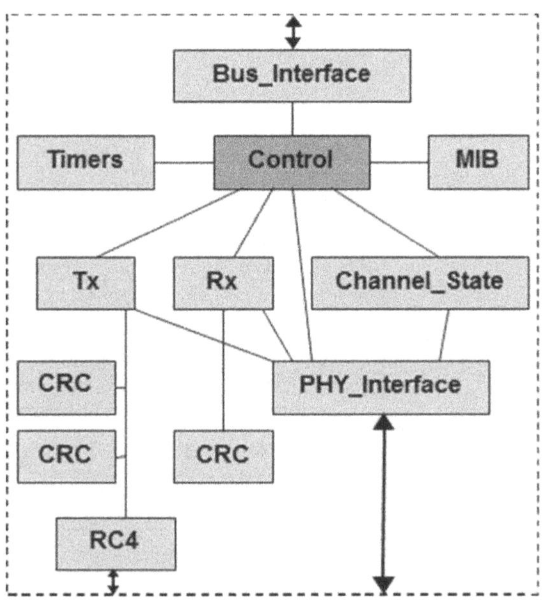

Fig. 4.4. MAC hardware accelerator

A single RTS frame is sent before each multicast transmission, while an RTS/CTS pair is generated before each unicast transmission. The RTS/CTS frames allow setting the optimal weights before the main frame transfer starts. They also define the duration of the following frame and acknowledgment (V-CS). The duration information from RTS/CTS frames is used to indicate the busy state until the end of frame transmission.

A system model has been developed and a C model of the protocol has been generated and compiled for the target processor. Simulations revealed that the system model and generated C code are not optimized in respect to timing requirements (which causes heavy processor overloads when executing the software). Therefore, some parts of the automatically generated C code have been replaced by a hand-optimized code. This speeds up the execution, but due to the tight timing requirements of the protocol, the processor running the software would have to be clocked at nearly 1 GHz in order to comply with the standard. This clock frequency (and energy consumption) is not affordable in mobile devices. Therefore, some of the timing-critical MAC functionality has to be implemented as a dedicated hardware. Based on profiling information from the C implementation and analysis of the real-time requirements specified in the standard, a hardware/software partitioning of the MAC architecture has been conducted. Among the timing-critical algorithms implemented in hardware are CRC calculation, RC4 encryption and decryption, as well as a large part of the frame reception process including the address filtering and acknowledgment generation. Rest of the functionality (which is not timing critical) has been implemented in software.

4.3 WLAN Baseband Model

We describe the WLAN digital baseband as a Matlab simulation model, which includes basic building blocks (such as scrambler/descrambler, encoder/decoder, IFFT/FFT, interleaver/deinterleaver, etc.) and testbenches for their verification. Such a model provides designers with a means for fast system simulation and development. The model's most important features are the following:

- It implements a scrambler/descrambler, which provides frequent transitions modifying periodic data or repetitive input to avoid very little change in the transmitted signal.
- It models the convolutional encoding, puncturing, depuncturing, and Viterbi decoding that combat the channel imperfections and correct the bit errors by adding redundancy to information packets before transmission.

- It includes interleaver/deinterleaver, which combats burst errors by reordering the bits at the sender end according to a particular scheme and restoring the original order at the receiver end.
- It also includes constellation mapper/demapper, which assigns individual data symbols to individual sub-carriers allowing multiple users to share the radio channel.
- It models pilot insertion, which mixes up dedicated pilot sub-carriers with data sub-carriers and pulse shaping, which reduces inter-carrier interference.
- It implements IFFT and FFT algorithms, which secure high data rates and avoidance of inter-symbol interference by partitioning the broadband into many orthogonal sub-carriers that transmit data in parallel.
- It includes time and frequency synchronization that corrects the symbol timing offset, carrier frequency offset, sampling clock offset, and carrier phase error.
- It implements a channel estimation, which provides information related to the signal distortion when it propagates through the channel.

Figure 4.5 presents a block diagram of the digital baseband Matlab model. White blocks denote the conventional IEEE802.11a baseband processing steps. On the other hand, gray blocks represent simulation testbenches including the input data modeling, performance analysis, and channel simulation. Chapter 5 describes specific Matlab models of the WLAN baseband processors and corresponding simulation testbenches.

A Matlab model can use the floating-point operations to implement the baseband blocks. Floating-point values are of a large dynamic range and high precision but require more hardware resources and energy compared to fixed-point values. Therefore, the floating-point toolbox of Matlab can be helpful to obtain an upper bound on the expected performance of digital baseband, but can be too complex for hardware implementation.

A simpler fixed-point Matlab model, which emulates the baseband signals and simulates the fixed-point hardware operations, is usually sufficient. The main properties of fixed-point variables are signedness (in the proposed model they are signed) and precision (the word and fraction lengths are in bits). Other properties associated to the fixed-point operations are the round mode (‘floor’ or ‘nearest’) and the overflow mode (‘wrap’ or ‘saturate’). The developed model can help to evaluate and modify these properties. This fixed-point model is bit-to-bit identical to the final hardware model.

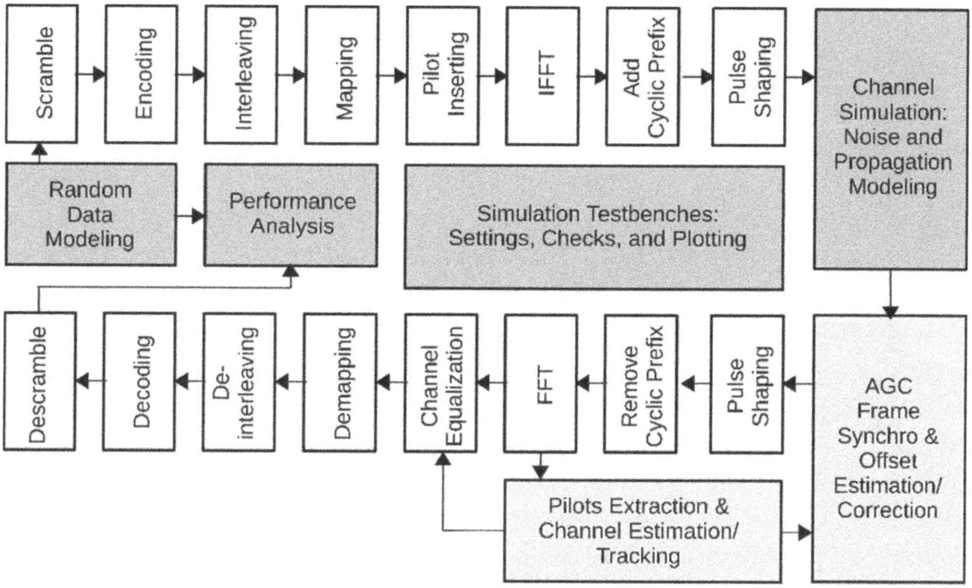

Fig. 4.5. Matlab model of an OFDM digital baseband

Figure 4.6 depicts an analog front-end capable of performing the intended OFDM operations for the IEEE802.11a standard. It is a direct conversion architecture, for both the transmitter and receiver. For the receiver chain, the incoming standard conform signals from the antennas have a signal power of -82 to -30 dBm. A bandpass filter (BPF) and a low noise amplifier (LNA) filter and amplify these signals. The following down-conversion path has to handle an effective dynamic range increased by approximately 12 dB. Thus, a passive down-conversion mixer driven by a phase locked loop (PLL) controls an 11 GHz voltage controlled oscillator (VCO). The last analogue stages incorporate variable gain and low pass filtering features, in order to set up optimal signal conditions for the receive analogue to digital converter (ADC). The transmit path (Tx) makes use of components similar to those the Rx such as parts of the baseband (BB), variable gain amplifier (VGA), and filter. An active up-conversion mixer driven by the PLL feeds its output signal to the Tx branch. The power amplifier (PA) and antenna switch form the last part of the Tx branch.

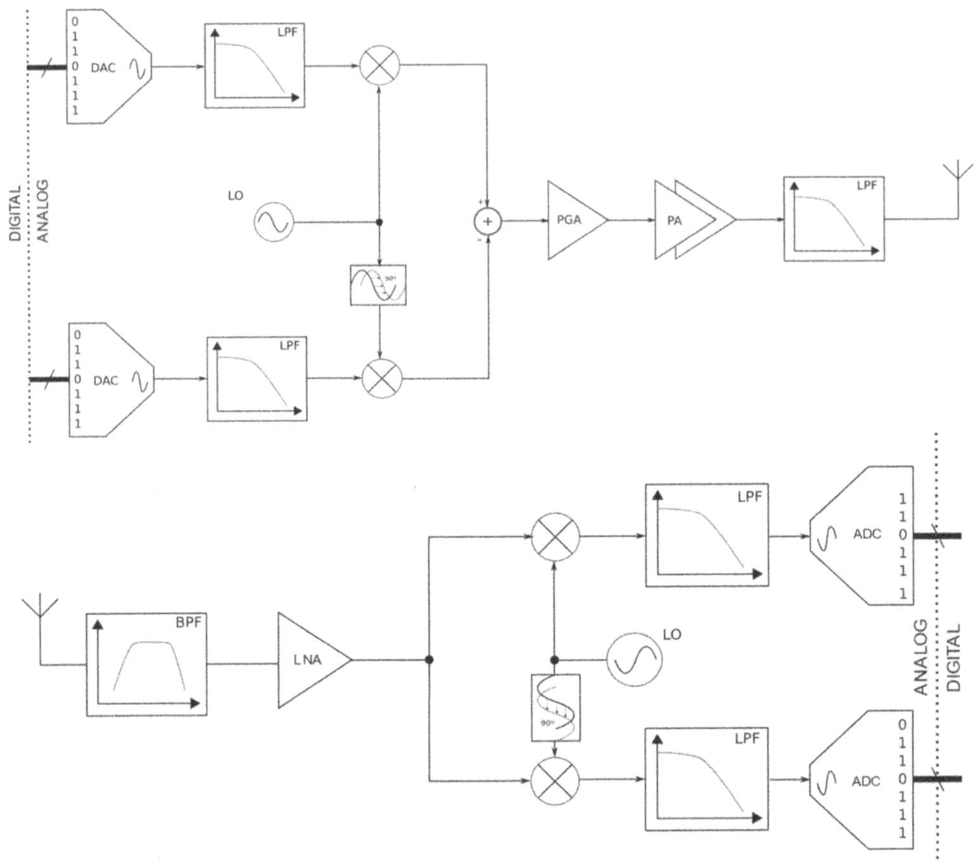

Fig. 4.6. Transmitter and receiver of a direct-conversion analog front-end

A fully differential topology of every building block secures the optimal signal-processing in RF domain. Virtual elimination of ground and supply nets reduces the risk and design time when assembling complex systems. Typical system-level models rely on Matlab™ Simulink [29]. Besides the "straightforward" modeling of high-level mathematical functions (gain, phase shifter, polynomial function, etc.), it provides a Communication System Toolbox that contains some valuable building blocks. Modeling the non-idealities of the front-end is cumbersome. The difficulty does not reside in the complexity of the architecture or the number of building blocks. The operation principles are easy to understand too. However, modeling the non-idealities of these building blocks with care is necessary to reach an accuracy sufficient for design optimization.

4.4 What is Next?

Next chapters summarize and describe models of digital (Chapter 5), analog and RF (Chapter 6), mixed-signal (Chapter 6), and embedded memory (Chapter 7) components and circuits typically used in an OFDM WLAN transceiver.

References

1. Stabler, E. P. (1970). System description languages, *IEEE Transactions on Computers*, C-19, pp. 1160–1173.
2. Smith, C., Frank, G. and Cuadrado, J. (1985). An architecture design and assessment system for software/hardware codesign, *Proc. 22nd Design Automation Conference*, pp. 417–424.
3. Gupta, R. and De Micheli, G. (1993). Hardware-software cosynthesis for digital systems, *IEEE Design & Test of Computers*, 10, pp. 29–41.
4. Mohanty, S. and Wilsey, P. A. (1995). System modeling, hardware-software codesign, and mixed modeling with hardware description languages, *Proc. European Design Automation Conference*, pp. 322–327.
5. Debardelaben, J. A. and Madisetti, V. K. (1996). Hardware/software codesign for signal processing systems - A survey and new results, *Proc. 29th Asilomar Conference on Signals, Systems and Computers*, pp. 1316–1320.
6. Rajesh, V. and Moona, R. (1999). Processor modeling for hardware software codesign, *Proc. 12th International Conference on VLSI Design*, pp. 132–137.
7. Saul, J. M. (1999). Hardware/software codesign for FPGA-based systems, *Proc. 32nd Hawaii International Conference on System Sciences*, pp. 1–10.
8. Slomka, F., Dorfel, M., Munzenberger, R. and Hofmann, R. (2000). Hardware/software codesign and rapid prototyping of embedded systems, *IEEE Design & Test of Computers*, 17, pp. 28–38.
9. Song, L., Parhi, K. K., Kuroda, I. and Nishitani, T. (2000). Hardware/software codesign of finite field datapath for low-energy Reed-Solomon codecs, *IEEE Transactions on Very Large Scale Integration Systems*, 8, pp. 160–172.
10. Keutzer, K., Malik, S., Newton, R., Rabaey, J. and Sangiovanni-Vincentelli, A. (2000). System level design: Orthogonalization of concerns and platform-based design, *IEEE Transactions on Computer-Aided Design of Circuits and Systems*, 19, pp. 1523–1543.
11. Baghdadi, A., Zergainoh, N.-E., Cesario, W. O. and Jerraya, A. A. (2002). Combining a performance estimation methodology with a hardware/software codesign flow supporting multiprocessor systems, *IEEE Transactions on Software Engineering*, 28, pp. 822–831.

12. Cote, C. and Zilic, Z. (2002). Automated SystemC to VHDL translation in hardware/software codesign, *Proc. 9th IEEE International Conference on Electronics, Circuits and Systems*, pp. 717–720.
13. Wolf, W. (2003). A decade of hardware/software codesign, *Computer*, 36, pp. 38–43.
14. Hosemann, M., Habendorf, R. and Fettweis, G. P. (2003). Hardware-software codesign of a 14.4 MBit - 64 state - Viterbi decoder for an application-specific digital signal processor, *Proc. IEEE Workshop on Signal Processing Systems*, pp. 45–50.
15. Jerraya, A. A. and Wolf, W. (2005). Hardware/software interface codesign for embedded systems, *Computer*, 38, pp. 63–69.
16. Wiangtong, T., Cheung, P. Y. K. and Luk, W. (2005). Hardware/software codesign: A systematic approach targeting data-intensive applications, *IEEE Signal Processing Magazine*, 22, pp. 14–22.
17. Sung, N. W. and Park, N. H. (2005). Hardware/software codesign of TDMA MAC applied to IEEE 802.16 subscriber station, *Proc. 16th IEEE Internat. Symposium on Personal, Indoor and Mobile Radio Communications*, pp. 2670–2674.
18. Monton, M., Portero, A., Moreno, M., Martinez, B. and Carrabina, J. (2007). Mixed SW/SystemC SoC emulation framework, *Proc. IEEE International Symposium on Industrial Electronics*, pp. 2338–2341.
19. Kim, Y., Kim, T., Park, E. K. and Tak, S. (2009). A novel hardware-software codesign of network protocol stacks on a SoC platform, *Proc. 18th Internat. Conference on Computer Communications and Networks*, pp. 1–6.
20. Bertels, K., Sima, V.-M., Yankova, Y., Kuzmanov, G., Luk, W., Coutinho, G. Ferrandi, F., Pilato, C., Lattuada, M., Sciuto, D. and Michelotti, A. (2010). HArtes: Hardware-software codesign for heterogeneous multicore platforms, *IEEE Micro*, 30, pp. 88–97.
21. Shalf. J., Quinlan, D. and Janssen, C. (2011). Rethinking hardware-software codesign for exascale systems, *Computer*, 44, pp. 22–30.
22. Khan, O. and Kundu, S. (2011). Hardware/software codesign architecture for online testing in chip multiprocessors, *IEEE Transactions on Dependable and Secure Computing*, 8, pp. 714–727.
23. Teich, J. (2012). Hardware/software codesign: The past, the present, and predicting the future, *Proceedings of the IEEE*, 100, pp. 1411–1430.
24. Sredojevic, R., Wright, A. and Stojanovic, V. (2013). Hardware-software codesign for embedded numerical acceleration, *Proc. 21st IEEE Annual International Symposium on Field-Programmable Custom Computing Machines*, pp. 169–172.
25. El-Moursy, M. A., Sheirah, A., Safar, M. and Salem A. (2014). Efficient embedded SoC hardware/software codesign using virtual platform, *Proc. 9th International Design and Test Symposium*, pp. 36–38.

26. Drozdenko, B., Zimmermann, M., Dao, T., Chowdhury, K. and Leeser, M. (2016). Modeling considerations for the hardware-software co-design of flexible modern wireless transceivers, *Proc. 26th International Conference on Field Programmable Logic and Applications*, pp. 1–4.
27. Drozdenko, B., Zimmermann, M., Dao, T., Chowdhury, K. and Leeser, M. (2018). Hardware-software codesign of wireless transceivers on Zynq heterogeneous systems, *IEEE Trans. on Emerging Topics in Computing*, 6, pp. 566–578.
28. Li, L., Saub, C., Fanni, T., Li, J., Viitanen, T., Christophe, F., Palumbo, F., Raffo, L., Huttunen, H., Takala, J. and Bhattacharyya, S. S. (2019). An integrated hardware/software design methodology for signal processing systems, *Journal of Systems Architecture*, 93, pp. 1–19.
29. https://www.mathworks.com/products/simulink.html
30. https://www.mentor.com/hls-lp/catapult-high-level-synthesis/
31. https://www.accellera.org/downloads/standards/systemc
32. https://standards.ieee.org/standard/1800-2017.html
33. https://www.mathworks.com/
34. https://www.synopsys.com/
35. https://www.cadence.com/
36. https://www.mentor.com/
37. Stamenkovic, Z., Miletic, E., Obrknezev, M. and Tittelbach-Helmrich, K. (2009). MAC protocol implementation in RF-MIMO WLAN, *Proc. 16th IEEE International Conference on Electronics, Circuits, and Systems*, pp. 303–306.

https://doi.org/10.1142/9789811210723_0005

Chapter 5

Digital Circuit Models

Zoran Stamenković
IHP - Leibniz-Institut für innovative Mikroelektronik
Frankfurt (Oder), Germany

Bojan Dimitrijević and **Mile Stojčev**
University of Niš
Niš, Serbia

5.1 Introduction

In this chapter, we describe the main digital components and circuits necessary to implement an OFDM WLAN transceiver. These circuits include configurable general-purpose processors, digital-signal processors, power-management units, debug-support units, buses, controllers, and peripherals. These components can be soft or hard intellectual property (IP) cores depending on the chosen design methodology and design flow.

A general-purpose processor (GPP), also commonly called a central processing unit (CPU), as a complex integrated circuit, consists of different building parts such as arithmetic-logic execution units (fixed and floating point), decoders, registers, comparators, and others. In principle, the GPP executes programs by performing the following steps:

1. Read an instruction from program memory and decode it,
2. Find any instruction-specified data needed to process the instruction,
3. Process (execute) the instruction, and
4. Write the results out (in destination locations).

Digital-signal processors (DSP) are the key components for implementing wireless communication algorithms. Most often, they perform the channel estimation and equalization, error detection/correction, encryption and decryption, interference suppression, multi-channel and multi-beam reception, synchronization, and other timing-critical functions.

Processors (GPP and DSP) communicate among themselves, with memory, as well as with other components via buses (system and peripheral), controllers (external memory, cache, and interrupt), and peripherals (timer, UART, GPIO, JTAG, etc.). Debug-support units help to test and debug the system and its components. Power-management units take care of the trade-off between system performance and energy consumption.

5.2 General Purpose Processors

A general-purpose processor [1] executes a sequence of instructions, where each instruction performs some primitive operation, such as subtracting two numbers, comparing two numbers, etc. An instruction has the binary form and is a sequence of one or more bytes. The instructions supported by a particular processor and their byte-level encodings are known as its instruction-set architecture (ISA). The ISA provides commands to the GPP, with aim to tell it what it needs to do. The ISA consists of instructions, data types, addressing modes, registers, memory architecture, interrupt and exception handling, and external and internal input-output. Different processor families, such as Intel [2], AMD [3], ARM [4], MIPS [5], LEON [6], and others, have different ISAs. A program compiled for one type of processor will not run on another. Within a single family, there are many different processor versions. Every new processor generation has employed larger and more complex circuits (multiport register files, complex cache hierarchies, multi-level bus hierarchy, reorder buffers, etc.) primarily intended to extract additional parallelism from a simple single-threaded program.

Processors have always been important in embedded systems design, but fundamental changes in technology and complexity of systems have made processors even more central. Especially in wireless applications, embedded protocols and data-types have become more sophisticated. Paradoxically, each new processor must support these new algorithms and functions, yet must generally be radically smaller and more power efficient than previous solutions. Resolving this paradox justifies a new approach to processor design – automatic generation of system-specific processors. Such a processor (configurable processor) can be tailored according to requirements of various applications and tasks. Configurations include one or more coprocessors, one or more caches, scratchpad memory, on-chip trace memory, on-chip buses, etc. Usually, the cache configuration attracts the highest attention. The configuration process supposes making decision on the size, associativity, and organization of the instruction and data cache. To select appropriate configurations and choose the power-optimal one for specific application is a difficult task [7-14].

A configurable processor core usually implements the architecture with a Memory Management Unit (MMU) interfacing between an Execution Unit and a Cache Controller. MMU can implement a simple Fixed Mapping Translation (FMT) or a more complex Joint Translation Lookaside Buffer (JTLB) mechanism. It performs virtual to physical address translation and provides attributes for the mapped regions. Figure 5.1 shows an example of the configurable core. It is the MIPS32 4KE classic core [15], which consists of required and optional blocks.

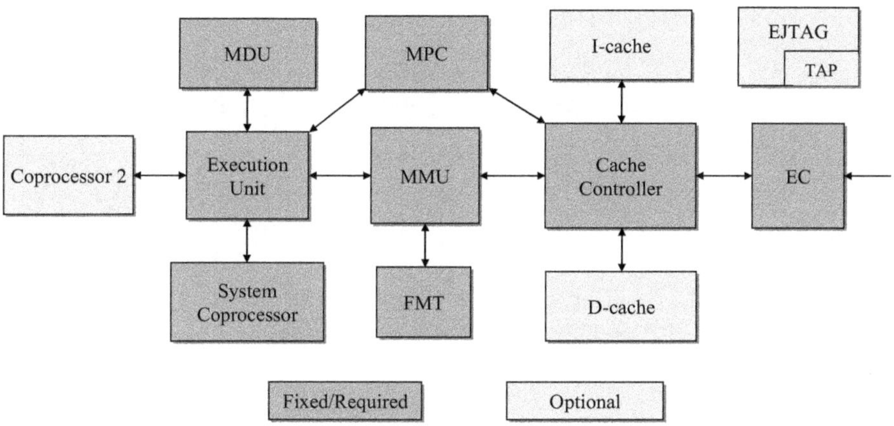

Fig. 5.1. MIPS32 4KE classic core block diagram

The Master Pipeline Control (MPC) block is responsible for recognizing and managing dependencies in the pipeline. This involves decoding each instruction and checking to see if there are interlocks that need to block the issue of the next instruction. MPC also contains the decoding necessary for issuing instructions to the proper blocks. Additionally, MPC is responsible for prioritizing and taking exceptions. The MPC block also receives the various stall conditions from the other functional blocks and manages the progression of the pipeline.

The Execution Unit implements fundamental 32-bit integer data manipulation functions of the 4KE classic core. In addition to data calculations, it contains logic for branch determination and branch target calculation and a load aligner. The integer functional units contained within the execution unit for performing integer operations include:

- Shared adder (Figure 5.2 presents a 4-bit carry look-ahead adder) for arithmetic operations, load/store address calculation, and branch target calculation;
- Shifter (store aligner);
- Logic unit for bitwise logical operations;
- Count leading hardware for CLO/CLZ instructions;

- Load aligner;
- Branch condition comparator;
- Trap condition comparator; and
- Next Fetch Address selection multiplexers.

It also contains the bypass multiplexers used to avoid stalls when executing the instruction streams where consumers of the results closely follow data-producing instructions.

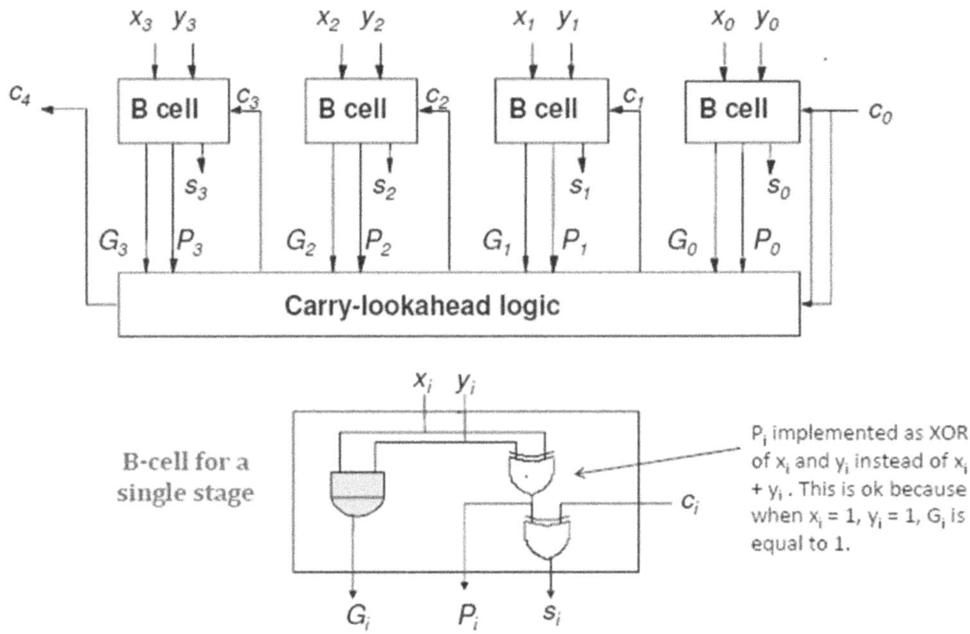

Fig. 5.2. A 4-bit carry look-ahead adder

To conserve area, the M4KE core uses an area-efficient iterative multiplier (Figure 5.3) rather than the low-latency 32x16 array (Figure 5.4) present in similar cores. An iterative one-bit shift-add happens every cycle. Multiplies are completed in 32 clocks; there is no early completion for 32x16 multiplies. Multiply accumulates and multiply subtracts take an additional two cycles to do the final accumulate step. Divides are also iterative, but do not have the data-dependent early-in optimization found in the high-performance Multiply/Divide Unit (MDU). All multiply and divide operations other than the general-purpose register targeted MUL instruction are non-blocking with respect to non-MDU instructions. The main pipeline can continue issuing instructions while a multiply is being done as long as no additional MDU instructions are encountered.

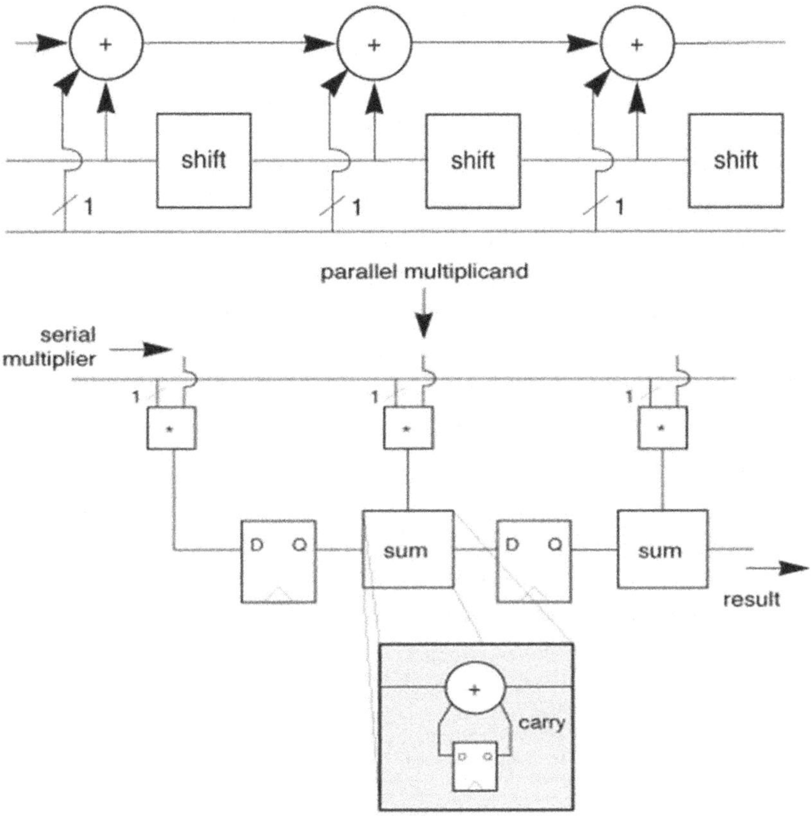

Fig. 5.3. A combinational iterative multiplier

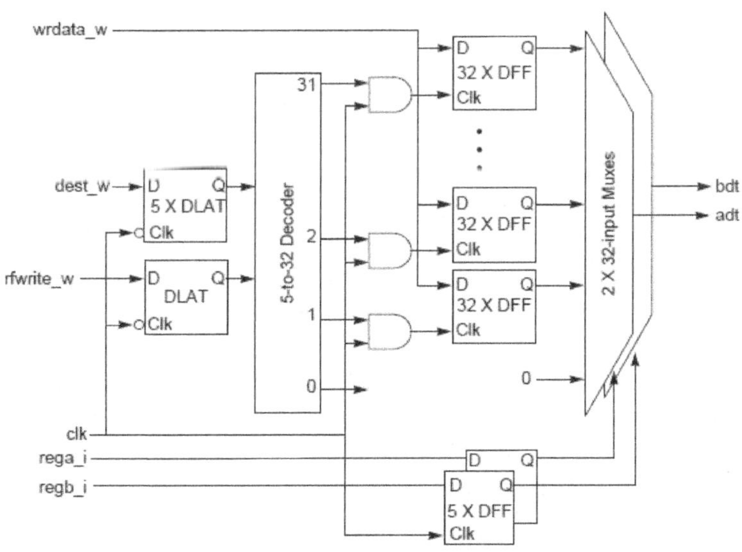

Fig. 5.4. An array-like multiplier used in serial-arithmetic operations

The Register File (RF) is a 31-entry by 32-bit register with 2 read ports and 1 write port. Register R0 always reads as zero and ignores writes. Therefore, the RF block does not implement it, which is why there are only 31 entries. As can be seen in Figure 5.5, a rising edge of the clock starts the register file read and write operations. If a write and read occur to the same address on the same clock edge, then the new write data flows through to the read port. Two synthesizable register file models are included with the 4KE core. One can select the right based on the methodology to handle gated clocks and automatic test-pattern generation.

The RF array consists of 31*32 (992) sequential elements that are composed of registers. Each of the 31 entries must be conditionally written with either gated clocks to each RF entry or a multiplexer in front of each flip-flop, which re-circulates the old value when an RF entry is not being written. The gated clock implementation will likely be smaller than the multiplexer implementation and consumes less energy since the effective clock switching capacitance is lower. Gated clocks impart certain requirements on the implementation methodology and scantest coverage as well. The read access time for the register file is about one third of the clock cycle time or faster to avoid affecting the maximum operating frequency.

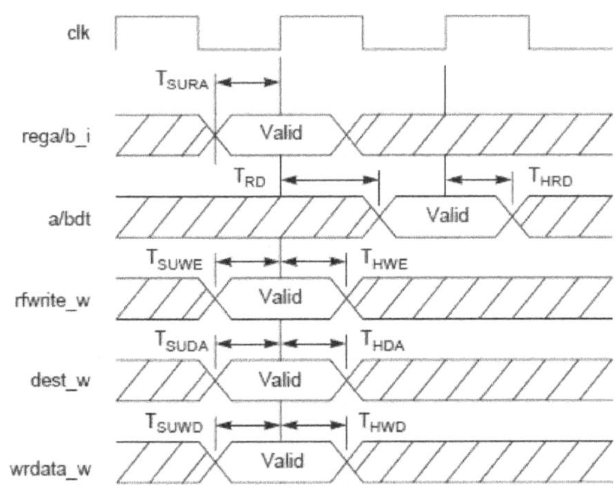

Fig. 5.5. Register file block diagram and timing

In this classic core, a Fixed Mapping Translation (FMT) module replaces the JTLB-based memory management unit. The FMT is smaller than the JTLB and easily synthesizable. The FMT provides a less general memory management mechanism than the JTLB. Like the JTLB, it performs virtual to physical address translation and provides attributes for the mapped regions. The FMT, however, does a direct address translation that is not configurable.

The Instruction Cache Controller (ICC) contains the logic to control accesses to the instruction cache. The cache arrays are located in the I-cache while the tag comparison logic and cache data multiplexers are located in the ICC module. The ICC generates the control signals needed to determine when to perform lookups, indexed reads, and fills to the I-cache block. The ICC also contains and manages the instruction-line fill buffer. Besides accumulating data to be written to the cache, instruction fetches that reference data in the line fill buffer are serviced via a bypass of that data or even data coming directly from the external interface. The ICC manages this bypass function and contains the multiplexers required to perform the bypass. The I-cache block sends static signals to configure the ICC. These signals indicate to the ICC the size and associativity of the instruction cache.

Like the ICC, the Data Cache Controller (DCC) manages the reading and writing of the data cache array (D-cache). It supports the same cache characteristics as ICC, although the instruction and data caches can be independently configured. In addition, the DCC block manages stores to the data cache. Since store data cannot be written to the cache until the cache lookup is done, the DCC maintains a buffer for the store data.

The External Control (EC) unit contains the logic to drive the external interface signals. Additionally, it contains the implementation of the 32-byte collapsing write buffer. The purpose of this buffer is to store and combine the write transactions before issuing them at the external interface. When using a write-through caching policy, the write buffer can significantly reduce the number of write transactions on the external interface as well as reduce the amount of stalling in the core due to issuance of multiple writes in a short period. When a write-back caching policy is used, the write buffer gathers the contents of a dirty cache line.

The System Coprocessor supports the virtual memory system and exception handling. It contains registers that hold information about the state of the processor to handle exceptions. These registers are Cause, Status, BadVA, EPC, ErrorEPC, and Context. Other miscellaneous control registers are also included in the coprocessor. WatchHi and WatchLo enable setting SW breakpoints on Instruction or Data addresses. TagLo and DataLo help in some of the cache operations for initializing and verifying the cache arrays. Count and Compare can play the role of a timer and generate periodic interrupts. TimerInt is an output of the 4KE core, which feeds back into the core to generate interrupts. Traditionally, this is done via muxing with Int[5]. PrID, Config, and Config1 registers contain information that allows software to determine the type of processor present and hardware resources available. The 4KE core supports the inclusion of an Enhanced JTAG (EJTAG) block that provides additional debugging capability. EJTAG provides mechanisms to allow debugging of software (including kernel code) without affecting the architectural state of the chip. Some of the EJTAG features spread outside of it and throughout the rest of the core. These include the CPZ registers DEPC, Debug, and DeSave, single step program execution, as well as support for the SDBBP and DERET instructions.

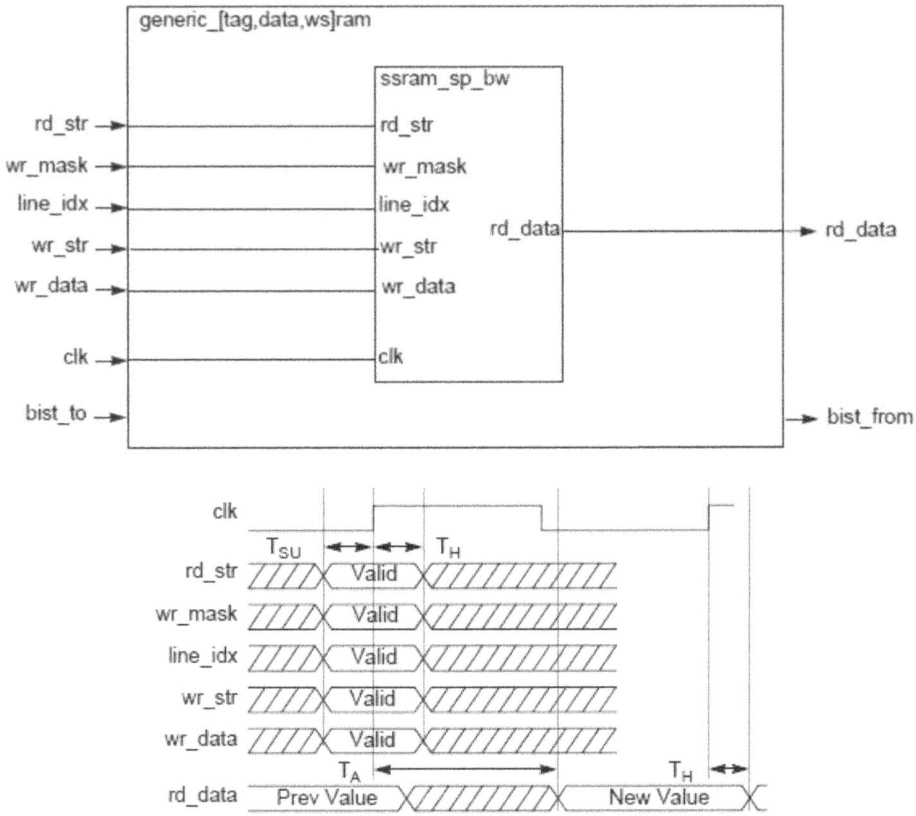

Fig. 5.6. Generic SRAM wrapper and timing

The next step is selection of SRAMs suitable for the desired technology. When choosing SRAMs for the caches, the implementation is often constrained by the characteristics of available memory generators [16, 17] for a specific technology. Generic wrappers from the cache-related SRAMs are necessary. The generic wrappers instantiate a parameterizable, single-port, synchronous, bit writeable memory model, which easies initial simulation of the cache modules. Figure 5.6 shows the ports and timing characteristics of this generic model. The actual SRAMs used for physical implementation must replace the wrapper module. Chapter 7 describes the functional characteristics of SRAMs used for the cache arrays. Figure 5.7 gives a general block diagram of the cache subsystem. Both the

I-cache and D-cache look similar, although their configurations (size and associativity) can be independently chosen.

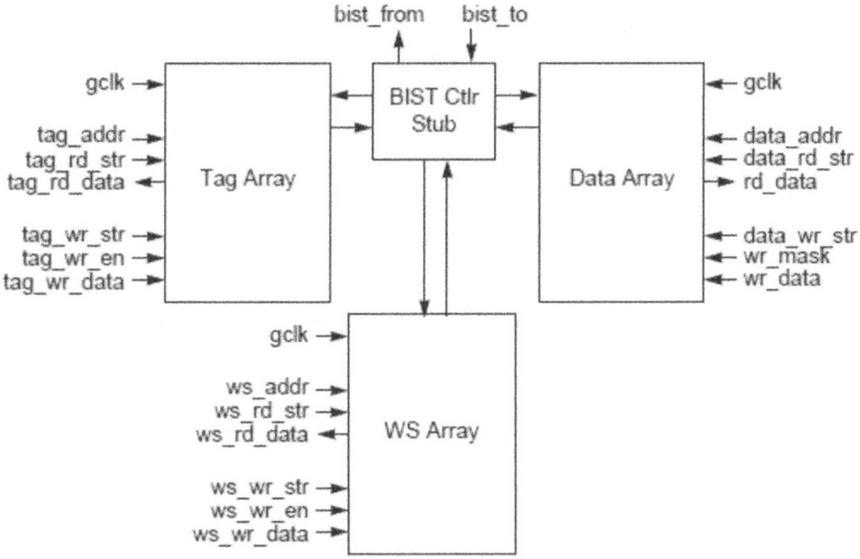

Fig. 5.7. Block diagram of the cache subsystem

Table 5.1 shows the possible cache sizes as a function of associativity and way size. The product of associativity and way size defines the total size of a cache. There are multiple cache configurations achieving the same total size that have different values for associativity and way size since the product of the two determines the total size. The total cache size always refers to the size of the data array. The size of the tag and way-select (WS) arrays, while related to the data array, is additional. The area of all three arrays determines the total cache area in an implementation.

Table 5.1. Cache size as a function of associativity and way size

Associativity	Way Size of Each Way				
	1 kB	2 kB	4 kB	8 kB	16 kB
Direct Mapped	1 kB	2 kB	4 kB	8 kB	16 kB
2-Way	2 kB	4 kB	8 kB	16 kB	32 kB
3-Way	3 kB	6 kB	12 kB	24 kB	48 kB
4-Way	4 kB	8 kB	16 kB	32 kB	64 kB

Table 5.2 shows the organization of the SRAM arrays required for a cache, based on the configuration choices for the associativity and way size. The tag and data array sizes are the same for both I-cache and D-cache. The WS array size is different for I-cache and D-cache due to the additional dirty bits needed in the D-cache. The I-cache WS array is not needed for direct-mapped caches.

The SRAM sizes shown in Table 5.2 assume a single monolithic entity for each array. This is the most efficient from an area perspective, but may not yield the fastest access time depending on the characteristics of the SRAM generator. Other physical organizations, which might affect speed, power and floor-planning considerations, are possible. It may be desirable to break the tag or data arrays into smaller individual physical arrays for each way. For example, the 1024x64 data array for an 8 kB, 2-way cache could be built from two 1024x32 SRAMs.

Table 5.2. Cache SRAM organization

Cache Size (kB)	Assoc. (Lines Per Set)	Way Size (kB)	Number of Sets	Tag RAM Size	Data RAM Size	I-Cache WS RAM Size	D-Cache WS RAM Size
1	1	1	64	64x24	256x32	No	64x1
2	1	2	128	128x24	512x32	No	128x1
	2	1	64	64x48	256x64	64x1	64x3
3	3	1	64	64x72	256x96	64x3	64x6
4	1	4	256	256x24	1024x32	No	256x1
	2	2	128	128x48	512x64	128x1	128x3
	4	1	64	64x96	256x128	64x6	64x10
6	3	2	128	128x72	512x96	128x3	128x6
8	1	8	512	512x24	2048x32	No	512x1
	2	4	256	256x48	1024x64	256x1	256x3
	4	2	128	128x96	512x128	128x6	128x10
16	1	16	1024	1024x24	4096x32	No	1024x1
	2	8	512	512x48	2048x64	512x1	512x3
	4	4	256	256x96	1024x128	256x6	256x10
24	3	8	512	512x72	2048x96	512x3	512x6
32	2	16	1024	1024x48	4096x64	1024x1	1024x3
	4	8	512	512x96	2048x128	512x6	512x10
64	4	16	1024	1024x96	4096x128	1024x6	1024x10

5.3 Digital Signal Processors

As promised in the previous chapter (System Models), we describe now Matlab models of the signal processing building blocks (or digital signal processors) of an OFDM WLAN transceiver. The models are based on theory and algorithms given in Part 1, Chapter 2 (Digital Baseband). However, unique functionalities of these signal processors and their models have to be verified by Matlab simulations using appropriate (custom) testbenches [18]. Therefore, each described model includes a simulation testbench. Exercising the models in corresponding test environment can also help to predict system performance and evaluate tradeoffs.

5.3.1 *Scrambler/descrambler*

In a digital communication system, it is common to meet long strings of 1's and 0's within the transmitted data what results in constant output levels. Tracking timing information is difficult to extract (additional synchronization information, i.e. pilot signals, is usually used) from such outputs because there will be no state-transition during these sequences, which may result in transmission errors at the receiver. Scrambler substantially reduces this trouble by detecting undesirable sequences of bits and inserting state transitions in a pseudo random behavior. If there is a long sequence of 1's, 0's will be pseudo randomly inserted in to the transmitted stream. It ensures that the probability of receiving a '1' is equal to the probability of receiving '0' and minimizes the probability of periodic or repetitive data transmission, which in turn makes clock recovery and synchronization processes easier for implementation. Simulation testbench includes a random number generator (RNG), which generates a test bit-stream for simulation. The RNG generates binary numbers with probability p using Bernoulli distribution with mean value 1-p and variance p(1-p). Values p = 0.3 was chosen since this probability provides a satisfactory increase in transitions. The RNG output feeds the input of the Matlab scrambler model. Parameters like scrambling polynomial and initial states are set according to the details given in Figure 5.3 of Part 1.

Figure 5.8 presents the code of a Matlab model and simulation testbench. The first entity defines the scrambler/descrambler functionality, while the second describes the testbench used for functional simulation.

```
% scrambling (descrambling) function
% y scrambled output
% s scrambler state
% x scrambler input
% n number of input/output bits
function [y, s]=m_scramble(x, s, n)
  m=uint8(144);                 % mask - polynomial x^7 + x^4 + 1
  t=uint8(0);
  for i=1:n
    s=bitshift(s, 1);             % shift left
    t=bitand(s, m);               % mask with m
    t=bitxor(bitshift(t,-4), t);  % parity check
    t=bitxor(bitshift(t,-2), t);
    t=bitxor(bitshift(t,-1), t);
    t=bitand(t,1);                % parity check done
    s=bitor(s,t);                 % set lsb
    y(i)=xor(x(i) ,t);            % xor to input
  end
end

% test script for scrambling (descrambling) function
n=32;
a(1:n)=[1 1 1 0 0 0 0 0 0 0 1 0 1 1 0 ...
        1 1 1 0 0 0 0 1 1 1 0 0 0 0 1 0]; % input signal for
scrambling
%a(1:n)=[1 1 1 1 1 1 1 1 1 1 1 1 1 1 1 ...
%        1 1 1 1 1 1 1 1 1 1 1 1 1 1 1 1]; % input all 1s
b(1:n)=0;                         % scrambled output
c(1:n)=0;                         % descrambled signal
s=uint8(255);                     % reset initial state
[b, s]=m_scramble(a, s, n);
s=uint8(255);                     % reset initial state
[c, s]=m_scramble(b, s, n);
a                                 % print input signal
b                                 % print scrambled signal
c                                 % print descrambled signal
c-a                               % difference
% optional signals ploting
x=1:0.1:n;
subplot(3, 1, 1);
plot(x,a(floor(x+0.5)),"linewidth",2)
ylabel('Signal');
axis([1 n -0.1 1.1]);
subplot(3, 1, 2);
plot(x,b(floor(x+0.5)),"linewidth",2)
ylabel('Scrembled');
axis([1 n -0.1 1.1]);
subplot(3, 1, 3);
plot(x,c(floor(x+0.5)),"linewidth",2)
xlabel('Time'); ylabel('Descrembled');
axis([1 n -0.1 1.1]);
```

Fig. 5.8. Matlab model of the scrambler/descrambler and simulation testbench

5.3.2 *Interleaver/deinterleaver*

The interleaving is a process of rearranging the samples of the transmitted signal so to spread bursts of errors over multiple code words. The simplest and most popular interleaving mechanisms is the block interleaving. The samples are first arranged to a matrix in a row-by-row manner. The reading is organized in a column-by-column manner. After deinterleaving, the error burst is effectively spread among different rows.

In the IEEE 802.11a standard for a single-stream mode, the interleaving algorithm includes two permutation processes. If more than one spatial stream exists, like in the IEEE 802.11n, a third permutation process, called frequency rotation, modulates additional spatial streams. The first permutation process modulates adjacent bits onto nonadjacent subcarriers. The second permutation maps adjacent bits alternatively onto less and more significant bits of the constellation. This is a way to avoid the long runs of low reliability bits (LSB). The IEEE 802.11a WLAN interleaver is a block interleaver, which size is equal to the number of bits in a single OFDM symbol and depends on the signal constellation. In the expression defining the first permutation:

$$j = (i\%16) * \left(\frac{N}{16}\right) + \lfloor i/16 \rfloor,$$

i denotes the bit index before the first permutation, j denotes the index after the first permutation, and k denotes the index after the second permutation. N is the number of bits in an OFDM symbol and N_{BPSC} is the number of bits per subcarrier.

The second permutation is defined by

$$k = s * \left\lfloor \frac{j}{s} \right\rfloor + \left[\left(j + N - \left\lfloor 16 * \frac{j}{N} \right\rfloor\right) \%s\right],$$

$$s = \max\left(\frac{N_{BPSC}}{2}, 1\right).$$

Figure 5.9 presents the Matlab code that models these permutation operations. The presented code has two sequences. The first sequence rearranges the input data (already written along the rows of a matrix memory) along the columns, i.e. permutes rows into columns. The second sequence permutes LSB and MSB (valid only for 16 QAM and 64 QAM). The number of columns in a matrix is 16 while the number of bits per row varies (192 or 288) and depends on the number of bits per OFDM symbol.

```
% interleaving function
% x input signal for interleaving
% y interleaved signal output
% b number of bits per symbol (1 for BPSK, 2 for QPSK,
% 4 for 16QAM or 6 for 64QAM)
% n number of bits to interleave
function [y] = m_interleaver(x, b, n)
  s=max(b/2,1);
  for i=1:n
% swapping rows and columns
    j=mod((i-1),16)*(n/16)+floor((i-1)/16);
% LSB to MSB permutation
    k=s*floor(j/s)+mod((j+n-floor(16*j/n)),s);
% copy input to output
    y(i)=x(k+1);
  end
end
% deinterleaving function
% x interleaved input signal
% y deinterleaved signal output
% b number of bits per symbol (1 for BPSK, 2 for QPSK,
% 4 for 16QAM or 6 for 64QAM)
% n number of bits to deinterleave
function [y] = m_deinterleaver(x, b, n)
  s=max(b/2,1);
  for i=1:n
% swapping rows and columns
    j=mod((i-1),16)*(n/16)+floor((i-1)/16);
% LSB to MSB permutation
    k=s*floor(j/s)+mod((j+n-floor(16*j/n)),s);
% copy input to output reversed
    y(k+1)=x(i);
  end
end
```

Fig. 5.9. Matlab model of the interleaver transposition operations

Simulation results given in Figure 5.10 verify the model correctness. The figure shows the positions of rearranged matrix elements (input matrix is of order 16x12) after the second permutation.

0	13	24	37	48	61	72	85	96	109	120	133	144	157	168	181
1	12	25	36	49	60	73	84	97	108	121	132	145	156	169	180
2	15	26	39	50	63	74	87	98	111	122	135	146	159	170	183
3	14	27	38	51	62	75	86	99	110	123	134	147	158	171	182
4	17	28	41	52	65	76	89	100	113	124	137	148	161	172	185
5	16	29	40	53	64	77	88	101	112	125	136	149	160	173	184
6	19	30	43	54	67	78	91	102	115	126	139	150	163	174	187
7	18	31	42	55	66	79	90	103	114	127	138	151	162	175	186
8	21	32	45	56	69	80	93	104	117	128	141	152	165	176	189
9	20	33	44	57	68	81	92	105	116	129	140	153	164	177	188
10	23	34	47	58	71	82	95	106	119	130	143	154	167	178	191
11	22	35	46	59	70	83	94	107	118	131	142	155	166	179	190

Fig. 5.10. Swapping the coded bits alternately onto LSB and MSB positions of the 16 QAM constellation

5.3.3 *Convolutional encoder and Viterbi decoder*

In digital communications, burst and random errors can be effectively corrected by convolutional encoding. The most popular algorithm for decoding convolutional codes is a Viterbi algorithm [19]. It provides maximum likelihood performance and is highly parallelizable. We have developed Matlab simulation models of a convolutional encoder, puncturing constituent, and Viterbi decoder following the IEEE 802.11a requirements. The functionality of each model has been verified.

The first model describes a convolutional encoder with puncturing (Figure 5.11). Depending on the selected value of a parameter r, the following configurations are possible: a) for $r = 0$, a convolutional encoder (2,1,7) without puncturing; b) for $r = 1$, a convolutional encoder (3,2,7) with puncturing; and c) for $r = 2$, a convolutional encoder (4,3,7) with puncturing.

The second model (Figure 5.12) describes a Viterbi decoder with depuncturing. The proposed decoder examines the received signal, computes the metric for each path, and selects the path with lowest metric. The Viterbi decoder implements a maximum likelihood estimator. In a hard decoding, before feeding in the received symbols to the decoder, they are quantized into two levels (zero and one). Afterwards, the Hamming distance is calculated. If a soft decoding is necessary, the received symbols are transformed into multiple-level signals (close to analog) corresponding to the log-likelihood ratio generated by a soft demapper. Then, the Euclidian distance is calculated. The presented Matlab code offers a possibility to select a code rate according to the chosen convolutional encoder (r = 0, 1, or 2). The value of a parameter h selects a decoder mode (for $h = 0$, soft and for $h = 1$, hard decision mode).

The third part of the Matlab code (Figure 5.13) is a simulation testbench, which verifies the functional correctness of the convolutional encoder and Viterbi decoder models.

```
% convolutional coding (217, 327, 437) function
% y coded bits out, x uncoded bits input, n number of uncoced bits
% r=0 code rate 1/2 code 217
% r=1 code rate 2/3 code 327
% r=2 code rate 3/4 code 437
function [y]=m_coder(x, r, n)
  g0=uint8(109);              % reversed bit order (133 octal)
  g1=uint8(79);               % reversed bit order (171 octal)
  s7=uint8(0);                      % 7 bits state
  m7=uint8(127);                    % 7 bits mask
  a=uint8(0);
  b=uint8(0);
  for i=1:n
    s7=bitshift(s7, 1);             % shift left
    s7=bitand(s7, m7);              % register size mask
    if x(i)>0
      s7=s7+1;                      % set input as lsb
    end
    a=bitand(s7, g0);               % mask with g0
    a=bitxor(bitshift(a,-4), a);    % parity check
    a=bitxor(bitshift(a,-2), a);
    a=bitxor(bitshift(a,-1), a);
    a=bitand(a,1);                  % parity check done
    b=bitand(s7, g1);               % mask with g0
    b=bitxor(bitshift(b,-4), b);    % parity check
    b=bitxor(bitshift(b,-2), b);
    b=bitxor(bitshift(b,-1), b);
    b=bitand(b,1);                  % parity check done
    if r==0                         % code rate 1/2 code 217
      y(i+i-1)=logical(a);          % set out 0
      y(i+i)=logical(b);            % set out 1
    end
    if r==1                       % code rate 2/3 code 327
% puncturing remove 1 of 4
      ia=floor((i-1)/2);        % ib 0 0 1 1 ... for i 1 2 3 4 ...
      ib=(i+i-2)-4*ia;      % ib 0 2 0 2 ... for i 1 2 3 4 ...
      y(ia*3+ib+1)=logical(a);      % set out 0
      ib=ib+1;              % ib 1 3 1 3 ... for i 1 2 3 4 ...
      if ib<3
        y(ia*3+ib+1)=logical(b);    % set out 1
      end
    end
    if r==2                         % code rate 3/4 code 437
% puncturing remove 2 of 6
      ia=floor((i-1)/3);% ib 0 0 0 1 1 1 ... for i 1 2 3 4 5 6 ...
      ib=(i+i-2)-6*ia;  % ib 0 2 4 0 2 4 ... for i 1 2 3 4 5 6 ...
      if ib<3
        y(ia*4+ib+1)=logical(a);    % set out 0
      end
      ib=ib+1;          % ib 1 3 5 1 3 5 ... for i 1 2 3 4 5 6 ...
      if ib<3
        y(ia*4+ib+1)=logical(b);    % set out 1
      end
      if ib>3
        y(ia*4+ib-1)=logical(b);    % set out 1
      end
    end
  end
end
```

Fig. 5.11. Matlab model of the convolutional encoder

```
% Viterbi decoding (217, 327, 437) function
% y decoded bits out
% x coded bits input
% h=1 hard input (bits 0 and 1 output from hard demapper)
% h=0 soft input (log likelihood ratio output from soft demapper)
% r=0 code rate 1/2 code 217
% r=1 code rate 2/3 code 327
% r=2 code rate 3/4 code 437
% n number of uncoded bits
function [y]=m_decoder(x, h, r, n)
  g0=uint8(109);            % reversed bit order (133 octal)
  g1=uint8(79);             % reversed bit order (171 octal)
  dl=40;                           % trelis max length
  pm(1:64)=0;                      % path metric
  npm(1:64)=0;                     % next path metric
% extended state up to 64 bits, lsb is last received bit
  es(1:64)=uint64(0);
  nes(1:64)=uint64(0);             % next extended state
  s7=uint8(0);                     % 7 bits state
  m7=uint8(127);                   % 7 bits mask
  m6=uint8(63);                    % 6 bits mask
  t=uint8(0);
  tpm=0;
% init extended states es and path metrics pm
  for i=1:64
    es(i)=uint64(i-1);
    if i>1;
      pm(i)=1000000000;
    else
      pm(i)=0;
    end
  end
% start decoding
  for i=1:2:n+n
    if r==0
      a=x(i);
      if h==1
        a=a+a-1;                 % convert to bipolar
      end
      b=x(i+1);
      if h==1
        b=b+b-1;                 % convert to bipolar
      end
    end
    if r==1
% depuncturing add 1 blank in 4
      ia=floor((i-1)/4); % ia 0 0 1 1 ... for i 1 3 5 7 ...
      ib=(i-1)-4*ia;     % ib 0 2 0 2 ... for i 1 2 3 4 ...
      a=x(ia*3+ib+1);              % get in 0
      a=a+a-1;                     % convert to bipolar
      ib=ib+1;           % ib 1 3 1 3 ... for i 1 2 3 4 ...
      b=0;                         % blank bit
      if ib<3
        b=x(ia*3+ib+1);            % get in 1
        if h==1
          b=b+b-1;                 % convert to bipolar
        end
      end
    end
    if r==2
% depuncturing add 2 blanks in 6
```

```
      ia=floor((i-1)/6);% ia 0 0 0 1 1 1 ... for i 1 3 5 7 9 11...
      ib=(i-1)-6*ia;    % ib 0 2 4 0 2 4 ... for i 1 2 3 4 9 11...
      a=0;                        % blank bit
      if ib<3
        a=x(ia*4+ib+1);           % get in 0
        if h==1
          a=a+a-1;                % convert to bipolar
        end
      end
      ib=ib+1;        % ib 1 3 5 1 3 5 ... for i 1 3 5 7 9 11...
      b=0;                        % blank bit
      if ib<3
        b=x(ia*4+ib+1);           % get in 1
        b=b+b-1;                  % convert to bipolar
      end
      if ib>3
        b=x(ia*4+ib-1);           % get in 1
        if h==1
          b=b+b-1;                % convert to bipolar
        end
      end
    end
    for j=1:64
% bit metric
      l=bitand(es(j), uint64(m6));  % short state l
      s7=uint8(bitshift(l, 1));     % shift left
      t=s7;                         % bit 0
      t=bitand(t, g0);              % mask with g0
      t=bitxor(bitshift(t,-4), t);  % parity check
      t=bitxor(bitshift(t,-2), t);
      t=bitxor(bitshift(t,-1), t);
      t=bitand(t,1);
% expected g0 parity for state (l<<1+0) for input bit 0
      if t>0
        bm0=-a;
      else
        bm0=a;
      end
      t=s7;                         % bit 0
      t=bitand(t, g1);              % mask with g1
      t=bitxor(bitshift(t,-4), t);  % parity check
      t=bitxor(bitshift(t,-2), t);
      t=bitxor(bitshift(t,-1), t);
% expected g1 parity for state (l<<1+0) for input bit 0
      t=bitand(t,1);
      if t>0
        bm0=bm0-b;
      else
        bm0=bm0+b;
      end
      t=s7+1;                       % bit 1
      t=bitand(t, g0);              % mask with g0
      t=bitxor(bitshift(t,-4), t);  % parity check
      t=bitxor(bitshift(t,-2), t);
      t=bitxor(bitshift(t,-1), t);
% expected g0 parity for state (l<<1+1) for input bit 1
      t=bitand(t,1);
      if t>0
        bm1=-a;
      else
        bm1=a;
```

```
      end
      t=s7+1;                          % bit 1
      t=bitand(t, g1);                 % mask with g1
      t=bitxor(bitshift(t,-4), t);  % parity check
      t=bitxor(bitshift(t,-2), t);
      t=bitxor(bitshift(t,-1), t);
% expected g1 parity for state (l<<1+1) for input bit 1
      t=bitand(t,1);
      if t>0
        bm1=bm1-b;
      else
        bm1=bm1+b;
      end
% path metrics
      tes=uint64(bitshift(es(j), 1));
      tes=bitor(tes, 1);
      k=bitand(tes, uint64(m6));
      k=k+1;
      tpm=pm(l+1)+bm1;
      if tpm<npm(k)
        nes(k)=tes;
        npm(k)=tpm;
      end
      tes=bitxor(tes, 1);
      k=bitand(tes, uint64(m6));
      k=k+1;
      tpm=pm(l+1)+bm0;
      if tpm<npm(k)
        nes(k)=tes;
        npm(k)=tpm;
      end
    end
    tpm=1000000000;
    tes=0;
% find minimal pm
    for j=1:64
      if npm(j)<tpm
        tpm=npm(j);
        tes=nes(j);
      end
    end
% set up es and pm
    for j=1:64
      pm(j)=npm(j)-tpm;
      es(j)=nes(j);
      npm(j)=1000000000;
    end
    j=(i+1)/2;
    j=j-dl;
    if j>0
      y(j)=bitget(tes, dl+1);
    end
  end
  for i=n-dl+1:n
    tes=bitshift(tes, 1);
    y(i)=bitget(tes, dl+1);
  end
end
```

Fig. 5.12. Matlab model of the Viterbi decoder

```
% test script for convolutional coding and convolutional hard decoding
n=48;
m=(n*3)/2;
x(1:n)=[0 1 1 1 0 0 0 0 1 1 1 1 1 1 1 0 ...
0 1 1 1 0 0 0 0 1 1 1 1 1 1 1 0 ...
0 1 1 1 0 0 0 0 1 1 1 1 1 1 1 0];   % input signal for scrambling
y(1:m)=0;                            % scrambled output
z(1:n)=0;                            % descrambled signal
x
%[y]=m_coder_327(x, n);
[y]=m_coder(x, 1, n);
y
%[z]=m_decoder_hard_327(y, n);
[z]=m_decoder(y, 1, 1, n);
z
z-x
```

Fig. 5.13. Matlab simulation testbench for functional verification of the convolutional encoder and Viterbi decoder

5.3.4 *Constellation mapper/demapper*

Modulators and demodulators in a digital communication system, such as OFDM WLAN 802.11a, require a constellation mapper and demapper. A constellation diagram is a representation of a digital modulation scheme in the complex plane. Constellation mappers (demappers) translate k-bit binary symbols to (from) a signal space of $\leq 2^k$ points. A constellation mapper takes a serial bit stream as input and segments the stream into k-bit symbols mapping them to coordinates of the signal constellation diagram. The coordinates of each point in a two-dimensional signal constellation diagram represent the baseband in-phase (I) and quadrature (Q) components that modulate the orthogonal IF carrier signals. At the transmitter side, the constellation mapper maps the incoming (scrambled, interleaved, and encoded) bits onto different sub-carriers. Different modulation techniques can be employed (BPSK, QPSK, and QAM) for different sub-carriers. At the receiver, the demapper simply extracts bits from the modulated symbols.

A high-order modulation scheme like QAM, offering high bandwidth efficiency, has large computational complexity. Therefore, in the low power receivers, low complexity demapping algorithms substitute the traditional soft-output demapping methods. Using a hard-decision strategy, these demappers only need to provide hard symbol-to-bit transformation via selecting the constellation point closest to the received symbol. If modulating bits are uncoded or algebraic coded (RS or BCH codes), the demapper output is typically hard. If modulating bits are coded with a convolutional, LDPC, or Turbo-Code encoder, the demapper output must be soft in order to reduce BER. Optimal soft-output demapping algorithms involve computationally complex functions such as logarithmic and exponential functions.

Figure 5.14 presents a Matlab model of the mapper, while Figure 5.15 shows Matlab models of the soft and hard demappers.

```
% mapping function
% x output I signal
% y output Q signal
% z input bits
% n input bits number
% b bits per output symbol
% 6 for 64QAM
% 4 for 16QAM
% 2 for QPSK
% 1 for BPSK
function [x, y]=m_mapper(z, b, n)
  bp=1;
  if b>1
    bp=b/2;
  end
  for i=1:b:n
% get next b/2 bits for I mapping
    for j=1:bp
      a(j)=z(i+bp-j);
    end
% I signal
    t=0;
    for j=1:bp
      t=t-(2^(j-1));
      if a(j)>0
        t=-t;
      end
    end
    x((i+b-1)/b)=t;
% get next b/2 bits for Q mapping
    for j=1:bp
      a(j)=z(i+b-j);
    end
% Q signal
    if b>1
      t=0;
      for j=1:bp
        t=t-(2^(j-1));
        if a(j)>0
          t=-t;
        end
      end
      y((i+b-1)/b)=t;
    else
      y(i)=0;
    end
  end
end
```

Fig. 5.14. Matlab model of the mapper

```
% demapping function
% x input I signal
% y input Q signal
% z output bits
% h=0 for soft demapper, h=1 for hard demapper
% n output bits number
% b bits per input symbol (6 for 64QAM, 4 for 16QAM, 2 for QPSK, 1 for BPSK)
function [z]=m_demapper(x, y, b, h, n)
  bp=1;
  if b>1
    bp=b/2;
  end
  for i=1:b:n
% I signal
    t=x((i+b-1)/b);
    if h==1                  % hard demapping, I component
      for j=bp:-1:1
        if t>0
          a(j)=1;
          t=-t;
        else
          a(j)=0;
        end
        t=t+(2^(j-1));
      end
% set next b/2 bits for I part
      for j=1:bp
        z(i+bp-j)=a(j);
      end
    else                     % soft demapping, I component
      for j=bp:-1:1
        if t>0
          c(j)=(-1-t)*(-1-t);
          a(j)=1;
          t=-t;
        else
          c(j)=-(1-t)*(1-t);
          a(j)=-1;
        end
        t=t+(2^(j-1));
      end
% set next b/2 bits for I part
      t=t*t;                 % distance to the closest constellation point
      for j=1:bp
        z(i+bp-j)=c(j)-t*a(j);
      end
    end
% Q signal
    if b>1
      t=y((i+b-1)/b);
      if h==1                % hard demapping, Q component
        for j=bp:-1:1
          if t>0
            a(j)=1;
            t=-t;
```

```
          else
            a(j)=0;
          end
          t=t+(2^(j-1));
        end
% set next b/2 bits for Q part
        for j=1:bp
          z(i+b-j)=a(j);
        end
      else                  % soft demapping, Q component
        for j=bp:-1:1
          if t>0
            c(j)=(-1-t)*(-1-t);
            a(j)=1;
            t=-t;
          else
            c(j)=-(1-t)*(1-t);
            a(j)=-1;
          end
          t=t+(2^(j-1));
        end
% set next b/2 bits for Q component
        t=t*t;              % distance to the closest constellation point
        for j=1:bp
          z(i+b-j)=c(j)-t*a(j);;
        end
      end
    end
  end
end
```

Fig. 5.15. Matlab models of the soft and hard demapper

5.3.5 *Pilot insertion and pulse shaping*

Aiming to make a coherent detection of each OFDM symbol immune to frequency offset and phase noise, the standard IEEE 802.11a defines four sub-carriers as pilot symbols. Pilot symbols are located at sub-carriers enumerated as -21, -7, 7, and 21. To avoid strong discrete lines in the OFDM spectrum, pilot symbols are BPSK modulated.

For modulation, the identical polynomial sequence to one for data scrambling is used. Initially, the sequence state register keeps all 1's. At the output, logical values 1 and 0 are transformed into -1 and 1 values of the pilot symbol.

This pilot insertion and modulation process applies for every OFDM symbol (starting from the SIGNAL symbol, DATA 1 symbol, DATA 2 symbol…). Figure 5.16 and Figure 5.17 describe a Matlab model and a simulation testbench of the pilot symbol generator, respectively.

```
% pilot generator function
% y pilot output
% s pilot state
% n number of output symbols
function [y, s]=m_pilot(s, n)
  m=uint8(144);                    % mask - polynomial x^7 + x^4 + 1
  t=uint8(0);
  for i=1:n
    s=bitshift(s, 1);              % shift left
    t=bitand(s, m);                % mask with m
    t=bitxor(bitshift(t,-4), t);   % parity check
    t=bitxor(bitshift(t,-2), t);
    t=bitxor(bitshift(t,-1), t);
    t=bitand(t,1);                 % parity check done
    s=bitor(s,t);                  % set lsb
    if t==0
      y(i)=1;                      % out pilot signal
    else
      y(i)=-1;                     % out pilot signal
    end
  end
end
```

Fig. 5.16. Matlab model of the pilot symbol generator

```
% testbench for pilot generator function
n=127;
a(1:n)=[0 0 0 0 1 1 1 0  1 1 1 1 0 0 1 0  1 1 0 0 1 0 0 1 ...
0 0 0 0 0 0 1 0  0 0 1 0 0 1 1 0  0 0 1 0 1 1 1 0  1 0 1 1 0 1 1 0
0 0 0 0 1 1 0 0 ...
1 1 0 1 0 1 0 0  1 1 1 0 0 1 1 1  1 0 1 1 0 1 0 0  0 0 1 0 1 0 1 0
1 1 1 1 1 0 1 0 ...
0 1 0 1 0 0 0 1  1 0 1 1 1 0 0 0  1 1 1 1 1 1 1];   % polynomial sequence
b(1:n)=0;                          % pilot symbol
s=uint8(255);                      % reset initial state
[b, s]=m_pilot(s, n);
b                                  % print pilot symbol
b - (1-a-a)                        % difference
% optional signal plotting
plot(x,b(floor(x+0.5)),"linewidth",2)
ylabel('Pilot');
axis([1 n -1.1 1.1]);
```

Fig. 5.17. Matlab simulation testbench for functional verification of the pilot symbol generator

Bearing in mind that OFDM systems use a band-limited channel, adjacent symbols may introduce the inter-symbol interference (ISI). In addition, time-variations of the channel corrupt orthogonality of the sub-carriers causing the inter-carrier interference (ICI). A low-pass filter is necessary to reduce such imperfections. One of the most effective solutions is a raised-cosine filter (RCF). The RCF satisfies the Nyquist criterion of suppressing the spectral distortion by multiplying the sampling rate. In practice, the RFC in OFDM systems splits into two identical parts, one at the transmitter and the other at the receiver side. Each of those is a square-root raised cosine filter (SRRCF). The total effect of two SRRCFs is equivalent to a single RCF.

Some common ISI suppression of 40 dB can be realized using an oversampling rate of 2 and spreading of 4 symbols on each side of the current symbol. A minimal oversampling rate of 2 is usually sufficient to reduce ISI. Higher oversampling rate values increase both the filter length (hardware complexity) and clock frequency.

Figure 5.18, Figure 5.19, and Figure 5.20 present the Matlab code of transmitter and receiver pulse-shaping filters, as well as a simulation testbench.

```
% Tx pulse shaping function
% x output I signal, y output Q signal
% a input I signal, b input Q signal
% os oversampling, ro roll-off factor
% 2*m+1 FIR filter length
% n input samples number
function [x, y]=m_pulse_shaper(a, b, os, ro, m, n)
% init filter waits
  tn=0.000001;
  ff(m+1)=(sin(pi*tn*(1-ro))+4*ro*tn*cos(pi*tn*(1+ro)))/(pi*tn*(1-
16*tn*tn*ro*ro));
  for i=1:m
    tn=i/os;
    rn=ro;
    if ((1-4*tn*ro)<0.0000001)
      rn=ro+0.0000001;
    end
    ff(m+1+i)=(sin(pi*tn*(1-rn))+4*rn*tn*cos(pi*tn*(1+rn)))/(pi*tn*(1-
16*tn*tn*rn*rn));
    ff(m+1-i)=ff(m+1+i);
  end
  for i=1:n
    for s=1:os
      k2=(i-1)*os+s;                  % k2 - oversampled out index
      x(k2)=0;                        % initial real part
      y(k2)=0;                        % initial imaginary part
      for j=1:(m+m+1)
        k1=((i-1)*os+s+m+1-j)/os;
        if k1<1
          k1=1;
        end
        if k1>n
          k1=n;
        end
        k1=round(k1);                 % k1 - input sample index
        x(k2)+=a(k1)*ff(j);           % summing real part
        y(k2)+=b(k1)*ff(j);           % summing imaginary part
      end
      x(k2)/=os;
      y(k2)/=os;
    end
  end
end
```

Fig. 5.18. Matlab model of the Tx pulse-shaping filter with oversampling

```
% Rx pulse shaping function
% x output I signal
% y output Q signal
% a input I signal
% b input Q signal
% us undersampling
% ro roll-off factor
% 2*m+1 FIR filter length
% n output samples number
function [x, y]=m_RX_pulse_shaper(a, b, us, ro, m, n)
% init filter waits
  tn=0.000001;
  ff(m+1)=(sin(pi*tn*(1-ro))+4*ro*tn*cos(pi*tn*(1+ro)))/(pi*tn*(1-
16*tn*tn*ro*ro));
  for i=1:m
    tn=i/us;
    rn=ro;
    if ((1-4*tn*ro)<0.0000001)
      rn=ro+0.0000001;
    end
    ff(m+1+i)=(sin(pi*tn*(1-rn))+4*rn*tn*cos(pi*tn*(1+rn)))/(pi*tn*(1-
16*tn*tn*rn*rn));
    ff(m+1-i)=ff(m+1+i);
  end
  for i=1:n
    for s=1:us
      x(i)=0;                         % init real part
      y(i)=0;                         % init imaginary part
      for j=1:(m+m+1)
        k1=(i-1)*us+s+m+1-j;
        if k1<1
          k1=1;
        end
        if k1>n*us
          k1=n*us;
        end
        x(i)+=a(k1)*ff(j);            % summing real part
        y(i)+=b(k1)*ff(j);            % summing imaginary part
      end
      x(i)/=us;
      y(i)/=us;
    end
  end
end
```

Fig. 5.19. Matlab model of the Rx pulse-shaping filter with undersampling

The inserts in Figure 5.20 show an impulse response of the pulse-shaping filter for two roll-off factor values (upper diagram) and an original, pulse shaped at the transmitter (with an oversampling rate of 4), and pulse shaped at the receiver (with an undersampling rate of 4) signals (lower diagram).

```
% simulation testbench for pulse-shaping verification
m=16;
os=4;
ro=0.25;
tn=0.000001;
f1(m+1)=(sin(pi*tn*(1-ro))+4*ro*tn*cos(pi*tn*(1+ro)))/(pi*tn*(1-
16*tn*tn*ro*ro));
for i=1:m
  tn=i/os;
  rn=ro;
  if ((1-4*tn*ro)<0.0000001)
    rn=ro+0.0000001;
  end
  f1(m+1+i)=(sin(pi*tn*(1-rn))+4*rn*tn*cos(pi*tn*(1+rn)))/(pi*tn*(1-
16*tn*tn*rn*rn));
  f1(m+1-i)=f1(m+1+i);
end
m=16;
os=4;
ro=0.8;
tn=0.000001;
f2(m+1)=(sin(pi*tn*(1-ro))+4*ro*tn*cos(pi*tn*(1+ro)))/(pi*tn*(1-
16*tn*tn*ro*ro));
for i=1:m
  tn=i/os;
  rn=ro;
  if ((1-4*tn*ro)<0.0000001)
    rn=ro+0.0000001;
  end
  f2(m+1+i)=(sin(pi*tn*(1-rn))+4*rn*tn*cos(pi*tn*(1+rn)))/(pi*tn*(1-
16*tn*tn*rn*rn));
  f2(m+1-i)=f2(m+1+i);
end
% optional signal plotting
subplot(2, 1, 1);
xx=-m:m;
plot(xx/os,f1(m+1+xx),xx/os,f2(m+1+xx),"linewidth",2)
ylabel('h(t)');
axis([-m/os m/os -0.5 1.5]);
m=16;
os=4;
ro=0.25;
n=16;
a(1:n)=[5, 5, 0, 7, 3 -2, 5, 0, 0, 0, 0, 7, 7, 7, 0, 0]
b(1:n)=0;
c(1:n)=0;
d(1:n)=0;
[x, y]=m_TX_pulse_shaper(a, b, os, ro, m, n);
[c, d]=m_RX_pulse_shaper(x, y, os, ro, m, n);
% optional signal plotting
subplot(2, 1, 2);
x1=os/2:n*os-os/2;
x2=1:n;
plot(x1,a(round(x1/os)),x1,x(x1),x1,c(round(x1/os)),"linewidth",3)
ylabel('h(t)');
axis([os/2 n*os+-os/2 -10 10]);
```

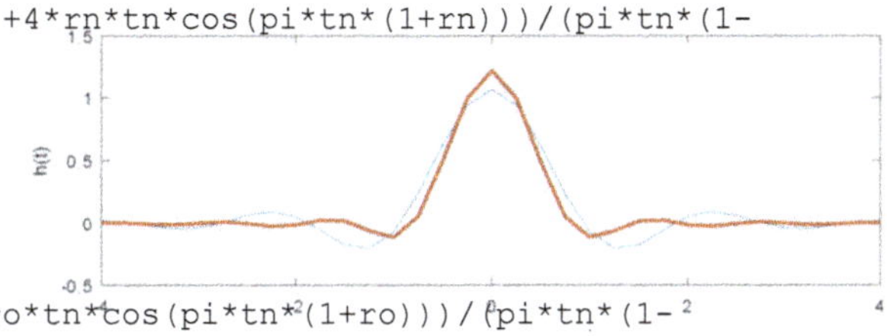

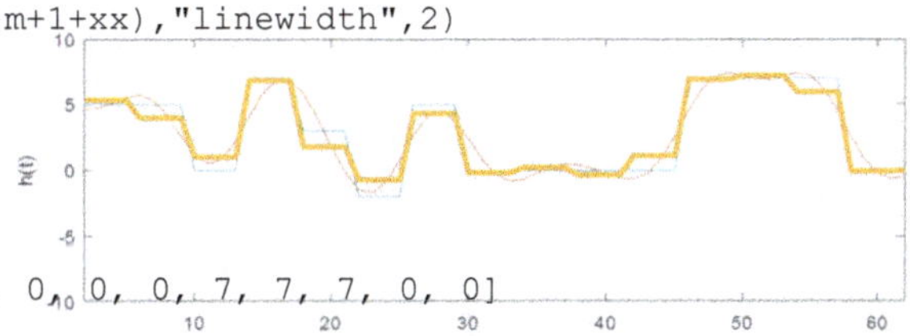

Fig. 5.20. Matlab simulation testbench for the pulse-shaping filter verification

5.3.6 *IFFT/FFT processor*

As we have already mentioned, the Fourier transform converts a time domain function into a frequency domain function while the inverse Fourier transform converts a frequency domain function into a time domain function. In practice, these functions are sampled at equally separated discrete points. The discrete Fourier transform (DFT) and the discrete inverse Fourier transform (IDFT) respectively are:

$$X(m) = \sum_{n=0}^{N-1} x(n)\, e^{\frac{2\pi jmn}{N}}, \quad m = 0,1,2,\ldots.,N-1 \qquad (5.3)$$

$$x(n) = \frac{1}{N}\sum_{m=0}^{N-1} X(m) e^{\frac{-2\pi jmn}{N}}, \qquad n = 0,1,2,\ldots.,N-1 \qquad (5.4)$$

where *n* represents the sampled points in the time domain, lower case *m* represents the sampled points in the frequency domain, and *N* is the number of sampled points.

Although the DFT is the most straightforward mathematical procedure for determining the frequency content of a time-domain sequence, it is terribly inefficient [20]. For instance, for computation of an *N*-point discrete Fourier transform (DFT) requires $(N - 1)^2$ complex multiplications and $N(N - 1)$ complex additions are needed. From the other hand, FFT is an algorithm, in which the original N-point sequence is decomposed into shorter sequence DFTs, thereby reducing the number of computations. For a radix-2 (in which the length of the decomposed sequence is two), the fundamental two-point transformation is known as a butterfly. $(N/2)\log_2 N$ butterfly operations are performed to complete FFT processing.

The FFT algorithms can be classified depending on the domains in which are calculated. If the length of the sequence is a 2^N value, the algorithm applied is called Radix-2. Other more efficient algorithms also exist, for instance the Radix-4, which can be applied when the input sequence length is a 4^N value.

Generally, FFT analyzes an input signal sequence by using a decimation-in-frequency (DIF) or decimation-in-time (DIT) decomposition to construct an efficiently computational signal-flow graph (SFG). Let note that an equivalent decimation-in-frequency FFT structure exists for each decimation-in-time FFT structure. It is important to note that the number of necessary multiplications to implement the decimation-in-frequency FFT algorithms is the same as the number necessary for the decimation-in-time FFT algorithms. In principle, there is no difference at all between both decimations, so the decision to take one or another depends only on the designer.

Radix-2 decimation-in-frequency (DIF) FFT algorithm

The most basic FFT algorithm is the Radix-2 decimation-in frequency algorithm [21]. This algorithm decomposes even and odd-indexed frequency samples shown mathematically in the following equation set:

$$\begin{aligned} X(2m) &= \sum_{n=0}^{N-1} x(n)W_N^{2mn}\ ,\quad m = 0,1,2,\dots,N-1 \\ &= \sum_{n=0}^{(\frac{N}{2})-1} x(n)W_N^{2mn} + \sum_{n=0}^{(\frac{N}{2})-1} x\left(n+\frac{N}{2}\right)W_N^{2m(n+\frac{N}{2})} \\ &= \sum_{n=0}^{(\frac{N}{2})-1} x(n)W_N^{2mn} + \sum_{n=0}^{(\frac{N}{2})-1} x\left(n+\frac{N}{2}\right)W_N^{2mn} \\ &= \sum_{n=0}^{(\frac{N}{2})-1} [(x(n) + x(n+\tfrac{N}{2}))\,W_{N/2}^{mn}] \end{aligned} \tag{5.5}$$

$$\begin{aligned} X(2m+1) &= \sum_{n=0}^{N-1} x(n)W_N^{(2m+1)n} \\ &= \sum_{n=0}^{(\frac{N}{2})-1} [(x(n) + W_N^{N/2}x(n+\tfrac{N}{2}))\,W_N^{(m+1)n}] \\ &= \sum_{n=0}^{(\frac{N}{2})-1} [(x(n) - x(n+\tfrac{N}{2}))\,W_N^{n}W_{N/2}^{mn}] \end{aligned} \tag{5.6}$$

Let

$a(n) = x(n) + x\left(n+\frac{N}{2}\right)$ and

$b(n) = x(n) - x\left(n+\frac{N}{2}\right)$.

Equations (5.5) and (5.6) can be more clearly written as two (N/2)-point DFT's or

$$X(2m)=\sum_{n=0}^{(\frac{N}{2})-1} a(n)W_{N/2}^{mn} \tag{5.7}$$

$$X(2m+1)=\sum_{n=0}^{(\frac{N}{2})-1} b(n)W_N^{n}W_{N/2}^{mn} \tag{5.8}$$

The decimation-in-frequency FFT algorithm decomposes the DFT by recursively splitting the sequence elements *X(k)* in the frequency domain into smaller and smaller sub sequences. The algorithm carries complex addition and complex multiplication operations exploiting the Radix-2 butterfly structure.

Figure 5.21 shows the flow graph of an 8-point FFT algorithm. This algorithm is referred as decimation-in-frequency (DIF) because the output sequence $X(k)$ is decomposed (decimated) into smaller subsequences, and this process continues through M iterations, where $N = 2^M$. The output $X(k)$ is complex with both real and imaginary components, and the FFT algorithm can accommodate either complex or real input values.

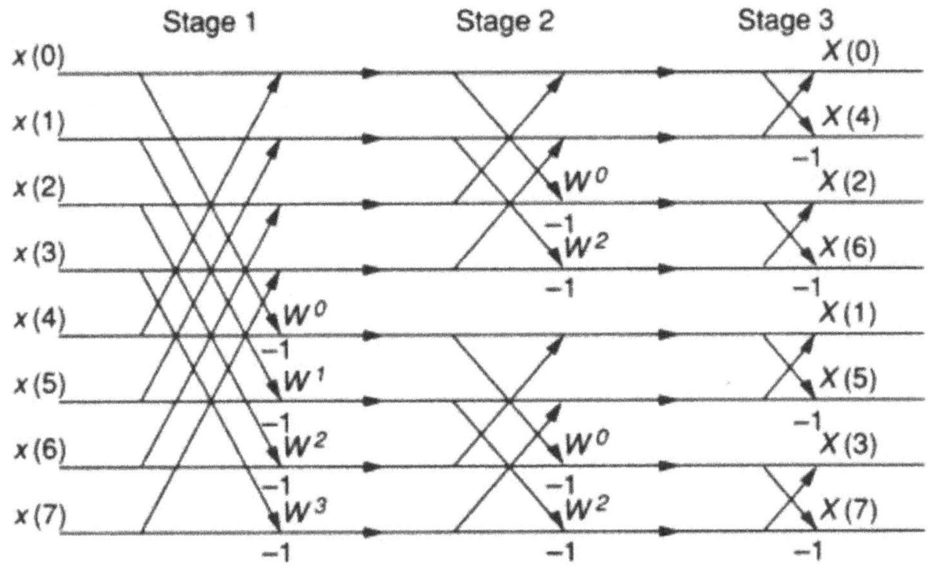

Fig. 5.21. Flow graph of an 8-point FFT using decimation-in-frequency

Radix-2 decimation-in-time (DIT) FFT algorithm

The decimation-in-time (DIT) is another process that decomposes the input sequence into smaller subsequences. Let the input sequence be decomposed into an even sequence and an odd sequence, or $x(0)$, $x(2)$, $x(4)$, . . . , $x(2n)$ and $x(1)$, $x(3)$, $x(5)$, . . . , $x(2n + 1)$ [21]. By applying Eq. (5.3) to both sequences on obtain:

$$X(m) = \sum_{n=0}^{\left(\frac{N}{2}\right)-1} x(2n)W_N^{2nm} + \sum_{n=0}^{\left(\frac{N}{2}\right)-1} x(2n+1)W_N^{2(n+1)m} \tag{5.9}$$

Using $W_N^2 = W_{n/2}$ in Eq. (5.9), we obtain

$$X(m) = \sum_{n=0}^{\left(\frac{N}{2}\right)-1} x(2n)W_{N/2}^{nm} + W_N^m \sum_{n=0}^{\left(\frac{N}{2}\right)-1} x(2n+1)W_{N/2}^{nm}, \tag{5.10}$$

which corresponds to two ($N/2$)-point DFT's. If we take

$$c(m) = \sum_{n=0}^{\left(\frac{n}{2}\right)-1} x(2n)W_{N/2}^{nm}, \tag{5.11}$$

$$d(m) = \sum_{n=0}^{\left(\frac{n}{2}\right)-1} x(2n+1)W_{N/2}^{nm}, \tag{5.12}$$

then $X(m)$ can be written as

$$X(m) = c(m) + W_N^m d(m). \tag{5.13}$$

Equation (5.13) is valid for $m > \left(\frac{N}{2}\right) - 1$. Using the symmetry property of the twiddle constant ($W_N^{m+N/2} = -W^m$), we have that

$$X(m + N/2) = c(m) - W_N^m d(m), \text{ for } m = 0, 1, 2, \ldots, \left(\frac{N}{2}\right) - 1 \tag{5.14}$$

Figure 5.22 presents the flow graph of an 8-point FFT using a decimation-in-time process. The input sequence is scrambled in the same manner as the output sequence $X(m)$ was scrambled during the decimation-in-frequency process. With the input sequence $x(n)$ scrambled, the resulting output sequence $X(m)$ becomes properly ordered. An FFT gives identical results using either the decimation-in-frequency (DIF) or the decimation-in-time (DIT) algorithm.

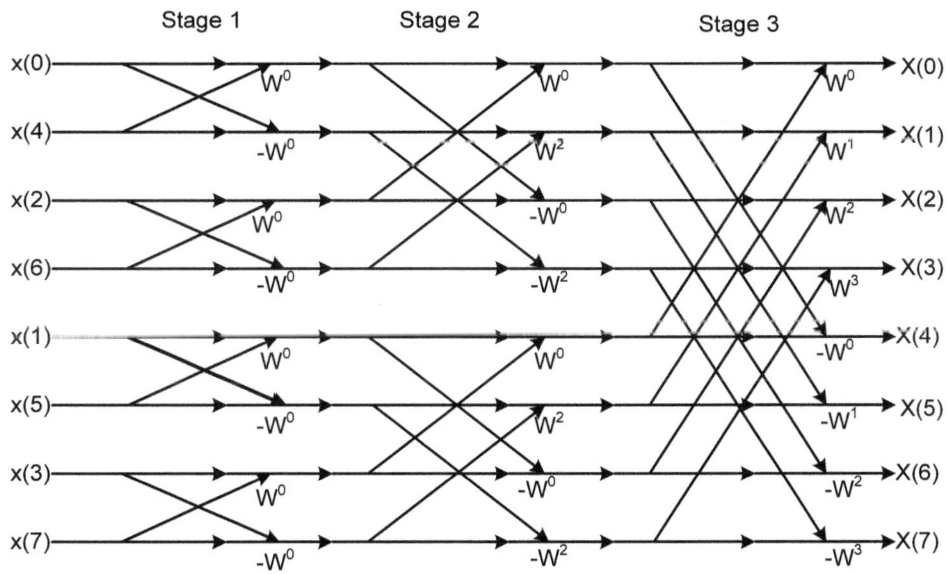

Fig. 5.22. Flow graph of an 8-point FFT using decimation-in-time

As mentioned previously, the inverse discrete Fourier transform (IDFT) converts a frequency-domain sequence $X(m)$ into an equivalent sequence $x(n)$ in the time domain [21]:

$$x(n) = \frac{1}{N}\sum_{m=0}^{N-1} X(m)e^{\frac{-2\pi jmn}{N}}, \ n = 0,1,2, \dots, N-1 \tag{5.15}$$

Comparing the expression (5.15) with the DFT definition in (5.1), one can use the FFT algorithm (forward) described previously to find the IFFT (reverse), with the following changes:

1. Add a scaling factor of $1/N$,
2. Replace W^{mn} by its complex conjugate W^{-mn}.

The Matlab tools offer an internally implemented FFT/IFFT processor model for selected N (2, 4… 1024… 8192). In addition, the Matlab users can always generate new *m*-files for user-dependent applications. Bearing in mind the aforementioned, but primarily guided by the goal to acquire better insight to model complexity and implementation details, a self-coded Matlab model (added function) of FFT/IFFT processor is presented in Figure 5.23. Figure 5.24 describes a Matlab simulation testbench.

```
% the Cooley-Tukey FFT algorithm
% a radix-2 decimation-in-time (DIT) FFT
% an in-place complex-to-complex FFT
% x output I signal, y output Q signal, a input I signal, b input Q signal
% d = 1 gives forward transform (FFT), d = 0 gives reverse transform (IFFT)
% 2^m points FFT / IFFT
function [x, y]=m_fft(d, m, a, b)
% calculate the number of points
  n=1;
  i=0;
  while i<m
    n=n+n;
    i=i+1;
  end
% do the bit reversal
  i2=n/2;
  j=0;
  for i=1:n
    if i<=j+1
      x(i)=a(j+1);
      y(i)=b(j+1);
      x(j+1)=a(i);
      y(j+1)=b(i);
    end;
    k=i2;
    while k<=j
      j=j-k;
      k=k/2;
    end;
    j=j+k;
  end;
% compute the FFT
  c1=-1.0;
  c2=0.0;
  l2=1;
  l=0;
  while l<m
```

```
    l1=l2;
    l2=l2+l2;
    u1=1.0;
    u2=0.0;
    j=0;
    while j<l1
      i=j;
      while i<n
        i1=i+l1;
        t1=u1*x(i1+1)-u2*y(i1+1);
        t2=u1*y(i1+1)+u2*x(i1+1);
        x(i1+1)=x(i+1)-t1;
        y(i1+1)=y(i+1)-t2;
        x(i+1)+=t1;
        y(i+1)+=t2;
        i=i+l2;
      end
      z=u1*c1-u2*c2;
      u2=u1*c2+u2*c1;
      u1=z;
      j=j+1;
    end
    c2=sqrt((1.0-c1)/2.0);
    if d==1
      c2=-c2;              % for forward transform only
    end
    c1=sqrt((1.0+c1)/2.0);
    l=l+1;
  end
% for forward transform only
  if d==1
    for i=1:n
      x(i)=x(i)/n;
      y(i)=y(i)/n;
    end
  end
end
```

Fig. 5.23. Matlab model of the Cooley-Tukey FFT/IFFT processor

```
% Simulation testbench for FFT/IFFT verification
m=6;
xx(1:64)=[0 1 1 1 8 0 0 0 1 1 1 1 1 1 1 0 0 1 1 1 0 0 0 0 1 1 1 1 1 1 1 0 0
1 1 1 0 0 0 0 1 1 1 1 1 1 1 0 0 1 1 1 0 0 0 0 1 1 1 1 1 1 1 0];
yy(1:64)=[0 1 1 1 0 0 5 0 1 1 1 1 1 1 1 0 0 1 1 1 0 0 0 0 1 1 1 1 1 1 1 0 0
1 1 1 0 0 0 0 1 1 1 1 1 1 1 0 0 1 1 1 0 0 0 0 1 1 1 1 1 1 1 0];
aa(1:64)=0;
bb(1:64)=0;
d=1;
[aa, bb]=m_fft(d, m, xx, yy)
d=0;
[xx, yy]=m_fft(d, m, aa, bb)
```

Fig. 5.24. Matlab simulation testbench for the FFT/IFFT processor verification

CORDIC rotator

Coordinate Rotation Digital Computer (CORDIC) is a simple processor designed to calculate mathematical, trigonometric and hyperbolic mathematical functions. CORDIC is a shift-add processor that can compute the discrete Fourier transform (DFT) rotating the vector in a plane [22-24]. It does not need multipliers, which reduces hardware size and energy consumption. CORDIC can operate in one of three configurations (linear, circular or hyperbolic) and two modes (rotation or vectoring). In the rotation mode, the input vector rotates by a specified angle, while in the vectoring mode, the input vector rotates to the *x*-axis while recording the angle of rotation required.

As Figure 5.25 presents, the vector (terminating in point x_{in}, y_{in}) rotates by the angle φ producing a new vector (terminating in point x_r, y_r). The following expressions describe the vector rotation:

$$x_r = A_{in} \cos(\alpha + \varphi) = x_{in} \cos\varphi - y_{in} yin\,\varphi$$

$$y_r = A_{in} \sin(\alpha + \varphi) = x_{in} \sin\varphi - y_{in} \cos\varphi \qquad (5.16)$$

where A_{in} is the vector modulus and α is the initial angle.

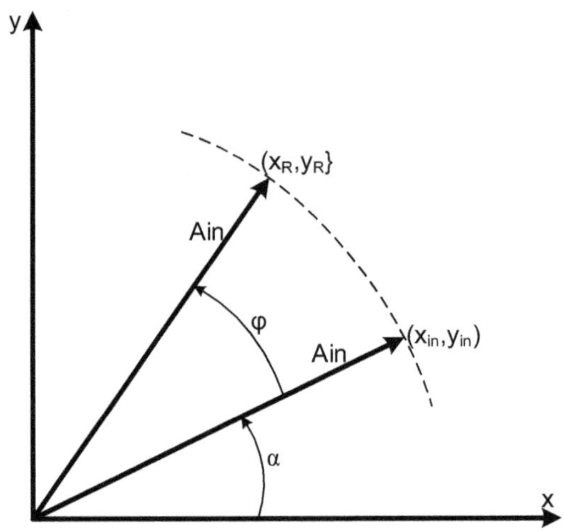

Fig. 5.25. Vector rotation

A matrix form of the rotation is as follows:

$$\begin{bmatrix} x_r \\ y_r \end{bmatrix} = \begin{bmatrix} \cos\varphi & -\sin\varphi \\ \sin\varphi & \cos\varphi \end{bmatrix} \begin{bmatrix} x_{in} \\ y_{in} \end{bmatrix} = ROT(\varphi) \begin{bmatrix} x_{in} \\ y_{in} \end{bmatrix} \qquad (5.17)$$

Solution of the equation (5.17) need four multiplications and two additions. In practical implementations, the CORDIC algorithm sequentially performs micro-rotations by elementary angles. In that sense, the angle φ can be decomposed into a sum of elementary angles φ_i:

$$\varphi = \sum_{i=0}^{\infty} \varphi_i \tag{5.18}$$

The term $ROT(\varphi)$ in the equation (5.17) is equal to

$$ROT\ (\varphi) = \prod_{i=0}^{\infty} ROT(\varphi_i) \tag{5.19}$$

For example, the product of $ROT\ (\varphi_1)$ and $ROT\ (\varphi_2)$ is equal to $ROT\ (\varphi_1 + \varphi_2)$, i.e., rotation through φ_1 then trough φ_2:

$$ROT\ (\varphi_1)\ ROT\ (\varphi_2) = \begin{bmatrix} \cos\varphi_1 & -\sin\ \varphi_1 \\ \sin\ \varphi_1 & \cos\ \varphi_1 \end{bmatrix} \begin{bmatrix} \cos\varphi_2 & -\sin\ \varphi_2 \\ \sin\ \varphi_2 & \cos\ \varphi_2 \end{bmatrix}$$

$$= \begin{bmatrix} \cos(\varphi_1 + \varphi_2) & -\sin\ (\varphi_1 + \varphi_2) \\ \sin\ (\varphi_1 + \varphi_2) & \cos\ (\varphi_1 + \varphi_2) \end{bmatrix} \tag{5.20}$$

A rotation matrix $ROT\ (\varphi) = ROT\ (\varphi_1 + \varphi_2)$ is a matrix that performs a rotation in the Euclidian space. The following equations describe the elementary rotation $ROT\ (\varphi_i)$:

$$x_r[i+1] = x_r[i]cos\ (\varphi_i) - y_r\ [i]\sin\ (\varphi_i)$$

$$y_r[i+1] = x_r[i]sin\ (\varphi_i) + y_r\ [i]\cos\ (\varphi_i) \tag{5.21}$$

Solutions of the equations (5.21) are still complex since multiplication operations are involved. Aiming to avoid multiplications, we can define the elementary angle φ_i (when a Radix-2 algorithm is used) by

$$\varphi_i = tan^{-1}(\delta_i(2^{-i})) = \delta_i \tan(2^{-i}) \tag{5.22}$$

where $\delta_i \in [-1, 1]$. By substituting the expression (5.22) into the equations (5.21), the latter obtain the following form:

$$x_r[i+1] = x_r[i] - \delta_i 2^{-1} y_r\ [i]$$

$$y_r[i+1] = y_r[i] + \delta_i 2^{-1} y x_r\ [i] \tag{5.23}$$

Now the equations (5.23) can be implemented using only addition and shift operations. A rotation step increases the length of vector A to

$$A[i+1] = \frac{A[i]}{\cos\varphi_i} = A[i](1 + tan^2\ \varphi_i)^{\frac{1}{2}}$$

As a real rotation does not change the vector length $A[i]$, it is necessary to eliminate the terms $(1 + tan^2\ \varphi_i)^{\frac{1}{2}}$ and choose the angle φ_i so that $\tan\varphi_i$ is a power of 2. Namely,

$$A[i+1] = \frac{A[i]}{\cos\varphi_i} = \left(1+ \delta_i^2 2^{-2i}\right) = \left(1+ 2^{-2i}\right)^{\frac{1}{2}} \qquad (2.24)$$

Since the CORDIC algorithm implements a sequence of micro-rotation steps, the total scaling factor S is equal to

$$S = \prod_{i=0}^{\infty}\left(1+ 2^{-2i}\right)^{\frac{1}{2}} \approx 1.6468$$

The scaling factor S is constant and independent of the rotation angle. With aim to decompose angle φ or to accumulate it, the following third recurrence equation is used:

$$z[i+1] = z[i] - \varphi_i = z[i] - \delta_i \, tan^{-1}(2^{-i}) \qquad (5.25)$$

Finally, the resulting iteration (called CORDIC micro-rotation) has the following mathematical form:

$$x_r[i+1] = x_r[i] - \delta_i 2^{-1} y_r\,[i]$$
$$y_r[i+1] = y_r[i] + \delta_i 2^{-1} y x_r\,[i]$$
$$z[i+1] = z[i] - \delta_i \, tan^{-1}(2^{-i}) \qquad (5.26)$$

Figure 2.26 presents the circuit implementation for one iteration.

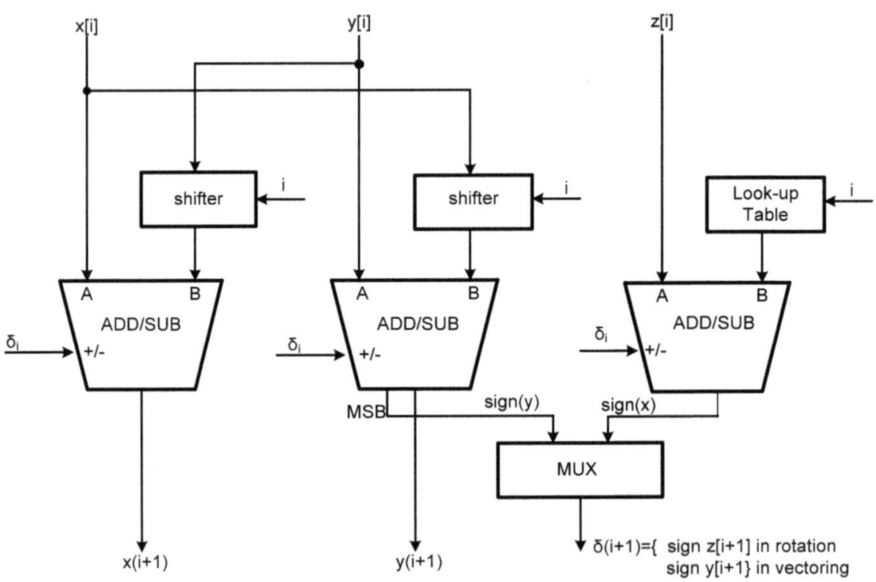

Fig. 5.26. Circuit performing one CORDIC iteration

The following set of equations describe the generalized CORDIC operations:

$$x_r[i+1] = x_r[i] - m\delta_i 2^{-1} y_r\,[i]$$

$$y_r[i+1] = y_r[i] + \delta_i 2^{-1} yx_r[i]$$

$$z[i+1] = z[i] - \delta_i \varphi_i$$

$$\varphi_i = \begin{cases} 2^{-i}, & linear \\ tan^{-1}(2^{-i}), & circular \end{cases}$$

$$\delta_i = \begin{cases} -sign(y_i), & for\ vector\ mode \\ sign\ (x_i), & for\ rotation\ mode \end{cases} \tag{5.27}$$

By choosing appropriate values for the m and φ_i parameters, we can select different coordinate systems. When m = 0, 1, -1, and the values of φ_i are $tan^{-1}(2^{-i})$, 2^{-i}, and $tanh^{-1}(2^{-1})$, the algorithm operates in linear, circular, and hyperbolic coordinate systems, respectively.

CORDIC-based Butterfly circuit

An FFT processor based on the CORDIC replicates the *sine* and *cosine* twiddle factors by CORDIC rotations and reduces the twiddle factor memories. The twiddle factor W_N is defined as

$$W_N = e^{\frac{-j2\pi}{N}} = \cos\frac{2\pi}{N} - jsin\frac{2\pi}{N}$$

Let A be a complex number

$$A = a_R + ja_I$$

When A is multiplexed by the twiddle factor, we have

$$AW_N^{mn} = (a_R + ja_I) * (\cos\frac{2\pi mn}{N} - j\sin\frac{2\pi mn}{N}) =$$

$$(a_R \cos\left(\frac{2\pi mn}{N}\right) + a_I \sin\left(\frac{2\pi mn}{N}\right)) + j\left(a_I \cos\left(\frac{2\pi mn}{N}\right) - a_R \sin\left(\frac{2\pi mn}{N}\right)\right) \tag{5.30}$$

By comparing expressions (5.16) and (5.30), we conclude that the real and imaginary part of (5.30) represents a circular rotation by angle $2\pi mn/_N$. The CORDIC algorithm is suitable for performing such a rotation. Bearing this in mind, we can write

$$X_i = (a_R \cos\left(\frac{2\pi mn}{N}\right) + a_I \sin\left(\frac{2\pi mn}{N}\right)),$$

$$Y_i = (a_I \cos(\frac{2\pi mn}{N}) - a_R \sin(\frac{2\pi nm}{N})),$$

where X_i is the real and Y_i is the imaginary part. For a Radix-2 length, the CORDIC equations are

$$x[i+1] = x[i] - \delta 2^{-i} y[i]$$
$$y[i+1] = y[i] + \delta 2^{-i} x[i]$$
$$z[i+1] = z[i] - \delta 2^{-i} tan^{-1}(2^{-i})$$

The initial conditions are

$$X[0] = a, Y[0] = b \text{ and } z[0] = \frac{2\pi mn}{N}.$$

Let us repeat that, for Radix-2, the scaling factor is $S = 1.64$. The exponential factor is realized by *sine* and *cosine* operations:

$$X = (a_R \cos\left(\frac{2\pi mn}{N}\right) + a_I \sin\left(\frac{2\pi mn}{N}\right))$$
$$Y = (a_I \cos(\frac{2\pi mn}{N}) - a_R \sin(\frac{2\pi nm}{N}))$$

Figure 5.27 shows the Butterfly computing circuit.

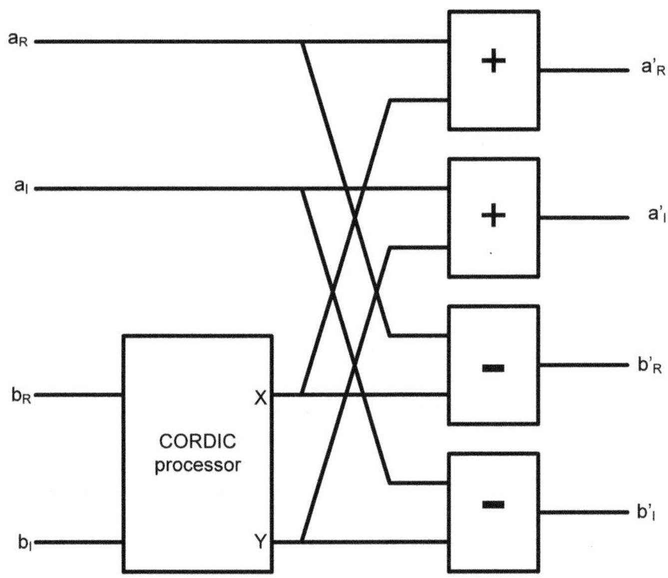

Fig. 5.27. Butterfly circuit

The Butterfly circuit performs the following computations:

$$a'_I = a_I + X, a'_R = a_R + Y, \; b'_I = a_I - X, \text{ and } b'_R = a_R - Y.$$

5.3.7 *Time and frequency synchronizer*

A synchronizer should detect an OFDM packet, store the timestamp of its arrival, and estimate the frequency offset between the received signal and reference clock. Detection of the OFDM packet relies on input-signal level variation (increase of the signal level) or change in auto-correlation of the input signal (signal repetition in the preamble). Similarly, cross-correlation between the input signal and defined preamble can identify the packet arrival timestamp. Finally, one can extract the frequency offset using either an auto-correlation or a cross-correlation function. We consider complex auto-correlation and cross-correlation functions.

Our Matlab model groups the three synchronization tasks in two entities. The first one deals with auto-correlation of the input signal (using short training symbols) to detect appearance of an OFDM packet. Afterwards, using the magnitude of the same auto-correlation function, the packet arrival time can be coarsely estimated. In addition, the argument of this function can be used for coarse frequency offset estimation. Figure 5.28 presents the corresponding Matlab code describing a simulation testbench, which includes the preparation of short and long training symbols. The insert in Figure 5.28 shows the real component of the ST and LT symbols of an OFDM packet. Figure 5.29 describes the synchronizer tasks. The insert in Figure 5.29 presents two curves: the signal amplitude at the auto-correlator output, which detects auto-correlation in ST symbols (fine line) and the cross-correlator output signal, which depicts accurate positions of ST symbols in time (bold line).

```
% simulation testbench preparing training symbols and adding noise
dsti=0;            % delay index for ST symbols starting
dlti=dsti+160;     % delay index for LT symbols starting
m=6;               % 2^m points FFT/IFFT
% real part of ST symbols in frequency domain
ast(1:64)=[0 ...                                          % null
         0  0  0 -1  0  0  0 -1  0  0  0  1  0 ... % channels 0  1 ... 13
         0  0  1  0  0  0  1  0  0  0  1  0  0 ... % channels 13 14 ... 26
         0 0 0 0 0 0 0 0 0 0 0 ...                 % null
         0  0  1  0  0  0 -1  0  0  0  1  0  0 ... % channels -26 -25...-14
         0 -1  0  0  0 -1  0  0  0  1  0  0  0]; % channels -13 -12 ... -1
% imaginary part of ST symbols in frequency domain
bst(1:64)=[0 ...                                          % null
         0  0  0 -1  0  0  0 -1  0  0  0  1  0 ... % channels  0  1 ... 13
         0  0  1  0  0  0  1  0  0  0  1  0  0 ... % channels 13  14 ... 26
         0 0 0 0 0 0 0 0 0 0 0 ...                 % null
         0  0  1  0  0  0 -1  0  0  0  1  0  0 ... % channels -26 -25...-14
         0 -1  0  0  0 -1  0  0  0  1  0  0  0]; % channels -13 -12 ... -1
% real part of LT symbols in frequency domain
alt(1:64)=[0 ...                                          % null
         1 -1 -1  1  1 -1  1 -1  1 -1 -1 -1 -1 ... % channels 1  2 ... 13
        -1  1  1 -1 -1  1 -1  1 -1  1  1  1  1 ... % channels 14  15 ... 26
         0 0 0 0 0 0 0 0 0 0 0 ...                 % null
         1  1 -1 -1  1  1 -1  1 -1  1  1  1  1 ... % channels -26 -25...-14
```

```
          1  1 -1 -1  1  1 -1  1 -1  1  1  1  1];  % channels -13 -12 ... -1

% imaginary part of LT symbols in frequency domain
blt(1:64)=[0 ...                                       % null
          0  0  0  0  0  0  0  0  0  0  0  0  0 ... % channels  0  1 ... 13
          0  0  0  0  0  0  0  0  0  0  0  0  0 ... % channels 14 15 ... 26
          0 0 0 0 0 0 0 0 0 0 0 ...                    % null
          0  0  0  0  0  0  0  0  0  0  0  0  0 ... % channels -26 -25...-14
          0  0  0  0  0  0  0  0  0  0  0  0  0];  % channels -13 -12 ... -1

% signal buffer
xx(1:dsti+320)=0;
yy(1:dsti+320)=0;

% plot1 and plot2 signals
plot1(1:dsti+320)=0;
plot2(1:dsti+320)=0;

% setup ST symbol
d=0;                            % perform IFFT
[xx(dsti+33:dsti+96), yy(dsti+33:dsti+96)]=m_fft(d, m, ast, bst);
c1=sqrt(13/6);                  % normalize power
for i=dsti+33:dsti+96
  xx(i)=xx(i)*c1;
  yy(i)=yy(i)*c1;
end
for i=dsti+97:dsti+160          % copy
  xx(i)=xx(i-64);
  yy(i)=yy(i-64);
end
for i=dsti+1:dsti+32            % add cyclic prefix
  xx(i)=xx(i+64);
  yy(i)=yy(i+64);
end

% reference ST symbol
for i=1:16
  rxx(i)=xx(dsti+32+i);
  ryy(i)=yy(dsti+32+i);
end

% setup LT symbol
d=0;                            % perform IFFT
[xx(dlti+33:dlti+96), yy(dlti+33:dlti+96)]=m_fft(d, m, alt, blt);
for i=dlti+97:dlti+160          % copy
  xx(i)=xx(i-64);
  yy(i)=yy(i-64);
end
for i=dlti+1:dlti+32            % add cyclic prefix
  xx(i)=xx(i+64);
  yy(i)=yy(i+64);
end

% add noise
sigma=1;
[xx, yy]=m_add_noise(sigma, 1, dsti+320, xx, yy);
xx
yy
```

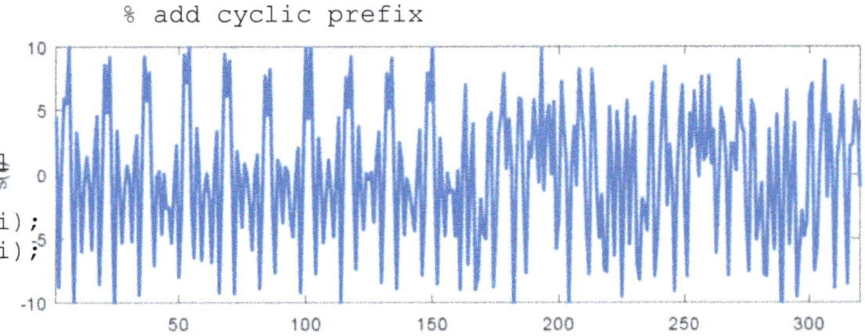

Fig. 5.28. Matlab model of the synchronizer simulation testbench

```
% start synchronizer
r=1;
[STD, CFO, plot1, plot2]=m_synchronizer(r, 20, dsti+320, xx, yy, rxx, ryy);
STD
CFO
% optional signals plotting
subplot(2, 1, 1);
tt=1:dsti+320;
plot(tt,xx(tt),"linewidth",2)
ylabel('s(t)');
xlabel('t');
axis([1 dsti+320 -10 10]);
subplot(2, 1, 2);
plot(tt,8*plot1(tt),tt,plot2(tt),"linewidth",2)
ylabel('s(t)');
xlabel('t');
axis([1 dsti+320 -10 10]);
% synchronizer function
% plot1 output auto-correlator, plot2 output cross-correlator
% a input I signal, b input Q signal
% n searching starting index n > 16
% m searching stopping index m > n+16
% xst input ST reference I signal, yst input ST reference Q signal
% r=1 reset internal buffers, r=0 continue searching
function [STD, CFO, plot1, plot2]=m_synchronizer(r, n, m, a, b, xst, yst)
  ca=0.0625;    % averaging coefficient for auto-correlation amplitude
  cb=0.0625;    % averaging coefficient for auto-correlation argument
  th1=0.7;      % threshold for plateau detection
  th2=0.5;      % threshold for coarse LT starting detection
  th3=0.5;      % latest repeated cross-correlation threshold
% cross-correlation range rca + rcb
  rca=64;       % start cross-correlator with rca previous samples
  rcb=32;       % check additional rcb samples
  if r==1
    sx(1:16)=0;
    sy(1:16)=0;
    sp(1:16)=0;
    si=1;
    sumx=0;
    sumy=0;
    sump=0;
    avr_sum=0;
    avr_sumx=0;
    avr_sumy=0;
  end
  STD=1;
  for i=n:m
% start auto-correlator
% next product [a(i)+jb(i)]*[a(i-16)-jb(i-16)]
    x=a(i)*a(i-16) + b(i)*b(i-16);
    y=b(i)*a(i-16) - a(i)*b(i-16);
    p=a(i)*a(i) + b(i)*b(i);
% remove oldest product from sum
    sumx=sumx-sx(si);
    sumy=sumy-sy(si);
    sump=sump-sp(si);
% add new product sum
    sumx=sumx+x;
    sumy=sumy+y;
    sump=sump+p;
% update auto-correlator register
```

```
    sx(si)=x;
    sy(si)=y;
    sp(si)=p;
    si=si+1;
    if si>16
      si=1;
    end
    if sump<0.00001
      sump=0.00001;
    end
    t_sum=sqrt(sumx*sumx+sumy*sumy)/sump;
    plot1(i)=t_sum;
    plot2(i)=0;
    avr_sum=(t_sum*ca)+avr_sum*(1-ca);  % average auto-correlation amplitude
    avr_sumx=(sumx*cb)+avr_sumx*(1-cb); % average auto-correlation argument
    avr_sumy=(sumy*cb)+avr_sumy*(1-cb); % average auto-correlation argument
    if avr_sum>th1      % plateau detection
      if t_sum<th2      % ST to LT transition
        CFO=atan2(avr_sumy,avr_sumx)/16;
        avr_sum=0;
        avr_sumx=0;
        avr_sumy=0;
        cdelay=i;
% generate CFO corrected signal in searching range rca+rcb+16
        for j=1:rca+rcb+17
          j1=j+cdelay-rca-17;
          cc=cos(CFO*j*50e-9);
          ss=sin(CFO*j*50e-9);
          cfx(j)=cc*a(j1)+ss*b(j1);
          cfy(j)=cc*b(j1)-ss*a(j1);
        end
% start cross-correlator from cdelay-rca to cdelay+rcb
        ccm=0;
        for j=17:rca+rcb+17
          ccx=0;
          ccy=0;
          for k=1:16
% cross-correlation product [a(j)+j*b(j)]*[xst(k)-j*yst(k)]
            ccx=ccx + cfx(j-k)*xst(17-k) + cfy(j-k)*yst(17-k);
            ccy=ccy + cfy(j-k)*xst(17-k) - cfx(j-k)*yst(17-k);
          end
          cct=(ccx*ccx+ccy*ccy)/100000;
          plot2(j+cdelay-rca-17)=cct;
          if STD==j+cdelay-rca-33
            if cct>ccm*th3      % prefer latest repeated cross-correlation
              ccm=cct;
              STD=j+cdelay-rca-17;
            end
          else
            if cct>ccm
              ccm=cct;
              STD=j+cdelay-rca-17;
            end
          end
        end
      end
    end
  end
end
```

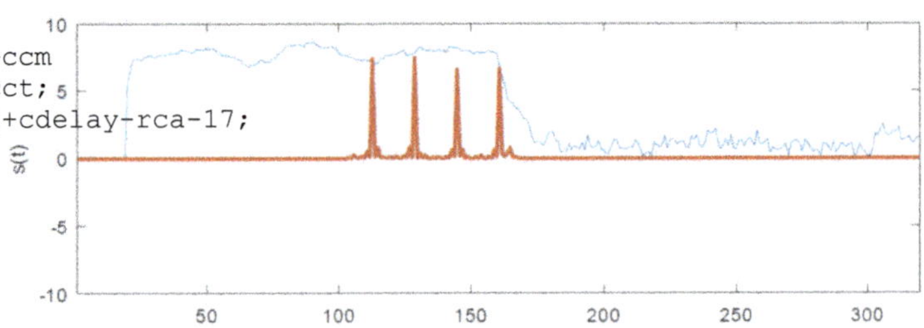

Fig. 5.29. Matlab model of the time and frequency synchronizer

5.3.8 *Channel estimator*

Channel estimator should collect information about the state of an OFDM channel and estimate the signal distortion introduced by this channel. The estimation process repeats periodically using known short and long training symbols. Taking into account typical values of the coherence time and coherence bandwidth of an OFDM channel, the standard IEEE 802.11a defines an adequate number of pilots (training symbols). As already mentioned in Part 1, Chapter 2 (Digital Baseband), the commonly used methods of channel estimation and tracking are a) time-domain windowing; b) frequency-domain interpolation; and c) coded decision-directed demodulation (CD3).

As lower complexity of the frequency-domain interpolation (comparing to the time-domain windowing complexity) secures a broader application field, we present an example of the interpolation in frequency-domain using the raised-cosine coefficients (RCI). These interpolation coefficients provide a good noise and aliasing attenuation, preserve most CIR information, and keep a high estimation accuracy. Figure 5.30 describes a Matlab model of the RCI channel estimator.

```
% channel estimator
m=64;
ro=0.66;
f1(1:m+m+1)=0;
%a(1:4)=[1, 1, 1, 1];
%b(1:4)=[1, 1, 1, 1];
a(1:4)=[1, 2, 1.6, 2.8];
b(1:4)=[1, 0.5, 2.5, 1.2];
tn=0.000001;
rn=ro;
f1(m+1)=(sin(pi*tn)/(pi*tn))*(cos(pi*rn*tn)/(1-(4*rn*rn*tn*tn)));
for i=1:m
  tn=4*i/64;
  rn=ro;
  if tn<0
    tn=-tn;
  end
  if (1-4*tn*ro)<0.0000001
    rn=ro+0.0000001;
  end
  if tn<0.0000001
    tn=tn+0.0000001;
  end
  f1(m+1+i)=(sin(pi*tn)/(pi*tn))*(cos(pi*rn*tn)/(1-(4*rn*rn*tn*tn)));
  f1(m+1-i)=f1(m+1+i);
end
% optional signals plotting
subplot(2, 1, 1);
xx=-m:m;
plot(xx/16,f1(m+1+xx),"linewidth",2)
ylabel('h(t)');
axis([-m/16 m/16 -0.5 1.5]);
x, y]=m_RC_interpolator(a, b, ro);
```

```
x
y
c(1)=x(27-21);
c(2)=x(27-7);
c(3)=x(27+7);
c(4)=x(27+21);
d(1)=y(27-21);
d(2)=y(27-7);
d(3)=y(27+7);
d(4)=y(27+21);
a
b
c
d
c-a
d-b
% optional signals plotting
subplot(2, 1, 2);
x1=1:53;
plot(x1-27,x(x1),x1-27,y(x1),"linewidth",2)
ylabel('h(f)');
axis([-26 26 -3 3]);

% RCI function
% x output I signal (channels state -26, -25, ... 26)
% y output Q signal (channels state -26, -25, ... 26)
% a input I signal (pilots channels state: -21, -7, 7, 21)
% b input Q signal (pilots channels state: -21, -7, 7, 21)
% ro roll-off factor
function [x, y]=m_RC_interpolator(a, b, ro)
  for j=1:53
    k=j-27;
% initial interpolator waits
    if k>21                        % channels 22, ... 26
      d(1)=(43-k)/22;                % distance to channel -21
      d(2)=1+(43-k)/22;              % distance to channel -7
      d(3)=1+(-21+k)/22;             % distance to channel 7
      d(4)=(-21+k)/22;               % distance to channel 21
    else
      if k>7                       % channels 8, ... 21
        d(1)=1+(21-k)/14;            % distance to channel -21
        d(2)=1+(-7+k)/14;            % distance to channel -7
        d(3)=(-7+k)/14;              % distance to channel 7
        d(4)=(21-k)/14;              % distance to channel 21
      else
        if k>-7                    % channels -6, ... 7
          d(1)=(-21-k)/14;           % distance to channel -21
          d(2)=(-7-k)/14;            % distance to channel -7
          d(3)=(7-k)/14;             % distance to channel 7
          d(4)=(21-k)/14;            % distance to channel 21
        else
          if k>-21                 % channels -20, ... -7
            d(1)=(21+k)/14;          % distance to channel -21
            d(2)=(-7-k)/14;          % distance to channel -7
            d(3)=1+(-7-k)/14;        % distance to channel 7
            d(4)=1+(21+k)/14;        % distance to channel 21
          else                     % channels -21, ... -26
            d(1)=(-21-k)/22;         % distance to channel -21
            d(2)=1+(-21-k)/22;       % distance to channel -7
            d(3)=1+(43+k)/22;        % distance to channel 7
            d(4)=(43+k)/22;          % distance to channel 21
```

```
          end
        end
      end
    end
    for i=1:4
      tn=d(i);
      rn=ro;
      if tn<0
        tn=-tn;
      end
      if (1-4*tn*ro)<0.0000001
        rn=ro+0.0000001;
      end
      if tn<0.0000001
        tn=tn+0.0000001;
      end
      ff(i)=(sin(pi*tn)/(pi*tn))*(cos(pi*rn*tn)/(1-(4*rn*rn*tn*tn)));
    end
% start interpolator
    x(j)=0;                       % initial real part
    y(j)=0;                       % initial imaginary part
    for i=1:4
      x(j)+=a(i)*ff(i);           % summing real part
      y(j)+=b(i)*ff(i);           % summing imaginary part
    end
  end
end
```

Fig. 5.30. Matlab model of the RCI channel estimator

5.4 System Bus

System bus architecture is among the top challenges in system-on-chip (SOC) design due to rapidly increasing operation frequency and growing chip size. In general, the performance of an SOC design heavily depends upon the efficiency of its bus structure. The balance of computation and communication tasks in any application fundamentally determines the delivered performance. It is possible to reach this balance using standard on-chip bus structures. Currently there are a few publicly available bus architectures from the leading manufacturers, such as CoreConnect from IBM [25], AMBA from ARM [26], and Wishbone from OpenCores [27]. These bus architectures usually correspond to the processor architectures to work with and require minimal extra interface logic. Figure 5.31 shows block diagrams of the AMBA high-performance and peripheral buses [28].

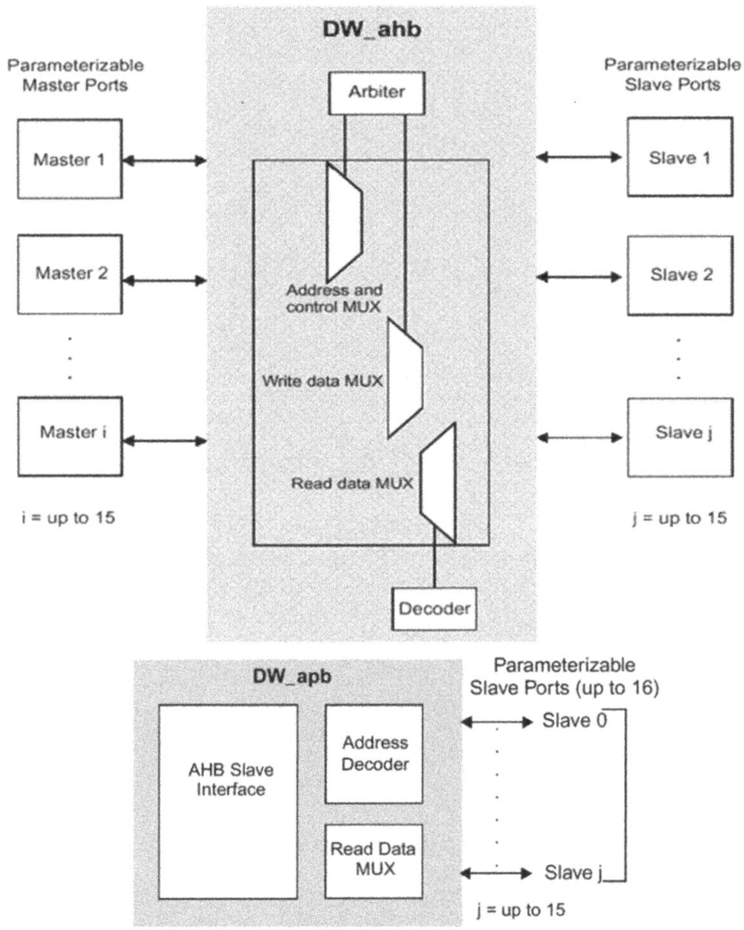

Fig. 5.31. Block diagram of a system bus

5.5 Memory Controller

A programmable memory controller usually controls the external memory bus. The controller shown in Figure 5.32 [28] acts as a slave on the system bus. Memory configuration registers help to program the memory controller. The memory bus provides a direct interface to PROM, memory mapped I/O devices and asynchronous static RAM (SRAM). The controller can decode and select several PROM banks, I/O banks, and SRAM banks. Therefore, there are numerous chip-select signals in the memory controller.

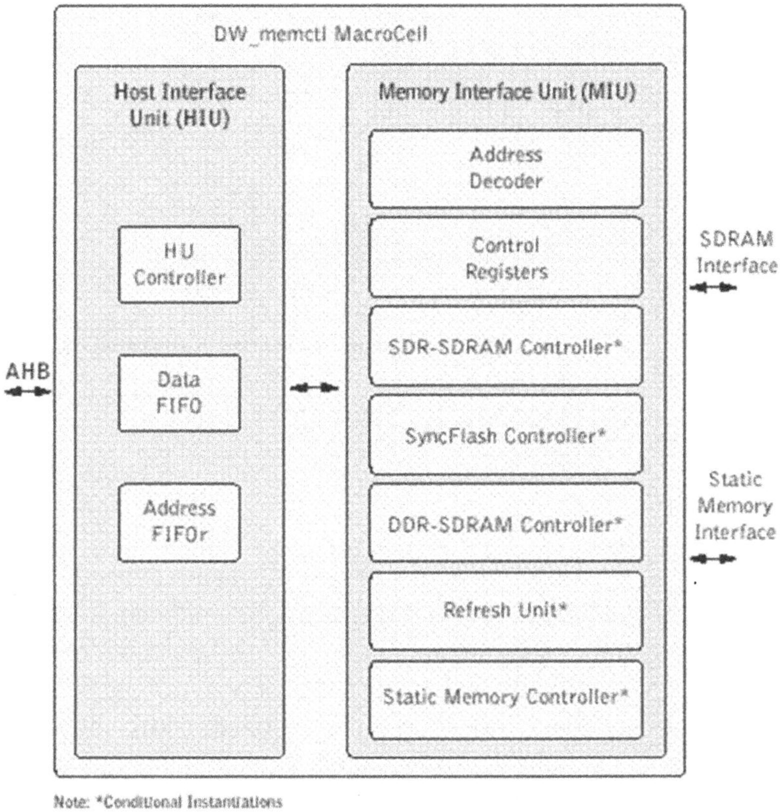

Fig. 5.32. Block diagram of a memory controller

5.6 Debug Support Unit

The debug support unit (DSU) takes the control during the processor debug mode. The DSU attaches to the system bus as a slave. Through the specific address space, any bus master can access the processor registers. The DSU control registers are accessible at any time, while the processor registers and caches are accessible when the processor has been in debug mode. In the debug mode, the processor pipeline stalls and DSU takes control over the processor. Figure 5.33 presents a functional block diagram of the JTAG (a simple debugger named after the Joint Test Action Group) with a test access port [29].

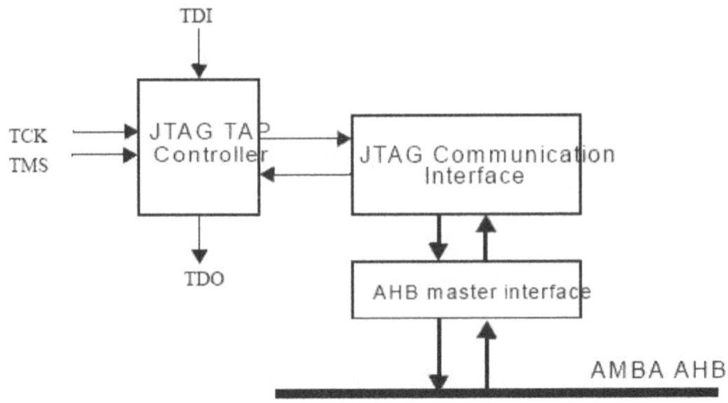

Fig. 5.33. JTAG debugger block diagram

5.7 Peripherals

Interrupt controller is used to prioritize and propagate interrupt requests from internal or external devices to the execution unit. The diagram in Figure 5.34 [30] illustrates the structure of an interrupt controller. All potential external interrupt sources, except one, are routed and prioritized resulting in a single interrupt request to the processor core. Two interrupt sources are treated differently by connecting directly to the processor and generating a machine check exception. These are the software watchdog timer and interrupt request zero. The software watchdog timeout indicates a severe software problem, typically an endless loop or runaway code, requiring the processor to stop normal operation and report or attempt to correct the error. The external interrupt request zero signal is reserved for extreme errors that also require similar processor action to the software watchdog timeout. The remaining external inputs can be masked. There is control over relative priorities between all these inputs. The port C can be selected to generate a request on any change in level. The inputs come from the event registers of the controllers. In some cases, the input can reflect numerous events. There is only one input from each controller, so if that controller is capable of reporting more than one event, then work is necessary to find the original cause. After receiving an interrupt request, the CPU initially recognizes it as an external interrupt request. However, the interrupt controller provides the means of determining which of the inputs generated the request.

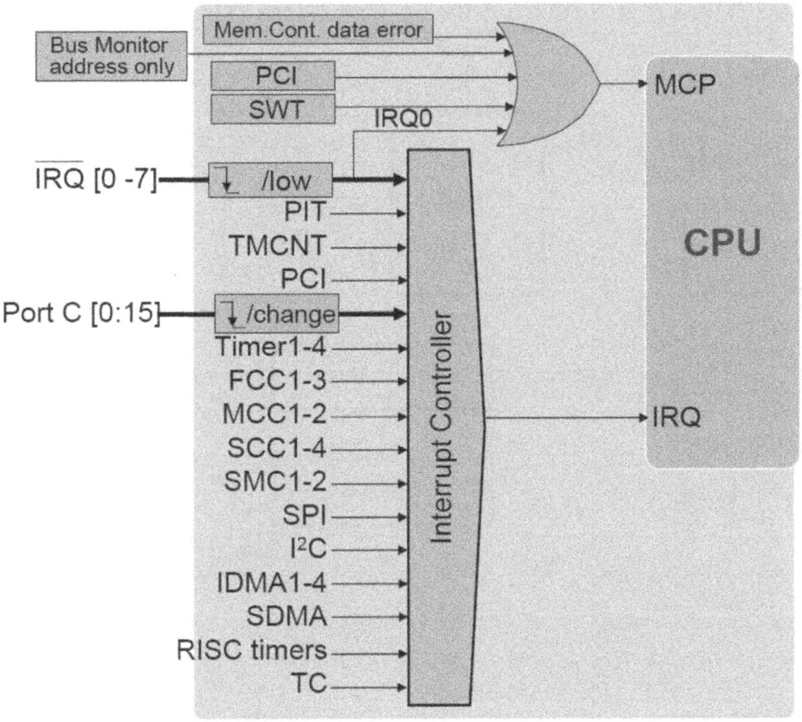

Fig. 5.34. Interrupt controller block diagram

Timer is a counter with several operating modes and capture/compare registers. It also supports multiple capture/compare operations, pulse generation, configurable outputs, asynchronous input and output latching, interval timing, and interrupts. Different components can generate interrupts: counters on overflow conditions and capture/compare registers. It can also send synchronization events to the direct-memory access (DMA) controller. The timer in Figure 5.35 [31] has two signaling modes and accepts an internal or an external clock source. The input and output pins (TINP and TOUT) can function as timer clock input and output or general-purpose input and output. With an internal clock, for example, the timer can signal an external A/D converter to start a conversion, or it can trigger the DMA controller to begin a data transfer. With an external clock, the timer can count external events and interrupt the processor after a specified number of events.

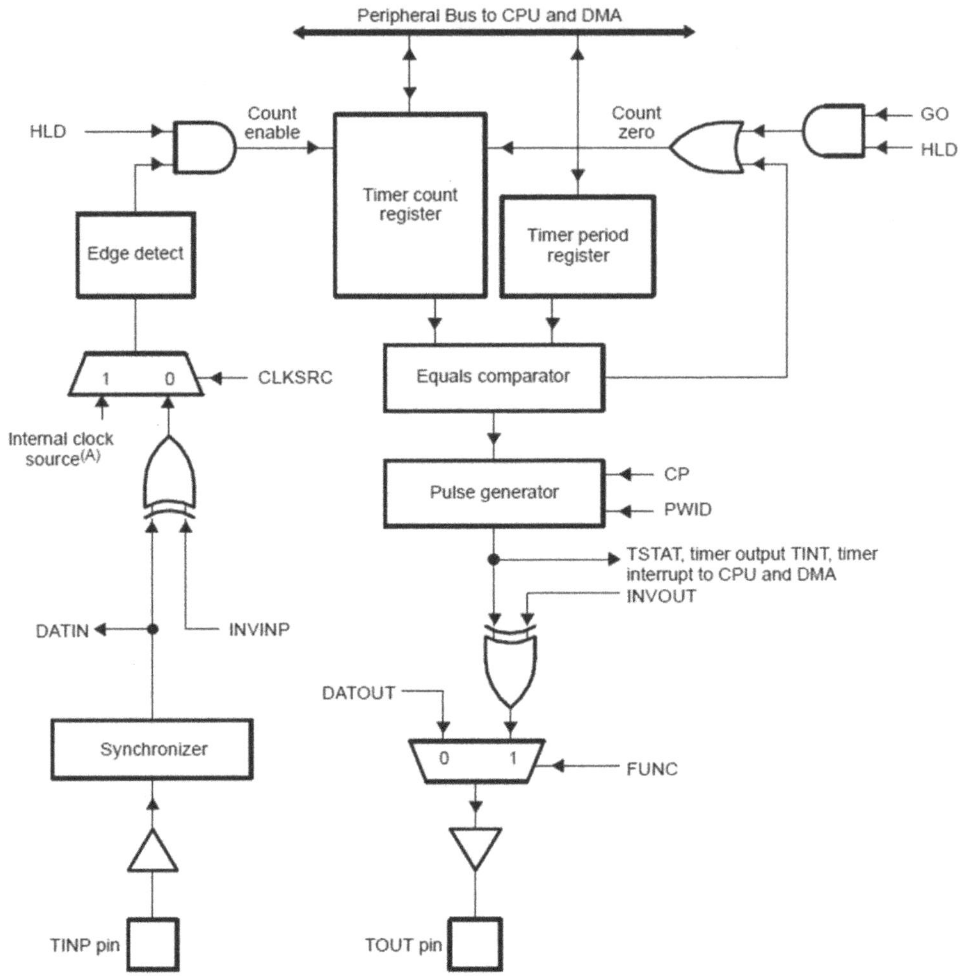

Fig. 5.35. Timer block diagram

Universal asynchronous receiver/transmitter (UART) supports communication with serial input/output devices [32]. Typically, the UART connects a central processor and a serial device. The processor can write to or read from the UART as a parallel port. To the serial device, the UART presents two data wires, one for input and one for output, which serially communicate data. The rate of data communication depends on the peripheral device. Some devices operate at a single clock speed and generate an internal clock. Other devices operate at multiple clock rates and get their clock input from the UART. Figure 5.36 shows a functional block diagram of the UART.

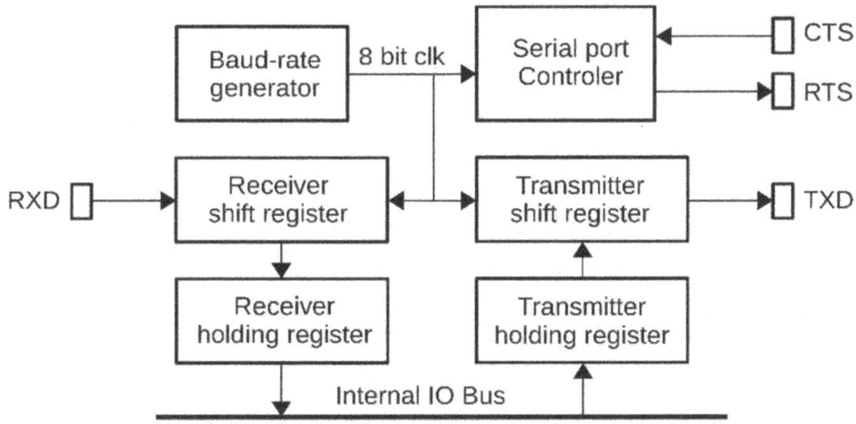

Fig. 5.36. UART block diagram

Parallel I/O port allows the processor core to interface to the user connectors. For each of the dedicated pins, the direction is programmable. The default direction is "input" after reset. In a typical multi-tasking environment, each process only owns some specific bits in an I/O register. Figure 5.37 presents a block diagram of the parallel I/O port described in [28].

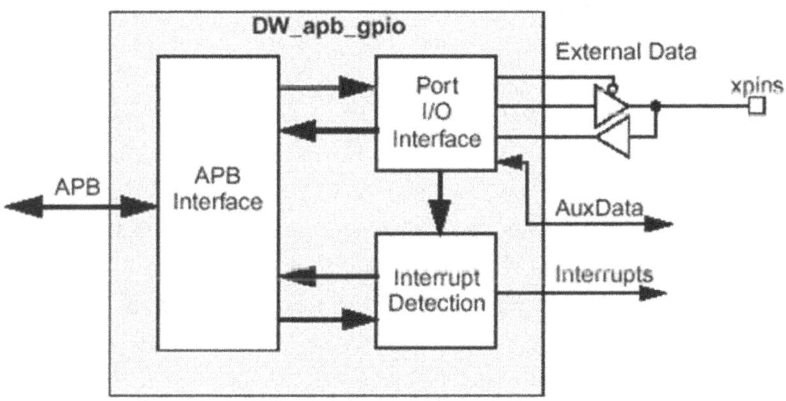

Fig. 5.37. I/O port block diagram

I²C controller supports a two-wire, bi-directional serial bus that provides a simple and efficient means of data exchange between devices. It is most suitable for applications requiring occasional communication over a short distance between many devices. The I²C standard [33] is a true multi-master bus including collision detection and arbitration that prevents data corruption if two or more masters attempt to control the bus simultaneously.

References

1. Hennessy, J. L. and Patterson, D. A. (2007) *Computer Architecture: A Quantitative Approach*, (Morgan Kaufmann, San Francisco, USA).
2. https://www.intel.com/content/www/us/en/products/processors.html
3. https://www.amd.com/en/ryzen
4. https://www.arm.com/products/silicon-ip-cpu
5. https://www.mips.com
6. https://www.gaisler.com/index.php/products/processors
7. Panda, P. R., Dutt, N. and Nicolau, A. (1999) *Memory Issues in Embedded Systems-On-Chip: Optimizations and Exploration*, (Kluwer Academic Publishers, Boston, USA).
8. Albonesi, D. H. (1999). Selective cache ways: on-demand cache resource allocation, *Proc. 32nd Annual International Symposium on Microarchitecture*, pp. 248–259.
9. Shiue, W. T. and Chakrabarti, C. (1999). Memory exploration for low power embedded systems, *Proc. 36th Design Automation Conference*, pp. 140–145.
10. Malik, A., Moyer, B. and Cermak, D. (2000). A low power unified cache architecture providing power and performance flexibility, *Proc. International Symposium on Low Power Electronics and Design*, pp. 241–243.
11. Givargis, T., Vahid, F. and Henkel, J. (2002). System-level exploration for Pareto-optimal configurations in parameterized SoC, *IEEE Trans. on VLSI Systems*, 10, pp. 416–422.
12. Banakar, R., Steinke, S., Lee, B.-S., Balakrishnan, M. and Marwedel, P. (2002). Scratchpad memory: a design alternative for cache on-chip memory in embedded systems, *Proc. 10th Intern. Symposium on Hardware/Software Codesign*, pp. 73–78.
13. Stamenkovic, Z., Vater, F. and Dyka, Z. (2003). A framework for selection of cache configurations for low power, *Proc. 4th International Workshop on IP-Based System-on-Chip Design*, pp. 137–140.
14. Kalla, P., Hu, X. S. and Henkel, J. (2005). A flexible framework for communication evaluation in SoC design, *Proc. 10th Asia and South Pacific Design Automation Conference*, pp. 956–959.
15. https://www.mips.com/products/classic
16. https://www.synopsys.com/dw/ipdir.php?ds=dwc_sram_memory_compiler
17. https://www.dolphin-technology.com/memory-compilers
18. Gopi, E. S. (2016) *Digital Signal Processing for Wireless Communication Using Matlab*, (Springer, New York, USA).
19. Sun Y. and Ding, Z. (2012). FPGA design and implementation of a convolutional encoder and a Viterbi decoder based on 802.11a for OFDM, *Wireless Engineering and Technology*, 3, pp. 125–131.

20. Kienle, F. (2014) *Architectures for Baseband Signal Processing*, (Springer, New York, USA).
21. Lyons, R. G. (2011) *Understanding Digital Signal Processing*, (Prentice Hall, Englewood Cliffs, USA).
22. Hu, Y. J. (1992). CORDIC-based VLSI architectures for digital signal processing, *IEEE Signal Processing Magazine*, 9, pp. 16–35.
23. Lakshmi, B. and Dhar, A. S. (2010). CORDIC architectures: A survey, *VLSI Design Hindawi*, pp. 1–19.
24. Tomar, R., Singh, P. K. and Raj, K. (2013). A review of CORDIC algorithms and architectures with applications for efficient designing, *International Journal of Scientific and Engineering Research*, 4, pp. 236–246.
25. CoreConnect, https://www.xilinx.com/Attachment/128bitPlbBus_v4_6.pdf
26. AMBA, https://www.arm.com/products/silicon-ip-system/embedded-system-design/amba-specifications
27. Wishbone, http://cdn.opencores.org/downloads/wbspec_b3.pdf
28. https://www.synopsys.com/designware-ip/soc-infrastructure-ip/designware-library.html
29. https://www.xjtag.com/about-jtag/what-is-jtag/
30. https://www.nxp.com/files-static/training_pdf/24819_PQ2_INT_CONTROL_WBT.pdf
31. https://www.ti.com/lit/ug/spru582b/spru582b.pdf
32. https://whatis.techtarget.com/definition/UART-Universal-Asynchronous-Receiver-Transmitter
33. https://www.ti.com/lit/an/slva704/slva704.pdf

https://doi.org/10.1142/9789811210723_0006

Chapter 6

AMS Circuit Budgets

Gildas Leger and **Antonio Gines**
Consejo Superior de Investigaciones Científicas
Instituto de Microelectrónica de Sevilla
Seville, Spain

6.1 Introduction

This chapter provides an overview of the design methodology of embedded transceivers. In part 1, we have seen how the main specifications of the communication standards can be translated into high-level requirements for both the transmitter and the receiver part of a transceiver. In this chapter we will take a step downward in the abstraction level and show how meeting these requirements depends on the correct balance between building blocks.

The high-level requirements for transceivers are expressions of the maximum errors that can be tolerated. For any architecture, these errors stem from two main degradation mechanisms: noise and distortion. Depending on the block where they show up, these degradation mechanisms can produce different effects. Noise in the PLL can give rise to clock jitter, while it will degrade the Noise Figure of an LNA. Distortion in a PA produces in-band spectral regrowth due to intermodulation but also significant RF power around harmonics of the carrier. Distortion is also the cause of receiver de-sensitization through compression. On top of these fundamental degradation mechanisms, other constraints may enter into play, like power consumption or silicon area. As a result of the complex interactions between all these parameters, the design of a transceiver is seldom a one-way procedure.

Despite it is possible to draw some analytical expressions of the transceiver error budgets, the complexity of the interactions between the different error sources suggest that the best way to proceed is to use a top-down design strategy and generate high-level models of the main building blocks with their first order non-idealities. Today's computational power allows running optimization in the high-level design space at low cost.

However, dropping the responsibility of the performance evaluation to the machine entails a risk of poor design choices. It is necessary for the designer to keep a phenomenological understanding of what is going on in the optimization loop. Most of the time, several iterations including architectural changes or refinements are necessary to strike the right balance.

6.2 From High-Level Specifications to Low(er)-level Requirements

6.2.1 *Block-level metrics*

For an analog designer, it is usually quite natural to separate non-idealities in two categories: noise and distortion. Noise is associated to a stochastic random process that is characterized by a spectral density which can vary with frequency — for instance, thermal noise has a flat density while flicker noise being higher at low frequency.

On the other hand, distortion is what makes a transfer function of a supposedly linear device deviate from a straight line. Typically, a smooth distortion is modelled by a low-order polynomial. Associated to these two categories we then find some low-level metric.

6.2.1.1 *Noise metrics*

Signal-to-Noise Ratio (SNR) simply relates the power of noise to the signal power:

$$SNR = \frac{P_{signal}}{P_{noise}} \quad \Leftrightarrow \quad SNR_{dB} = 10log_{10}\left(\frac{P_{signal}}{P_{noise}}\right) = P_{signal,dBm} - P_{noise,dBm} \qquad (6.1)$$

This relative metric is actually of much more interest than the bare noise power since a signal is practically detectable only if its power is above the noise level.

The Noise Figure, on the other hand, expresses the amount of noise added by a device:

$$NF_{dB} = 10\log_{10}\frac{SNR_{in}}{SNR_{out}} = 10\log_{10}\frac{P_{in} / P_{noise}}{SNR_{out}} \tag{6.2}$$

For instance, if the block under consideration is the complete receiver, P_{in} is the input signal power and P_{noise} is the antenna thermal noise in the bandwidth BW of interest (assuming matching, $R_S \approx R_{in}$),

$$P_{noise} = 4KTR_S \cdot \left(\frac{R_S}{R_S + R_{in}}\right)^2 \cdot BW \approx KT \cdot BW \tag{6.3}$$

The sensitivity threshold at room temperature can thus be derived as follows,

$$Sensitivity\big|_{dBm} = \underbrace{-174\big|_{dBm/Hz} + NF\big|_{dB} + 10\log_{10}(BW)}_{\text{Integrated Thermal Noise Floor } (TNF)} + SNR_{\min}\big|_{dB} \tag{6.4}$$

The standard assumes a typical Noise Figure (NF) of 10dB with 5dB additional margin for dealing with the implementation errors, such as antenna insertion-loss, non-linearity or IQ impairments. Particularizing the sensitivity expression for this assumption, we can obtain the minimum *SNR* for the receiver as,

$$SNR_{\min}\big|_{dB} = Sensitivity\big|_{dBm} + 174\big|_{dBm/Hz} - 10\log_{10}(BW) - 15dB \tag{6.5}$$

This equation shows that if, for a given modulation scheme, the signal-to-noise ratio is to be maintained, the sensitivity must be scaled by 3dB by doubling the bandwidth. This trend is precisely the one observed in the standard requirement (see Table 3.4 of Part 1). To assure compatibility with the different modulation schemes, the worst case should be evaluated according to the higher order modulation, that is 256-QAM with 5/6 code rate, for which the sensitivity ranges between −57dBm and −48dBm depending on the bandwidth, leading in all cases to a minimum *SNR* of 29dB.

This estimation is based on NF limit of 10dB. In practice, this feature is easily achievable in the current state-of-the-art, as shown in Table 6.1, taking advantage of the difference to improve sensitivity with respect to the minimum requirement obtaining more competitive figure of merit (FoM). Actually, in [1], NF values of 2.9 and 3.2 dB are reported for the 2.4GHz and 5GHz bands, respectively. This improvement allows reducing the sensitivity from –51dBm to –65.4dBm for the higher modulation with 80MHz bandwidth.

Table 6.1. Specifications of some state-of-the-art receivers compliant with the IEE 802.11

		JSSCC18 [1]	ISSCC17 [2]	JSSCC17[3]	CICC15 [4]	RFIC14 [5]	ISSCC14 [6]
Standards		4x4 11 abgn/ac/ax	4x4 11 abgn/ac	2x2 11 abgn/ac	1x1 11 abgn/ac	2x2 11 abgn/ac	3x3 11 abgn/ac
Technology node		28nm	40nm	40nm	55nm	55nm	40nm
TX EVM (dB)	**2.4G**	-42.1 (n,64QAM) -42.5 (ax,40M,1KQAM)	NA	-40 (20M, Floor)	NA	-38 (20M, Floor)	-41 (HT40)
	5G	-38.4 (ac,80M,256 QAM) -38.1 (ax,80M,1KQAM)	-36.5 (ac,80M,MCS9)	-38 (20M, Floor)	-37.8 (80M, Floor)	-37.5 (80M, Floor)	-37
RX sensitivity (dBm)	**2.4G**	-78.4 (g,54M)	-77 (LG,54M)	-78.3 (54M)	-77.5 (LG,54M)	-77.5 (LG,54M)	NA
	5G	-64.2 (ax,40M,1kQAM)	-62 (ac,80M,MCS9)	-66 (MCS 9)	-63.5 (80M, MCS9)	-62.5 (ac,80M, MCS9)	NA
RX NF (dB)	**2.4G**	2.9	NA	2.9	3	3.8	3
	5G	3.2	NA	4.5	3.7	4	4.3
RF power consumption (mW)	**TX 2.4G**	844 (4SS+1LO, -5dBm)	3863 (4SS+1LO, 21dBm)	1460 (2SS+1LO, 20dBm)	705 (1SS+1LO, 20.5dBm)	1588 (2SS+1LO, 20dBm)	1080 (3SS+1LO, full SoC)
	TX 5G	832 (4SS+2LO, -5dBm)	4164 (2SS+2LO, 21dBm)	1750 (2SS+2LO, 18dBm)	996 (2SS+1LO, 18.5dBm)	1722 (2SS+1LO, 17.5dBm)	1170 (3SS+1LO)
	RX 2.4G	354 (4SS +1LO)	297 (4SS +1LO)	179 (2SS +1LO)	84 (1SS +1LO)	303 (2SS +1LO)	1170 (3SS+1LO, full SoC)
	RX 5G	447 (4SS +2LO)	474 (4SS +2LO)	243 (2SS +2LO)	156 (1SS +1LO)	317 (2SS +1LO)	2080 (3SS+1LO, full SoC)
Image rejection ratio (post cal, dB)		-64 (TX @5MHz)	-61 (TX @5MHz)	NA	-45 (RX)	-45 (RX)	NA
Area (mm^2)		12	11.4	8.6	3.4	7.7	21.5 (full SoC)

6.2.1.2 *Distortion metrics*

The smooth non-linearity in RF blocks is usually represented by the 1dB compression point and the Third-order Input-referred Intercept Point IIP3 (IP3 would be the same intercept point referred at the output). These metrics can easily be related to the coefficients of a polynomial approximation of the transfer function.

If a third-order non-linearity model, $y(t) = \alpha_1 x(t) + \alpha_2 x(t)^2 + \alpha_3 x(t)^3$, is considered with a sinusoidal stimulus of amplitude A_{in}, the effective gain at the tone frequency will be reduced by the effect of the third order coefficient [7],

$$Gain = \alpha_1 + \frac{3}{4}\alpha_3 A_{in}^2 \quad \text{with} \quad \alpha_1\alpha_3 < 0 \tag{6.6}$$

This reduction of the effective gain for high power inputs is known as a compressive behavior. Hence, the associated metric is the 1dB compression point, which is the amplitude at which the gain is reduced by 1dB with respect to its small-signal behavior. It can be evaluated as,

$$A_{1dB} = \sqrt{0.145|\alpha_1 / \alpha_3|} \tag{6.7}$$

In addition, polynomial-like non-linearity generates intermodulation components when two tones are applied at the input. Figure 6.1 shows this effect for a third-order non-linear model. If the interference tones at the frequency ω_1 and ω_2 are placed at the adjacent and non-adjacent channels, the intermodulation product at ($\omega_{in} = 2\omega_1 - \omega_2$) will lay precisely on the channel of interest corrupting the wanted information, the error signal being,

$$IM \text{ at } \omega_{in} - 2\omega_1 - \omega_2 : \quad \underbrace{\frac{3}{4}\alpha_3 A_1^2 A_2}_{A_{IM}} \cdot \cos(\omega_{in} t) \tag{6.8}$$

The amplitude at which the power of the fundamental, $|\alpha_1 A_{in}|$ become equal to the power of the intermodulation component, $|3/4\alpha_3 A_{in}^3|$, is called *Input-Referred Third Intercept Point* (IIP3),

$$A_{IIP3} = \sqrt{4/3 \cdot |\alpha_1 / \alpha_3|} \tag{6.9}$$

the value of which is related to the 1-dB compression point in term of amplitude and power according to Eq. (6.7), as follows,

$$A_{IIP3} \approx \sqrt{4/0.46} \cdot A_{in,1dB} \quad \rightarrow \quad IIP3\Big|_{dBm} \approx P_{1dB}\Big|_{dBm} + 9.6dB \qquad (6.10)$$

Notice that this equivalence only holds for a strict third-order polynomial model and would have to be corrected for higher-order ones. Actually, for higher order models, equivalent n^{th}-order intercept point metrics can be considered.

In practical applications, the RF power in adjacent channels will seldom be a pure tone, but rather a distributed random-like signal. However, the IIP3 still gives some insight on the amount of power that will leak into the channel of interest and that is why it is usually used as a block-level metric for high-level sizing.

For baseband circuits like ADCs or DACs, the static non-linearity may be characterized by the Integral Non-Linearity and the dynamic one by the Total Harmonic Distortion (THD). In communication applications, the latter is of much more useful than the former. As its name indicates, the THD is simply the ratio of the fundamental power over the sum of its harmonics power, usually expressed in decibels.

6.2.1.3 *Signal definition*

In a practical situation, the radio signal will be a modulated signal significantly different from a sinusoidal one. In spite of the description limitations, referring to the performance for a sinusoidal input has been proven to be a quite powerful theoretical tool for the analysis of more general cases. Actually, it is usually sufficient to consider the *peak-to-average power ratio* (PAR or PAPR) as a safety margin.

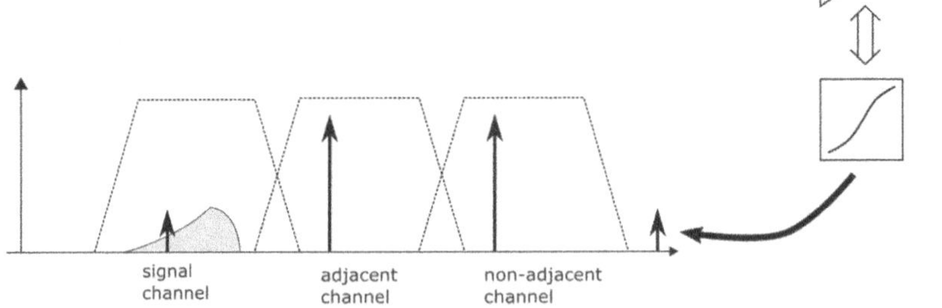

Fig. 6.1. Intermodulation products caused by two tones in the adjacent and non-adjacent channels

The Peak to Average Power Ratio can be computed from the complex envelope of the baseband signal as,

$$PAR = \frac{\rho_{peak}^2}{\overline{\rho^2}} \tag{6.11}$$

where ρ is the modulus of that envelope signal.

For OFDM modulation schemes, the direct computation of PAR is not straightforward, since the baseband complex envelope is the sum of subcarriers with random (but discrete) amplitudes. It comes that the probability density function of the envelope modulus tends to a Rayleigh distribution, but with a deterministic maximum. However, it appears that the absolute peak value is quite unlikely and may lead to overestimate the PAR. A probabilistic approach is thus usually followed to determine the peak value over an OFDM frame duration as a random variable and take the peak value at the 99.9% quantile. For an OFDM modulation with a 64 QAM scheme, a 20MHz band and 48 data subcarriers out of 64, simple numerical simulations show that the PAR is around 11dB. As a first order approximation, the PAR (in dB) is often taken equal to: $2 \ln N$, where N is the sum of the data and pilot carriers [8].

The requirements for the input power [9] can be maintained without lack of generality,

$$P_{1dB} > P_{in} + PAR \tag{6.12}$$

The previous equation can be particularized for the 802.11 standard, as shown in Table 6.2, considering the worst case OFDM modulation in terms of non-linearity[1] with negligible insertion loss at the antenna front-end. For completeness, the specification of the *Input Third Intercept Point* (IIP3) according to the in-band non-linearity, later introduced, is also reported.

[1] The linearity requirements for DSSS modulation scheme are more relaxed since constructive addition between different carriers cannot occur as in ODFM.

Table 6.2. In-band Non-linearity Requirements derived from IEEE 802.11bag/n/ac standard at maximum input power

Bandwidth (BW) (MHz)	Total Subcarriers (DataCarriers)	PAR (dB)	Maximum P_{in} (dBm)	In-band Non-Linearity Thresholds	
				P_{1dB} (dBm)	IIP3(dBm)
20 MHz	48+4	9.0	-30	-21.0	-11.4
20 MHz	52+4	9.1	-30	-20.9	-11.3
		9.1	-20*	-10.9	-1.3
40 MHz	108+6	9.8	-30	-20.2	-10.6
80 MHz	234+8	10.4	-30	-19.6	-10.0
160 MHz	468+16	10.9	-30	-19.1	-9.5

* HT-OFDM (802.11n) in the 2.4GHz Radio Band.

Notice that the power level associated to the 1-dB compression point shows similar values between different configurations, being the HT-OFDM (802.11n) in the 2.4GHz radio Band the unique outlier due to the more restrictive constrains in terms of input power. In any case, these specifications are again easily achievable in current state-of-art. Thus, in reference [3], P_{1dB} values of 5.5dBm and 7.0dBm were reported at 2.4GHz and 5GHz, respectively.

6.2.2 *Error sources*

Providing the details of transmitter design methodology is not an easy task, even at a system level of abstraction. Other books in the literature dedicate complete chapters to explain the non-idealities of the building blocks and how they can be related to specifications at higher abstraction levels. Our purpose here is not to replicate these studies but rather to give a clear concept of the different dependencies in a transmitter budget. For that purpose, we will rely on the diagram shown in Figure 6.2. The error sources for the transmitter and the receiver have a lot in common since their functions are symmetrical, but their implications can be quite different.

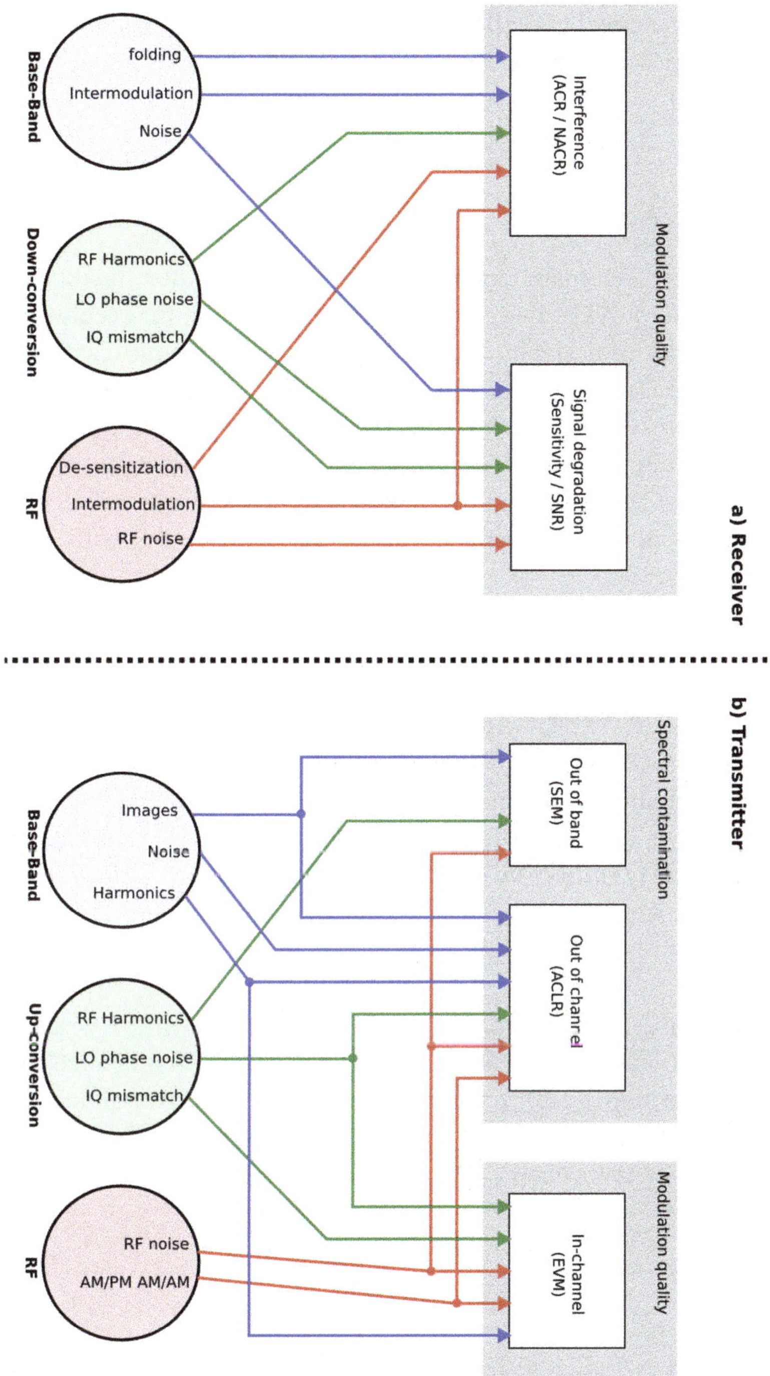

Fig. 6.2. Link between the three main error budgets and the transceiver functions

Unfortunately, the direct classification of performance degradation mechanisms in noise and distortion contributions is not as straightforward as it may seem. Indeed, the distortion of some building blocks can appear as noise-like contribution at system-level.

6.2.2.1 *The receiver case*

For the receiver, the main concern is the de-modulation quality. While the transmitter has to provide a controlled spectral emission, the receiver has to deal with a given non-ideal signal. This signal can exhibit significant RF power in adjacent channels or even outside the RF band and the amount of RF power in the channel of interest can vary to a great extent due to the attenuation of the transmission medium. Hence, we can still consider two categories for the non-ideal effects: those that are linked to unwanted RF signals and that will degrade the de-modulation through an interference process, and those that directly affect the RF signal to be converted.

RF section

The LNA and associated RF filter has to implement sufficient programmable gain to handle both small and large RF signal situations. This must be achieved with a low noise contribution (this is usually the main design challenge) and maintaining a decent linearity. The issue of linearity is not only related to the degradation of the wanted signal (just like for the transmitter) but also to a phenomenon called de-sensitization, where a strong out-of-channel signal saturates the LNA such that the effective gain is significantly reduced for the potentially low-power desired signal.

The minimum specification on the system linearity can be derived from the adjacent and non-adjacent requirements in the standard. As detailed in Chapter 3 of Part 1 (see Section 3.2.2), for OFDM scheme the receiver should guarantee a correct transmission, i.e. BER < target, in presence of undesired aggressors with relatively high power levels (P_1, P_2),

$$\begin{aligned} P_1\big|_{dBm} &= Sensitivity\big|_{dBm} + 3dB + ACR\big|_{dB} \quad \text{with} \quad P_1 = A_1^2/2 \\ P_2\big|_{dBm} &= Sensitivity\big|_{dBm} + 3dB + NACR\big|_{dB} \quad \text{with} \quad P_2 = A_2^2/2 \end{aligned} \tag{6.13}$$

where ACR and NACR are the *Adjacent Channel Rejection* and *Non-Adjacent Channel Rejection* parameters, respectively.

In this situation, the power associated to the intermodulation product P_{IM} should be significantly lower than the reference input level, that is, 3 dB above the sensitivity for the OFDM case. If the inter-modulation product is selected to be equal to integrated thermal-noise power (P_{noise}) in Eq. (6.3), the total error power will be incremented by a factor 2 (i.e. +3dB), being in the limit for the minimum SNR. In practice, a safety margin (Δ) should be budget to account for additional errors sources.

$$P_{IM} = A_{IM}^2 / 2 = \alpha_1^2 P_{noise} / \Delta \tag{6.14}$$

where α_1^2 takes into account the noise amplification at the output.

Normalizing both expression sides by the power of the fundamental harmonic $\alpha_1 A_{in}^2 / 2$, the dependence on the IIP3 and the SNR can be explicitly shown as,

$$\left(\frac{3}{4} \frac{\alpha_3}{\alpha_1} \frac{A_1^2 A_2}{A_{in}} \right)^2 = \left(\frac{1}{A_{IIP3}^2} \frac{A_1^2 A_2}{A_{in}} \right)^2 < \frac{2P_{noise}}{A_{in}^2 \cdot \Delta} = \frac{1}{SNR \cdot \Delta} \tag{6.15}$$

Table 6.3. Requirements of the out-band *Input Third Intercept Point* (IIP3) based on the IEEE 802.11 specifications for the OFDM transmission (Δ = 12dB) at low input power levels

Modulation Scheme	Code Rate	Minimum SNR (dB)	Out-band Input Third Intercept Point (IIP3) in dBm					
			802.11a/g BW = 20 MHz	802.11n/ac BW (MHz)				802.11ac BW = 80+80 MHz
				20	40	80	160	
BPSK	1/2	4.0	-20.5	-37.5	-34.5	-31.5	-28.5	-33.0
BPSK*	3/4	5.0	-20.5	-	-	-	-	-
QPSK	1/2	7.0	-20.5	-37.5	-34.5	-31.5	-28.5	-33.0
QPSK	3/4	9.0	-20.5	-37.5	-34.5	-31.5	-28.5	-34.0
16-QAM	1/2	12.0	-20.5	-37.5	-34.5	-31.5	-28.5	-33.0
16-QAM	3/4	16.0	-20.5	-37.5	-34.5	-31.5	-28.5	-33.0
64-QAM	2/3	20.0	-20.5	-37.5	-34.5	-31.5	-28.5	-33.0
64-QAM	3/4	21.0	-20.5	-37.5	-34.5	-31.5	-28.5	-33.0
64-QAM	5/6	22.0	-	-37.5	-34.5	-31.5	-28.5	-33.0
256-QAM	3/4	27.0	-	-37.5	-34.5	-31.5	-28.5	-33.0
256-QAM	5/6	29.0	-	-37.5	-34.5	-31.5	-28.5	-33.0

which expressed in decibels results,

$$IIP3\big|_{dBm} > P_1\big|_{dBm} + \frac{1}{2}\left(P_2\big|_{dBm} - P_{in}\big|_{dBm} + SNR\big|_{dB} + \Delta\big|_{dB}\right) \tag{6.16}$$

or equivalently considering Eq. (6.13) in the form,

$$IIP3\big|_{dBm} > sensitivity\big|_{dBm} + ACR\big|_{dBm} + \frac{1}{2}\left(NACR\big|_{dBm} + SNR_{\min}\big|_{dB} + \Delta_{dB} + 9\right) \tag{6.17}$$

Table 6.3 shows the IIP3 minimum requirements derived from this expression for IEEE 802.11 OFDM transmission at low input power levels (a safety margin Δ = 12dB is assumed). The consistence of values and independence on the modulation scheme are again a consequence that the standard itself was thought to facilitate the operation between the multiple operation modes. Notice that these values are quite relaxed when compare to the in-band requirement of Table 6.2. In any case, the overall IIP3 threshold is several tens of decibels below what it is feasible in current state-of-the-art — for instance, IIP3 = +17.7dBm can be achieved in a 1V 28nm CMOS process [10].

In general, direct conversion receivers are more susceptible to second-order non-linearity [11] than for third order counterpart. Figure 6.3 shows this effect when two out-bands interferes are closely spaced in frequency. If a second order model is assumed, $y(t) = \alpha_1 x(t) + \alpha_2 x(t)^2$, the output for a two tones inputs at ω_1 and ω_2 will generate an undesired inter-modulation product (IM2) at $\omega_2 - \omega_1$,

$$y(t) = y_{DC} + \alpha_2 \cdot A_{beat}{}^2 \cos(\omega_2 - \omega_1)t + \ldots \tag{6.18}$$

called *Beat* tone, which could lay on the channel of interest after down-conversion if the mixer does not provide enough port isolation.

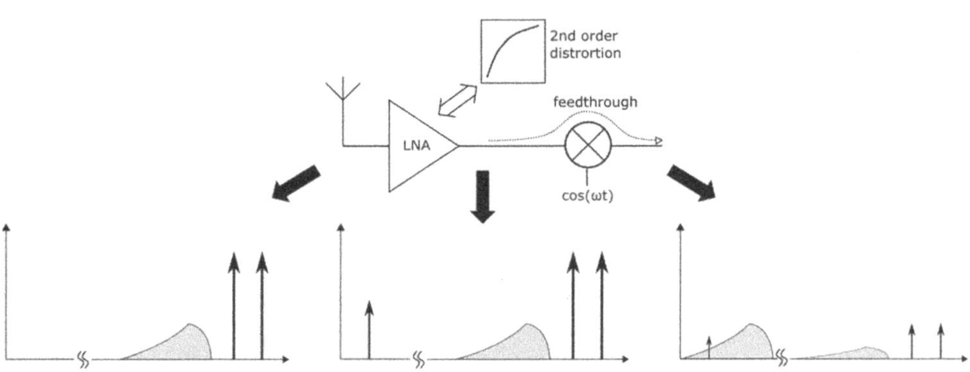

Fig. 6.3. Effect of even-order non-linearity on the direct conversion receiver

Defining R_{FH} as the feedthrough rejection factor in this path and G_{mixer} as the mixer gain, the *Second-order Intercept Point* (A_{IIP2}) can be defined as the input amplitude for which the IM2 and the primary input are equal, in the form,

$$R_{FH}\left|\alpha_2\right|A_{IIP2}^{\ 2}=G_{mixer}\left|\alpha_1\right|A_{IIP2} \quad\rightarrow\quad A_{IIP2}=\frac{G_{mixer}}{R_{FH}}\cdot\left|\alpha_1/\alpha_2\right| \tag{6.19}$$

Under the same considerations as for the third-order intermodulation product, the power associated to IM2 should be below the integrated noise after down-conversion, which taking into account the broad-band nature of RF front-end in the receiver, can be approximately scaled by the mixer gain without significant filtering,

$$P_{IM2}=A_{IM2}^{\ 2}/2=\alpha_1^{\ 2}G_{mixer}^{\ 2}P_{noise}/\Delta \tag{6.20}$$

Again, normalizing by the input power, $\alpha_1 G_{mixer}^{\ 2}A_{in}^{\ 2}/2$, the dependence on the IIP2 and the SNR becomes,

$$\left(\frac{R_{FH}}{G_{mixer}}\frac{\alpha_2}{\alpha_1}\frac{A_{beat}^{\ 2}}{A_{in}}\right)^2=\left(\frac{1}{A_{IIP2}}\frac{A_{beat}^{\ 2}}{A_{in}}\right)^2<\frac{2P_{noise}}{A_{in}^{\ 2}\cdot\Delta}=\frac{1}{SNR\cdot\Delta} \tag{6.21}$$

Table 6.4. Requirements of the out-band *Input Second Intercept Point* (IIP2) based on the IEEE 802.11 specifications for the OFDM transmission (Δ = 12dB)

Modulation Scheme	Code Rate	Minimum SNR (dB)	Out-band Input Second Intercept Point (IIP2) in dBm					
			802.11a/g BW = 20 MHz	802.11n/ac BW (MHz) 20	40	80	160	802.11ac BW = 80+80 MHz
All modes		4.0 to 29.0	12.0	-8.0	-5.0	-2.0	1.0	-5.0
BPSK*	3/4	5.0	12.0					
QPSK	1/2	7.0	12.0	-8.0	-5.0	-2.0	1.0	-5.0
QPSK	3/4	9.0	12.0	-8.0	-5.0	-2.0	1.0	-5.0
16-QAM	1/2	12.0	12.0	-8.0	-5.0	-2.0	1.0	-5.0
16-QAM	3/4	16.0	12.0	-8.0	-5.0	-2.0	1.0	-5.0
64-QAM	2/3	20.0	12.0	-8.0	-5.0	-2.0	1.0	-5.0
64-QAM	3/4	21.0	12.0	-8.0	-5.0	-2.0	1.0	-5.0
64-QAM	5/6	22.0		-8.0	-5.0	-2.0	1.0	-5.0
256-QAM	3/4	27.0		-8.0	-5.0	-2.0	1.0	-5.0
256-QAM	5/6	29.0		-8.0	-5.0	-2.0	1.0	-5.0

$$IIP2\big|_{dBm} > 2\,P_{beat}\big|_{dBm} - P_{in}\big|_{dBm} + SNR\big|_{dB} + \Delta\big|_{dB} \tag{6.22}$$

that under the worst-case conditions in which each of the two interferes contribute –3dB below the maximum non-adjacent channel power, it will lead to,

$$IIP2\big|_{dBm} > sensitivity\big|_{dBm} + 2NACR\big|_{dB} - 6dB + SNR_{\min}\big|_{dB} + \Delta\big|_{dB} \tag{6.23}$$

from which the results in Table 6.4 are estimated ($\Delta = 12$dB).

These results should be considered as a pessimistic scenario, since they assume a broad-band characteristic of the RF front-end. Although this approach is relatively accurate for the 5GHz transmissions (based on low quality factors, Q, to avoid frequency roll-off with high channel bandwidth), it could over-estimate the noise budget for the 2.4GHz band in which narrow band structures are preferred. Actually, in [12] a less pessimistic 0dBm constrain was determined based on model simulations. In any case, the requirements in the table are easily achievable in the current state-of-the-art. For instance in [3], IIP2 values of +35dBm and +43dBm were reported at 2.4GHz and 5GHz, respectively, taking advantage of the extra margin to develop more competitive devices.

Down-conversion and mixing

The mixer stage is usually passive and can even be implemented by the sampling function of the ADC in some cases. Here, the major concerns are similar to those of the transmitter: IQ mismatch that will cause bad image reconstruction, and LO noise that will modulate the desired signal. IQ mismatch can be managed to suitable levels by calibration, but phase noise can be a severe design constrain.
As introduced in Chapter 3 of Part 1 (Section 3.2.2), the maximum error in the frequency synthesizer should be below 20-25ppm depending on the operation band and modulation scheme to assure that the different messages between sub-carriers do not interfere each other.

Regardless of the stability at the peak frequency, the spectral purity of the frequency itself, characterized by the phase noise, is one of the main characteristic in the frequency synthesizer. The double-side[2] power density function associated to the phase noise referred to the carrier frequency, $S_{DS}(f)$, typically has, as shown in Fig. 6.4., two different regions depending on the cut-off frequency (f_{coff}) of the PLL loop, typically around 200kHz [1], [3]. Below f_{coff}, it remains relatively constant, $S_{\text{coff}} = S_{DS}(f < f_{\text{coff}})$. Above the cutting frequency, the phase noise is equal

[2] This option is preferred in the bibliography with respect to the single-side version for historic reasons related to the experimental characterization with spectral analyzers.

to the free-running VCO contribution, which assuming a thermal noise model, becomes,

$$S_{DS}(f) = \frac{k}{f^2} \quad \text{with} \quad k = S_c f_c \tag{6.24}$$

which is often characterized by the integrated phase noise in the band: $[f_1, f_2] = [a_1 BW, a_2 BW]$, in the form,

$$P_{\theta(f_1,f_2)} = P_{\theta(a_1,a_2)} = \int_{f_1}^{f_2} \frac{k}{f^2} \cdot df = k\left(\frac{1}{f_1} - \frac{1}{f_2}\right) = \frac{k}{a_1 a_2 BW} \quad ; \quad f_{coff} < f_1 < f_2 \tag{6.25}$$

The errors in the mixing frequency will be translated to two different effects in the down-conversion process. First, the tails of $S_{DS}(f)$ produce that the interferers at the channel are partially translated to baseband by the mixers. The contribution of both the adjacent and non-adjacent to the error power (P_{PN}) will be dominated by the non-adjacent contribution due to its stronger signal level,

$$P_{PN} = 2 \cdot (P_{in} \cdot ADR \cdot P_{\theta(0.5,1.5)} + P_{in} \cdot NADR \cdot P_{\theta(1.5,2.5)}) \approx 2 \cdot P_{in} \cdot NADR \cdot P_{\theta(1.5,2.5)} \tag{6.26}$$

where the factor 2 comes from the double-side band contribution.

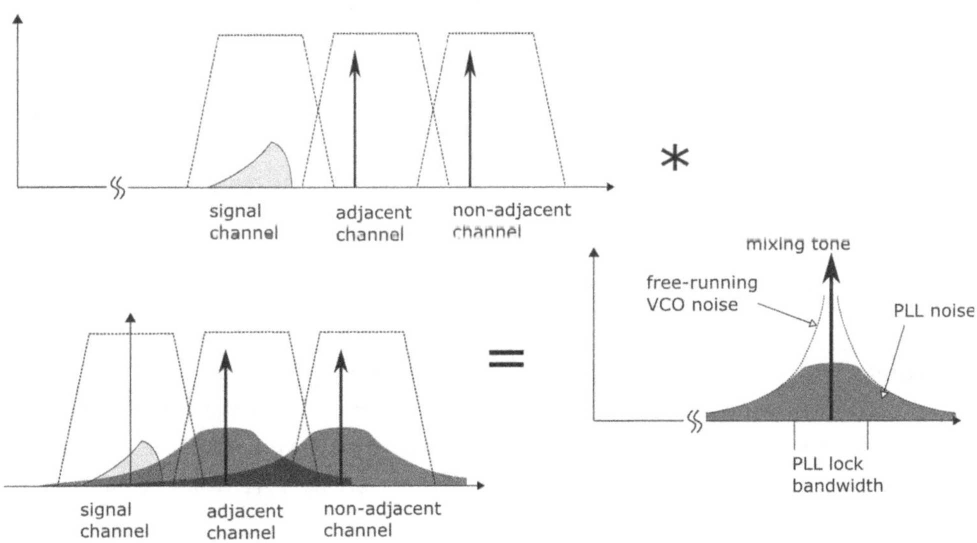

Fig. 6.4. Phase noise characteristic of the frequency synthesizer and its effect of on the adjacent and non-adjacent channel interferers

Following the same theoretical development than for the IIP3 and IIP2, this error power should be below the thermal noise at minimum input power, leading to an upper limit on the phase noise, given by:

$$P_{\theta(1.5,2.5)} < \frac{1}{2 \cdot NACR \cdot SNR \cdot \Delta} \quad \text{and} \quad k = 1.5 \cdot 2.5 \cdot BW \cdot P_{\theta(1.5,2.5)} \tag{6.27}$$

which can be equivalently expressed according to Eq. (6.24), as follows,

$$S_{DS}(f)\big|_{dB/Hz} < 10\log_{10}\left(\frac{BW}{f^2}\right) - NACR\big|_{dB} - SNR\big|_{dB} + 10\log_{10}\left(\frac{1.5 \cdot 2.5}{2}\right) - \Delta\big|_{dB} \tag{6.28}$$

Table 6.5 shows the phase noise requirements evaluated with this expression at 10MHz frequency offset from the carriers. The parameter Δ should account for +3db extra safety margin, since both the transmitter and the receiver paths contribute to the phase noise in the transmission, that is, following the same criteria than for the other source of errors, Δ = 12dB+3dB. The worst case corresponds, as usual to the 802.11a/g, for which a minimum phase noise of −125.3dB/Hz@2.4GHz[3] is imposed at 10MHz. Alternatively, the phase noise specification is often expressed in rms degrees derived from the integrated phase noise as,

$$\theta_{rms(f_1,f_2)} = \sqrt{2P_{\theta(f_1,f_2)}} \cdot \frac{360^\circ}{2\pi} \tag{6.29}$$

using a typical integration bandwidth between 10kHz up to 10MHz. In this situation, expression (6.25) should be modified to account for the VCO phase noise rejection close to the carrier due to the PLL loop high-pass feature,

$$P_{\theta(f_1,f_2)} = \frac{k}{f_{off}^{\;2}}(f_{coff} - f_1) + k\left(\frac{1}{f_{coff}} - \frac{1}{f_2}\right) \quad \text{when} \quad f_1 < f_{coff} < f_2 \tag{3.30}$$

Assuming a typical cutting frequency f_{coff} of 0.3MHz [1], [3], a worst case value, θ_{rms} = 1.12°, is equivalently obtained in our example. From the point of view of out-band interferer, the specifications are relatively relaxed. Thus, in [3] a phase noise uncertainly of 0.2° and 0.27° were reported at 2.4Hz and 5.5GHz, the

[3] Assuming as usual that the 2.4GHz frequency is derived from a common VCO working a twice the frequency followed by a frequency divider-by-2, the phase noise can be relaxed by +6dB at the 5GHz-band.

S_{DS}(10MHz) taking −133dB/Hz and −138dB/Hz, respectively. Even better figures of merit are shown in [2] with 0.105°@2.4Ghz and 0.25°@5.95GHz.

Table 6.5. Minimum phase noise power density function (dB/Hz) at 10MHz offset from the carrier based on the IEEE 802.11 specifications for the OFDM transmission ($\Delta = 12 + 3$ dB)

Modulation Scheme	Code Rate	Minimum SNR (dB)	Phase noise power density function at 10MHz offset (dB/Hz)					
			802.11a/g BW = 20 MHz	802.11n/ac BW (MHz) 20	40	80	160	802.11ac BW = 80+80 MHz
BPSK	1/2	4.0	-125.3	-115.3	-112.2	-109.2	-106.2	-103.2
BPSK*	3/4	5.0	-125.3	-	-	-	-	-
QPSK	1/2	7.0	-125.3	-115.3	-112.2	-109.2	-106.2	-103.2
QPSK	3/4	9.0	-125.3	-115.3	-112.2	-109.2	-106.2	-103.2
16-QAM	1/2	12.0	-125.3	-115.3	-112.2	-109.2	-106.2	-103.2
16-QAM	3/4	16.0	-125.3	-115.3	-112.2	-109.2	-106.2	-103.2
64-QAM	2/3	20.0	-125.3	-115.3	-112.2	-109.2	-106.2	-103.2
64-QAM	3/4	21.0	-125.3	-115.3	-112.2	-109.2	-106.2	-103.2
64-QAM	5/6	22.0	-	-115.3	-112.2	-109.2	-106.2	-103.2
256-QAM	3/4	27.0	-	-115.3	-112.2	-109.2	-106.2	-103.2
256-QAM	5/6	29.0	-	-115.3	-112.2	-109.2	-106.2	-103.2

The safety margin with respect to values in Table 6.5 of these implementations are not exclusively justified by the more restrictive target in terms of sensibility, but it is also motivated by the impact of the phase noise on the wanted message signal itself. Actually, even if the out-of-band components are negligible, the errors in the mixing frequency will generate a self-interference which could affect the BER if not properly handled. This effect, which should be evaluated by simulation [13], implies greater constrains in the phase noise with θ_{rms} values below 1° to assure standard compliance.

Baseband function

The baseband section of the receiver is formed essentially by an ADC and by an anti-aliasing filter. The first unavoidable impact of using an ADC is precisely the aliasing effect associated to the sampling function: all the spectral content around multiples of the sampling frequency will be folded into the first Nyquist zone. The anti-aliasing filter is built to limit the amount of folded power but cannot be perfect. The amount and nature of RF components that can be superimposed to the desired signal due to this folding must thus be evaluated.

On top of that, the ADC has to provide sufficient dynamic range and resolution to handle the minimum and maximum down-converted signal, which will define the thermal and quantization noise budget. The analog and mixed-signal section of a receiver compliant with the IEEE 802.11 standard should be able to operate under different input signal levels, ranging from the lower sensitivity of –82dBm (for the BPSK modulation scheme), up to a maximum power of –30dBm in both 2.4/5.0GHz. Moreover, if compatibility with the high-throughput mode in HT-OFDM is needed, the upper limit extends up to –20dBm in the 2.4GHz band (see Table 3.5 in Chapter 3 of Part 1).

Taking into account the reference (R) at the back-end ADC, the optimum gain G_{ch} in the channel should be sized to cover the complete ADC full-scale (FS =2R in differential), that is: $|V_{out}| = G_{ch}|V_{in}| \leq R$, or equivalently in terms of power,

$$P_{out}\big|_{dBm} = G_{ch}\big|_{dB} + P_{in}\big|_{dBm} \leq 10\log_{10}\left(\frac{R^2}{PAR \cdot R_s} \cdot 1000\right) \tag{6.30}$$

where $PAR \in [9, 10.9]$ dB is the *peak-to-average power ratio* introduced above, and R_s is the antenna impedance.

As numerical example, considering $R = 0.5$V and $R_s = 50\Omega$, the ADC maximum output power would be between –3.9dBm and –2dBm; while the gain should vary approximately from 80dB at minimum input power down to 16.1dB at maximum power. In practice, some extra margins should be allocated at both sides to deal with insertion loss due to antenna, band filter and RF switches, channel variations, etc. These considerations, in addition to dealing with multiple modulation schemes, make necessary to include a quite flexible programmable gain scheme for achieving the best performance. In this selection, extra bits than strictly necessary in the ADC could be considered to maintain the target SNR when the excursion at the ADC does not achieve the maximum available amplitude.

6.2.2.2 *The transmitter case*

The system-level specifications of the transmitter greatly depend on the communication standard. However, they can usually be grouped in two concepts: spectral contamination and modulation quality. The first concept deals with the unwanted part of the transmitted signal spectrum, which can act as an interferer for other users. Within this category, we will find the interferences out of the specified RF band and the interferences within the same RF band. The former could impact users of other applications while the latter could impact users of the same application, operating in a different channel. As we will see further, there is a reason why these two types of interferences are specified separately: they cannot rely on the same corrective actions. The second concept is self-explanatory: modulation quality is defined by the fidelity of the transmitted signal to the intended one. All the errors that impact the spectral content within the selected channel will alter the transmitted signal and cannot be corrected at the receiver side. Typical specifications for modulation quality are EVM or BER, which can be related to SNR (in the wide acceptation of "noise"). Out-of-band impact is usually specified by SEM (Spectral Emission Masks) which define the acceptable power in a given bandwidth at different offsets of the RF band. Out-of-channel noise is usually worst in the closest channel and is thus specified by ACLR (Adjacent Channel Leakage Ratio).

These three errors budgets (out-of-band, out-of-channel and in-channel) are contributed by the different building blocks of the transmitter. In the first stage of the system-level design, the transmitter architecture is not yet fixed, and it is thus difficult to detail the different contributions from an unknown set of building blocks. However, we can still provide some insight regarding the main functions that these building blocks should provide, which can be simply related to the direct conversion architecture seen in Chapter 3 of Part 1 (Section 3.3.1).

Baseband function

The baseband section is formed essentially by the DAC and its reconstruction filter. The main impact of these blocks is not related to some non-idealities but to its principle of operation: The data comes in form a discrete-time digital signal. As a consequence, its spectrum not only contains the desired low-frequency signal but also its copies around the multiples of the sampling frequency. For an ideal block, these copies would be replicated to the infinite and thus affect out-of-band and out-of-channel errors. In-channel error is not impacted since the sampling frequency is necessarily higher than the channel bandwidth. These signal images have to be properly attenuated by a reconstruction filter.

In order to properly size the requirements of that filter, the aperture effect must be taken into account. Indeed, the hold function performed by most DAC architectures inherently implements a cardinal sine filtering function. This *sinc* function is null at multiples of the sampling frequency and the worst attenuation of the image is provided at the edge of the signal bandwidth (BW/2) and can easily be computed as:

$$A_{app} = \frac{\sin\left(\pi\left(1 - BW/\left(2f_s\right)\right)\right)}{\pi\left(1 - BW/\left(2f_s\right)\right)} \tag{6.31}$$

It appears that the higher the sampling frequency, the larger the inherent attenuation. This is thus a first argument to set the DAC sampling frequency as high as possible. Then the reconstruction filter is also much easier to design for higher sampling frequencies since the required additional attenuation can be reached with a lower-order filter. Unfortunately, low-level design constraints limit the practical value of DAC sampling frequency to a few hundreds of MHz.

The baseband noise contribution is also band-limited by the reconstruction filter and should thus not affect the far out-of-band part of the SEM. However, since practical implementations of the reconstruction filter will not achieve perfect filtering in the channel bandwidth, a fraction of the noise may still impact the near out-of-band part of the SEM. It will also clearly affect the out-of-channel error and the in-channel error. That being said, the baseband of the transmitter is always operated close to Full-Scale and the impact of the quantization noise contribution can be made almost negligible for reasonable values of the resolution.

The same can be said about the spectral content corresponding to the distortion of the base-band blocks. This distortion will produce an error whose bandwidth is broader than that of the signal and that will essentially affect the out-of-channel and the in-channel errors. A sufficient linearity for the DAC must thus be ensured. The effects of the non-linearity depend greatly on its structure. If the static transfer function is smoothly varying, the dominating effects will be similar to those of the RF blocks: a marked spectral regrowth and an intermodulation component in the channel. If the static transfer function is more "jittery", like the one obtained from segmented architectures, the impact of non-linearity can almost be assimilated to a noise contribution (uncorrelated with the signal) spreading over the entire Nyquist bandwidth.

The operation frequency of the digital circuitry can also be chosen to minimize the impact in particularly restricted bands. Indeed, synchronous logic generate a significant amount of dynamic power (current surges) at multiples of the clock frequency, which could leak in the transmitted RF signal through substrate coupling, package parasitics, etc.

Up-conversion and mixing

The up-conversion function is carried out by the mixers and the Local Oscillator (LO). An ideal mixer should perform a perfect multiplication and would not contribute to any error budget in first order. However, practical designs normally rely on chopper-based mixers: these are reasonably linear on the low-frequency input but the LO input uses square waves. Differential implementations cancel the signal images around the even harmonics, but a significant amount of RF power is generated around the odd harmonics. This will obviously impact the out-of-band error budget.

The phase noise of the LO is narrowband, but the multiplication in the time domain corresponds to a convolution in the frequency domain. As a result, the spectrum of the base-band signal will be broadened, which impacts both the out-of-channel error and the in-channel error. Finally, just like for the receiver, any mismatch (in amplitude and phase) between the In-phase and Quadrature channels results in an imperfect cancellation of the unwanted signal image at the summing node. Since the signal image is symmetrical to the desired signal, this IQ mismatch only affects in-channel error. In practice, the IQ mismatch is handled by system-level calibration procedures that allow to finely trim the gain, offset and timing of the IQ channels both in the transmitter and the receiver.

RF section

The RF section is comprised by the power amplifier and eventually a programmable gain amplifier (also sometime called pre-power amplifier). The RF broadband noise of these blocks will affect the three error budgets. The harmonic distortion would affect the out-of-band contribution but for normal implementations, its contribution is much lower than that of the chopper-based mixers and will be handled by the same filters.

However, the non-linearity of the RF section also degrades the amplitude modulation (AM/AM) and the phase modulation (AM/PM). The former is typically linked to smooth distortion with no memory effect, that can be modelled by a low-order polynomial, while the latter is linked to memory effects. These non-linearities will impact both the in-channel error budget and the out-of-channel error budget through spectral regrowth. For instance, the maximum achievable SNR in the presence of odd-order AM/AM distortion for an amplitude modulated signal can be written as

$$SNR = \frac{\left(1 - \frac{P}{IIP3}(\Gamma_4 + 1)\right)^2}{\left(\frac{P}{IIP3}\right)^2 \left((\Gamma_6 + 1) - (\Gamma_4 + 1)^2\right)} \tag{6.32}$$

where P is the power of the input signal of the RF block, IIP3 is its input third order intercept point. The Γ_{2k} terms correspond to the high-order moments of the signal and thus depends on the modulation scheme. For an OFDM modulation scheme, the signal is shown to follow a Rayleigh distribution and the moments can be approximated by,

$$\Gamma_{2k} = k! - 1 \tag{6.33}$$

This expression is derived considering that all the inter-modulation distortion power produced by the non-linearity lies in the channel, which is an approximation. Indeed, the 3rd order intermodulation distortion also contains a component that corresponds to 3rd order self-convolution of the signal spectrum. It can be shown that the bandwidth of that perturbation is 3 times the bandwidth of the original signal. The higher the order of the non-linearity, the larger the spectral regrowth.

Dynamic non-linearity (for instance caused by the non-linear gate-to-source parasitic capacitor in a MOS transistor) can cause both amplitude and phase distortion. It can be shown that the AM/PM contribution is similar to that of AM/AM distortion if no memory effect is involved. Indeed, the phase distortion component is orthogonal to the amplitude distortion component and thus the power simply sums up. The presence of memory effect, though, alters the symmetry of the distortion side-lobes around the carrier frequency.

Other sources of errors (not mentioned in Figure 6.2) are:

- Carrier leakage (occurs at the mixers through RF coupling or by the presence of an undue offset in the BB signal). This can be a stringent concern for some architectures and/or communication standards and could be contemplated in the error budgets (typically the in-channel and out-of-band errors). However, for communication standards based on OFDM, the center sub-carrier remains unused. To some extent, the base-band circuitry can thus afford some offset and the mixers can also afford some carrier leakage. By the way, the amount of tolerated carrier leakage is specified separately in IEEE 802.11 standard [14]. Notice that, although carrier leakage will probably be the main cause of center frequency leakage it is not necessarily the only one. Since the specification considers the ratio of power transmitted in the central subcarrier relative to the total transmitted power, it will integrate all the possible contributors.
- Spectral flatness within a channel is also usually specified, particularly for OFDM. It imposes that the average power transmitted on all the

subcarriers should be similar (with some upper and lower margins). For instance, this limits the possibility to use aggressive filtering for limiting spectral regrowth. Base-band circuitry will also alter the flatness, for instance through the aperture effect of the DAC. However, spectral flatness can usually be handled in the digital domain trough proper equalization.

- LO injection pulling is a phenomenon that is particularly difficult to put into equations. The idea is that the presence of significant amount of power around the LO frequency (i.e. in the +/- 20% vicinity) can produce significant phase glitches. It is shown in [15] that injection pulling by a sinusoidal signal in a PLL produces two symmetrical sidebands at $\pm\Delta\omega$, where $\Delta\omega$ is the difference between the locked frequency and the injection frequency. The sideband attenuation shows a bandpass behavior. If the injection signal is not a pure sinewave, deriving analytically the impact on the LO spectrum is cumbersome. However, in first approximation, we can suggest dividing the injection spectrum into narrowband slices and apply the bandpass transfer function to these slices as if they were pure sinewaves. Thus, for a flat spectrum injector (like the signal generated by an OFDM scheme) we can expect to see a broadening of the LO spectrum with a bandpass behavior, as conceptually sketched in Figure 6.5.

The choice of the LO generation architecture is part of the frequency planning. The most straightforward approach consists in synthesizing a signal at the double of the LO frequency and use a divider by two. The main benefit of this approach is that it naturally offers an easy mean to generate the I and Q phases with no duty-cycle mismatch. However, it may be challenging for direct conversions transceivers operating in high-frequency bands to keep a reasonable power consumption. Indeed, the 5GHz band would require a synthesized frequency and a divider operating in the range of 10GHz. In addition, a certain amount of RF power which may cause pulling may still be present around the 10GHz due to the second-order harmonics of the RF blocks. That being said, this would be a second-order effect much more manageable by proper isolation techniques at layout level than for the direct LO synthesis.

Another possibility is based on Single Side-Band LO generation. The idea is that the I and Q phases are generated by mixing lower frequencies, in a way similar to the heterodyne architecture, but applied to the LO generation only. Such a scheme is actually a practical implementation of the well-known trigonometric equations:

$$\begin{aligned}\cos(\omega_1+\omega_2)&=\cos(\omega_1)\cos(\omega_2)-\sin(\omega_1)\sin(\omega_2)\\ \sin(\omega_1+\omega_2)&=\cos(\omega_1)\sin(\omega_2)+\sin(\omega_1)\cos(\omega_2)\end{aligned} \tag{6.34}$$

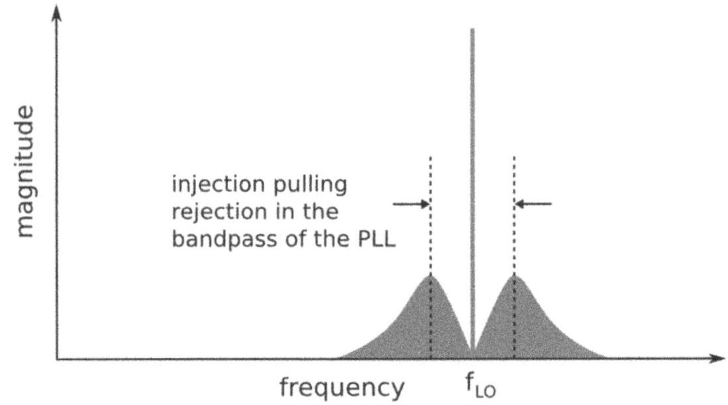

Fig. 6.5. Bandpass behavior of LO pulling for synthesizer based on a PLL. This spectrum shape is only indicative of what would be obtained for an aggressor of flat power around the LO frequency

Taking $\omega_1 = 2\omega_c/3$ and $\omega_2 = \omega_1/2$, we obtain the I and Q phases at the carrier frequency ω_c. Unfortunately, this scheme considers perfect mixing with pure sine-waves, which in practice is not feasible. Common mixers can be made linear on one of their two inputs, but not on both. As a result, a number of intermodulation components will show up in the spectrum and corrupt it. These components have thus to be filtered out.

6.3 Budgeting the Transceiver

The error budgets presented in this section can be partially estimated through analytical expressions, whose parameters relate to the block-level performance. Providing a valuable explanation of these expressions would require entering into the details of all the performance degradation mechanisms and cannot be done in a few pages. The interested reader is referred to other textbooks that address analog RF sizing in much more detail [7].

Instead, we will describe a top-down bottom-up methodology that has been successfully applied to the design of RF transceivers. For the sake of simplicity, we will take the receiver as an example, but a similar approach can be followed for the transmitter.

Optimization of the Error Budget

Budgeting the receiver is more complex than budgeting the transmitter, essentially because the RF signal is not controlled: the non-idealities are not restricted to those introduced by the block design, but some already come with the signal acquired at the antenna. In particular, the receiver has to deal with a wide variation of wanted signal power in the presence of potentially large interferers.

Budgeting the receiver for a given architecture consists in distributing noise and distortion error budget among the different stages to satisfy the target performance, while minimizing power consumption. The key block specifications are the average power consumption of each building block ($P_{AVG,i}$), the noise figure (NF_i), the 2nd and 3rd order intercept points ($IIP_{3,i}$, $IIP_{2,i}$), the power gains (G_i) and the internal power levels (P_i) at the input of the different stages in the receiver queue, {STG_1=Band Filter, STG_2= LNA, ... , STG_{NSTG}=ADC} in the conceptual example of Figure 6.6. Note that $P_1 \equiv P_{in}$ is the input power available at the antenna port.

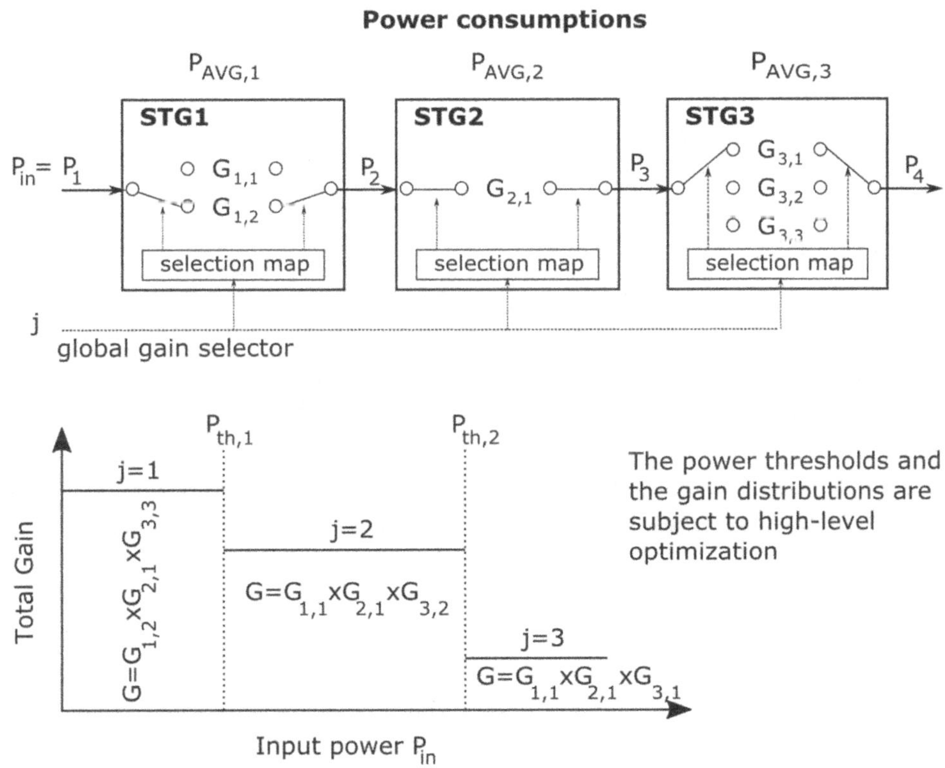

Fig. 6.6. Conceptual example of a RF receiver with N_G=3 different gain modes, $j \in [1, 3]$

In addition, the architecture shall implement some form of gain programmability and this programmability has also to be properly distributed among the stages. We thus have to find the adequate gain modes that satisfy the requirements while maintaining feasible signal levels. The stage specifications must thus be considered in the different gain modes and the input power thresholds for the gain mode selection have to be properly defined. The manual search of such a large design space is a tedious task and though expert knowledge can guide the designer towards a working solution, such an implementation will hardly be optimum in terms of area and power efficiency.

In [16], a guided synthesis procedure for RF receiver channels was proposed. It relied on semi-analytical expressions that did not incorporate gain as a design parameter, limiting the introduction of gain-programmability modes. As alternative, a systematic procedure for the synthesis of RF receivers can be found in [17]. The synthesis is formulated as an optimization problem with multiple constrains, generally described in the form,

$$\text{find}$$
$$\underset{\sim}{x} = \left\{ N_G,\ \left\{ P_{th,j} \right\}_{j=1 \ldots N_G-1},\ \left\{ G_{i,k_j} \right\}_{j=1 \ldots N_G},\ \left\{ P_{AVG,i} \right\}_{i=1 \ldots NSTG} \right\}$$
$$\text{which minimizes: } P_{AVG}(\underset{\sim}{x}) = \sum_{i=1}^{N_{STG}} P_{AVG,i} \tag{6.35}$$
$$\text{constrained by:}$$
$$NF(P_{in}) \le NF^{spec}(P_{in})\ ;\ IIP_3(P_{in}) \ge IIP_3^{spec}(P_{in})\ ;\ IIP_2(P_{in}) \ge IIP_2^{spec}(P_{in})$$

where N_G is the number of gain modes, $P_{th,j}$ are the input power thresholds for the different modes and $P_{AVG,i}$ are the power consumption of the different stages. This procedure minimizes the total power consumption while fulfilling the target noise and distortion specifications of the receiver chain expressed as the total NF^{spec} and intermodulation products (IIP_3^{spec}, IIP_2^{spec}) as function of the input signal power (P_{in}).

A top-down bottom-up design flow to address this optimization problem is depicted in Figure 6.7.

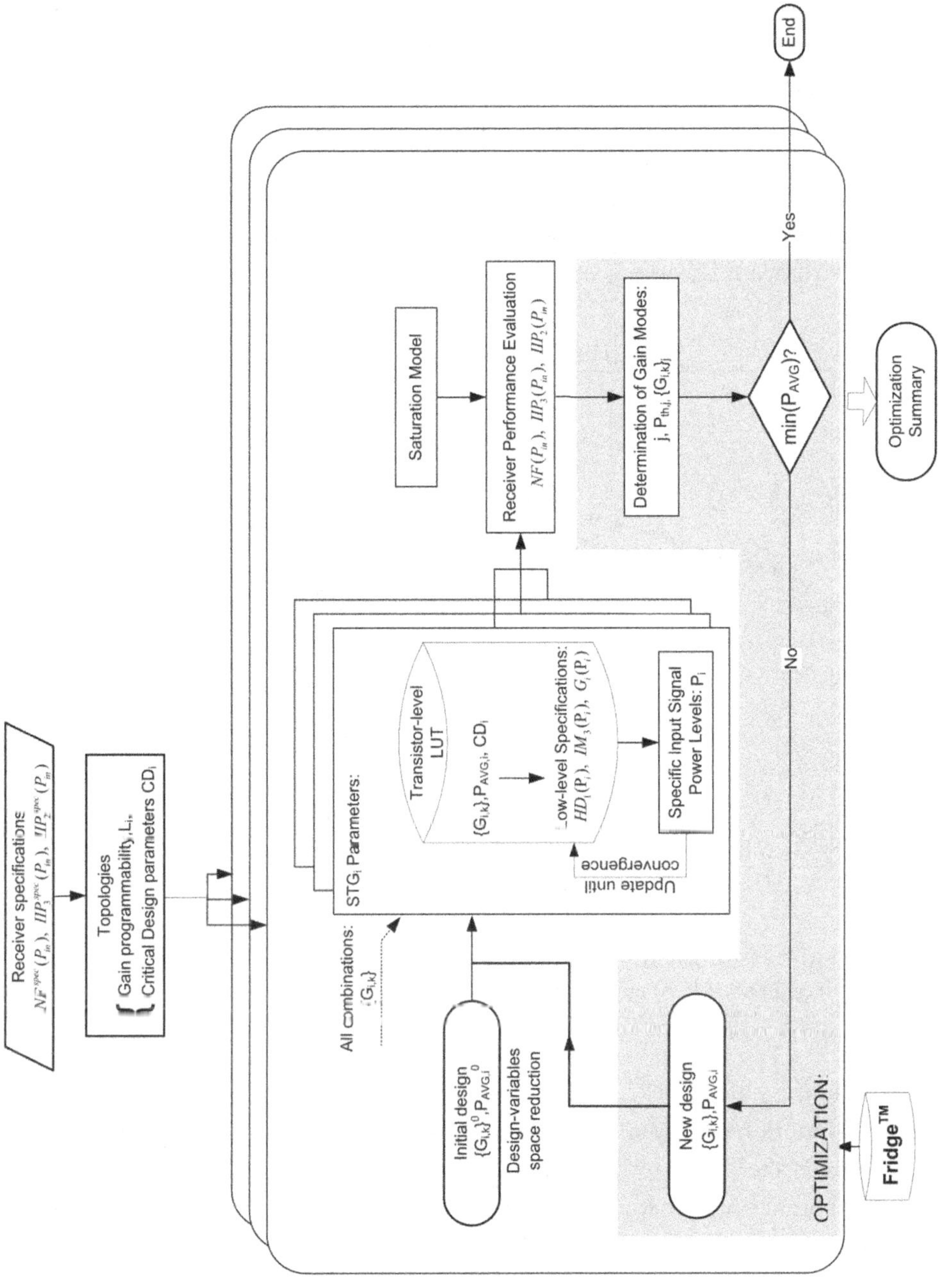

Fig. 6.7. Diagram of the top-down bottom-up optimization flow

- The Bottom-up part: Technological Look-Up Tables (LUTs) are generated for the different building blocks from transistor-level simulations. The idea consists in having a library of blocks that are electrically characterized. These LUTs simply gather the block-level performance metrics as a function of the input power level: Power consumption, Noise Figure, 2nd and 3rd order Input-Referred Intercept Points, 1dB Compression Point, effective gain but also eventually the Total Harmonic distortion, the Signal-to-Noise Ratio or any other relevant feature.

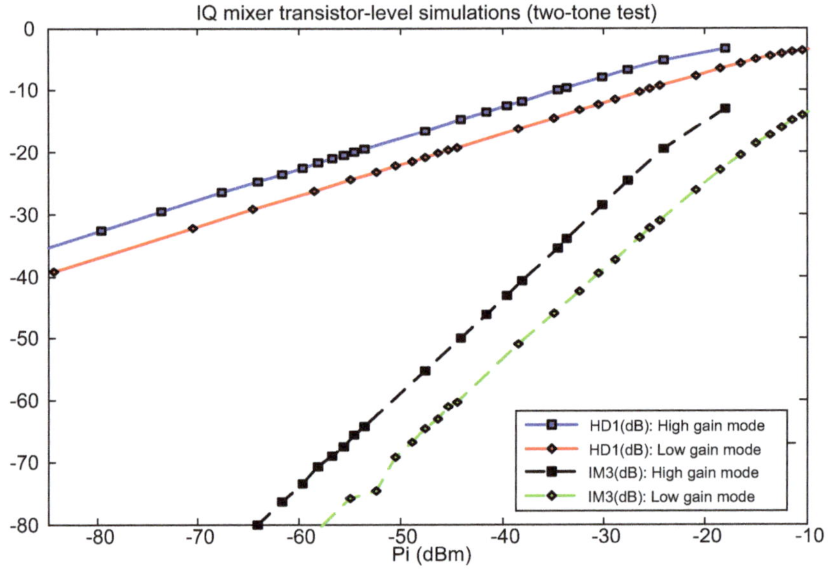

Fig. 6.8. Transistor-level simulations of HD_1 and IM_3 versus input signal power of an implementation of the IQ-mixers in a 1.2 V 90 nm CMOS process

Generating the LUTs for all the blocks and for different levels of power consumption or gain modes can be quite time-consuming if only interpolation between the characterized cases is allowed. Figure 6.8 shows an example of the simulations that can be used to build a LUT: The transistor-level simulations of a two-tone test for an implementation of IQ-mixers in a 1.2-V 90-nm CMOS process. Details on the amplitude of one of the tones HD_1 and the corresponding intermodulation product IM_3 for two possible gain values are depicted. It can be seen how the IQ mixer incurs in compression for the higher signal levels, particularly in high-gain modes.

In order to relax the characterization effort for building the LUTs, analytical scaling expressions can eventually be used, which are specific to a particular block realization. For instance, it is possible for some blocks to predict how will scale the noise figure as a function of the input power level and the power consumption: $NF_i = f_i(\mathrm{P}_i, P_{AVG,i})$.
Adding the information of the effective gain may also predict the value of the different effective intercept points: $IIP_{x,i} = g_{x,i}(\mathrm{P}_i, P_{AVG,i}, G_{i,k})$. Indeed, the IIPs are in principle small-signal metrics and should thus be independent of the input signal level. However, instead of considering higher-order metrics to reflect the more complex non-linearity at high signal levels, it is possible to simply consider a piecewise affine approximation of the low-order IIP3 using the local derivative of the intermodulation products to extimate and input-dependent IIP3 (this effective IIP3 would take into account the curvature of both the HD1 and the IMD3 curves in Figure 6.8.). The development of these expressions relies on circuit analysis and is out of the scope of this chapter. However, it is intuitive to most designers that the noise can be reduced incrementing power consumption, and that for a given input signal power $\mathrm{P_i}$, the input referred intermodulation products ($\mathrm{IIP_{3,i}}$, $\mathrm{IIP_{2,i}}$) have an inverse relationship with gain since non-linearities increase with the output signal level.
Similar analytical expressions may be developed for other performance metrics. In any case, it is advisable to perform a sanity check and verify by simulation that the predicted values are correct at some points of the design space. Once the LUTs (and the analytical expressions) have been generated, it becomes possible to predict the performance of all the building blocks at different input levels, power consumption levels and gain levels.

- The Top-down part is actually an optimization loop. The first section of this loop is the topology generator, which distributes the gain programmability feature to the different stages and assigns the number of gain levels to the selected blocks. This step requires some collaboration form the block designers to guarantee the feasibility of the selection. In the conceptual example of Figure 6.6, the first and third stages are selected to be programmable, with the first stage having 2 gain levels and the third stage 3 gain levels. This topology generator could also eventually contemplate some critical design ($\mathrm{CD_i}$) parameters that actually condition some implementation trade-offs like for instance the biasing conditions (g_m/I_D) of some blocks.
  The topology generator thus presents a reduced number of candidate architectures to the second section that implements the loop itself. The design

variables here are reduced to the power consumption of each block ($P_{AVG,i}$) and the gain of each block ($G_{i,k}$) which also eventually depends on the programmable gain mode (for those blocks that do present gain programmability, as proposed by the topology selection). The first step approximation determines the ideal signal levels at the input of each block and the LUTs are used to obtain the block-level performance parameters for the current selection of design variables. Since the effective gains affect the signal level of following blocks, a few iterations are necessary for the approximation to converge to a consistent value. This task has to be performed covering the allowed signal range at the antenna (from the sensitivity threshold to the maximum power) and for all the gain configurations allowed by the topology.

The next step (step 2), computes the receiver-level performance metrics from the converged block LUTs. The well-known Frii's formula is used to combine the block NFs to get the receiver NF which is annotated as a function of the block gains and the block power consumptions (i.e. the design variables of the optimization loop), and the input power level (P_I only).

$$NF = 1 + (NF_1 - 1) + \frac{(NF_2 - 1)}{G_1} + \frac{(NF_3 - 1)}{G_1 \cdot G_2} + \ldots + \frac{(NF_m - 1)}{G_1 \cdot G_2 \cdot \ldots G_{(m-1)}} \tag{6.36}$$

Similar expressions are used for the 2nd and 3rd order Intercept Points:

$$\frac{1}{a_{IIP_3}^2} = \frac{1}{a_{IIP_{3,1}}^2} + \frac{A_1^2}{a_{IIP_{3,2}}^2} + \frac{A_1^2 \cdot A_2^2}{a_{IIP_{3,3}}^2} + \ldots + \frac{A_1^2 \cdot A_2^2 \cdot \ldots \cdot A_{(m-1)}^2}{a_{IIP_{3,m}}^2} \tag{6.37}$$

$$\frac{1}{a_{IIP2}} = \frac{1}{a_{IIP2,1}} + \frac{A_1}{a_{IIP2,2}} + \frac{A_1 \cdot A_2}{a_{IIP2,3}} + \ldots + \frac{A_1 \cdot A_2 \cdot \ldots \cdot A_{(m-1)}}{a_{IIP2,m}} \tag{6.38}$$

where $a_{IIPx,i}$ represents the input referred amplitude of the intermodulation products (in Volts rms), and A_i is the linear voltage gain of block i.

Notice that the NF formula does not require impedance matching. In addition, the SNR degradation associated to the baseband blocks has to be translated into an effective noise figure (which is not the usual way to describe the noise performance from a baseband designer viewpoint).

In order to avoid the selection of gain values that would lead to unrealistic signal ranges, a soft saturation penalty is added to the optimization loop as a correction factor. The block designers define a safety range for their implementation and the effective gain of the block is artificially degraded beyond this range. The idea is that such a degradation will automatically penalize the NF and IIPx performance of the receiver and these designs will thus be discarded rapidly by the optimization routine as inefficient.

Once the receiver performance metrics over the full input signal range are available for all the gain modes, the next step consists in determining the optimum distribution (and effective number N_G) of the gain modes and the associated power thresholds $P_{th,j}$. For that, the sections that do not meet the receiver specification constrains are discarded and among the remaining sections, those with the lowest total power consumption are selected. The crossing points of the total power consumption curves thus define the thresholds, as conceptually represented in Figure 6.6. In this conceptual example, only 3 gain modes would actually be selected among the 6 possible combinations.

The optimization loop then updates the block gain values and their power consumptions and iterates the process. In [17], the whole synthesis loop is developed in MATLAB™ and the optimization tool is an engine that processes performance results and generates new iterations seeds following a simulated annealing algorithm [18]. However, other optimization algorithms like genetic algorithms could be envisioned.

The optimization for each topology finishes upon convergence of the minimum power consumption subject to the condition that all the specifications are met. This process usually takes less than half an hour for each topology but if convergence cannot be achieved, a maximum number of iterations can be defined to stop the loop.

Ideally, a transistor-level simulation of the receiver in the chosen configuration has to be performed as a sanity check, to verify the quality of the predicted performance. If this is not feasible, block models that incorporate all the relevant non-ideal effects have to be developed to simulate the complete receiver at the lowest feasible abstraction level.

As an example, Figure 6.9. shows the sensitivity simulation for a ZigBee receiver whose design followed the explained procedure [17]. The figure shows the Chip Error Rate (CER) as a function of the input power, and the evolution of the total receiver gain is superimposed to illustrate the transition points between the different configurations. This simulation not only includes the noise effects but also the distortion in the receiver blocks and phase noise in the LO. The CER is below the standard 6.89% for input powers above -88dBm, with a Noise Figure of 21dB, showing that the chosen distribution of specifications among receiver building blocks meets the standard requirements.

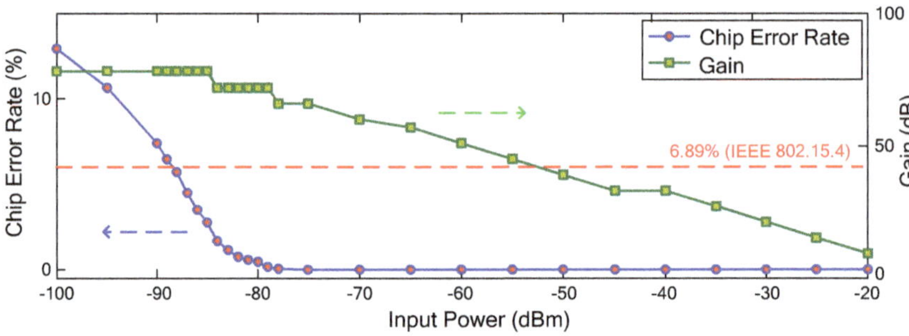

Fig. 6.9. System simulation results in the receive path: sensitivity performance

References

1. Kawai, S. *et al.* (2018). An 802.11ax 4x4 high-efficiency WLAN AP transceiver SoC supporting 1024-QAM with frequency-dependent IQ calibration and integrated interference analyzer, *IEEE Journal of Solid-State Circuits*, 53, pp. 3688–3699.
2. Chen, T.M. *et al.* (2017). An 802.11ac dual-band reconfigurable transceiver supporting up to four VHT80 spatial streams with 116fs_{rms}-jitter frequency synthesizer and integrated LNA/PA delivering 256QAM 19dBm per stream achieving 1.733Gb/s PHY rate, *Proc. IEEE International Solid-State Circuits Conference, ISSCC*, pp. 126–127.
3. Yan, S.T. *et al.* (2017). An 802.11a/b/g/n/ac WLAN transceiver for 2x2 MIMO and simultaneous dual-band operation with +29 dBm P_{sat} integrated power amplifiers, *IEEE Journal of Solid-State Circuits*, 52, pp. 1798–1813.
4. Chung, Y.H. *et al.* (2015). A dual-band 802.11abgn/ac transceiver with integrated PA and T/R switch in a digital noise controlled SoC, *Proc. IEEE Custom Integrated Circuits Conference, CICC*, pp. 1–8.
5. Chen, T.M. *et al.* (2014). A 2x2 MIMO 802.11 abgn/ac WLAN SoC with integrated T/R switch and on-chip PA delivering VHT80 256QAM 17.5dBm in 55nm CMOS, *Proc. IEEE Radio Frequency Integrated Circuits Symposium*, RFIC, pp. 225–228.
6. He, M. *et al.* (2014). A 40nm dual-band 3-stream 802.11a/b/g/n/ac MIMO WLAN SoC with 1.1Gb/s over-the-air throughput, *Proc. IEEE International Solid-State Circuits Conference, ISSCC*, pp. 350–351.
7. Razavi, B. (2011) *RF Microelectronics*, 2nd Ed. (Prentice Hall Press, USA).
8. Dinur, N. and Wulich, D. (2001). Peak-to-average power ratio in high-order OFDM, *IEEE Transactions on Communications*, 49, pp. 1063–1072.
9. Leung Lai Kan, L. *et al.* (2007). A 1-V 86-mW-RX 53-mW-TX single-chip CMOS transceiver for WLAN IEEE 802.11a, *IEEE Journal of Solid-State Circuits*, 42, pp. 1986–1998.

10. Shirane, A. *et al.* (2017). A low voltage 0.8V RF receiver in 28nm CMOS for 5GHz WLAN, *Proc. 43rd IEEE European Solid-State Circuits Conference, ESSCIRC*, pp. 328–331.
11. Zhang, P. *et al.* (2003). A 5-GHz direct-conversion CMOS transceiver, *IEEE Journal of Solid-State Circuits*, 38, pp. 2232–2238.
12. Cai, K.Z. and Zhang, P. (2004). The effects of IP_2 impairment on an 802.11a OFDM direct conversion radio system, *Microwave Journal*, 47, p. 22-26.
13. Vassiliou, I. *et al.* (2003). A single-chip digitally calibrated 5.15~5.825-GHz 0.18-μm CMOS transceiver for 802.11a wireless LAN, *IEEE Journal of Solid-State Circuits*, 38, pp. 2221–2231.
14. IEEE Std 802.11™-2016, IEEE Standard for Information technology—Telecommunications and information exchange between systems—Local and metropolitan area networks—Specific requirements—Part 11: Wireless LAN Medium Access Control, p. 3534, 2016.
15. Razavi, B. (2004). A study of injection locking and pulling in oscillators, *IEEE Journal of Solid-State Circuits*, 39, pp. 1415–1424.
16. Sheng, W., Emira, A. and Sanchez-Sinencio, E. (2006). CMOS RF receiver system design: a systematic approach, *IEEE Transactions on Circuits and Systems I*, 53, pp. 1023–1034.
17. Gines, A. et al. (2015). *Mixed-Signal Circuits*, ed. Noulis, Chapter 7: Design of an energy-efficient ZigBee transceiver, (CRC-Press, USA).
18. Medeiro, F. *et al.* (1994). A statistical optimization-based approach for automated sizing of analog cells, *Proc. IEEE/ACM International Conference on Computer-Aided Design, ICCAD*, pp. 594–597.

https://doi.org/10.1142/9789811210723_0007

Chapter 7

Embedded Memories

Hassen Aziza
Aix-Marseille Univ, Univ Toulon, CNRS, IM2NP
38 rue Joliot Curie, F-13451, Marseille, France

7.1 Introduction

Memory is an essential component of today's electronic systems. It is used in any equipment using a processor such as computers, smart phones, tablets, digital cameras, entertainment devices, global positioning systems, automotive systems, etc. Moreover, the unprecedented growth in Internet of Things (IoT) devices across all industry verticals continuously generates a massive amount of data which increases the demand for even more physical space for memory. Alternatively, memory technology plays an important role in security as it stores sensitive information and cryptographic keys [1]. In the same way, memory technology can also be used for the generation of basic security primitives such as Physically Unclonable Functions (PUFs) and True Random Number Generators (TRNGs) [2].

Memories constituted 20% of the semiconductor market for the last 30 years and are expected to increase in the coming years [3]. Generally, semiconductor memory technologies can be split into two categories: volatile and non volatile. Volatile memory does not retain data when power is turned off; conversely, non-volatile memory retains data once power is turned off. Static Random Access Memory (SRAM) and Dynamic Random Access Memory (DRAM) fall into the volatile category, while "Flash" which is the short form of "flash electrically erasable programmable read-only memory" is categorized as a Non Volatile Memory (NVM) technology.

Today's dominant memory technologies are DRAM and Flash, for performance and storage purposes. Both technologies are reaching a state of development that presents scaling issues [4]. The DRAM offers very high endurance (approximately 10^{14} cycles) but it is volatile. Flash is nonvolatile, but its endurance is limited (approximately 10^6 cycles), and its operation access frequency is very slow, with program/erase times in the order of microseconds to milliseconds. Moreover, Flash devices need high voltages for storage operations, thus it is hardly compatible with ultra-low power applications. Storage Class Memory (SCM) is seen as a disruptive storage technology as it is aimed to change both storage and application architectures. This technology has been identified for a memory system that could stay in-between DRAM and hard drives, both in term of latency and costs [5]. SCM combines the benefits of solid-state memory, such as high performance and robustness, with the archival capabilities of conventional hard disk magnetic storage. Such a device requires a high-density nonvolatile memory technology that can be manufactured at a very low cost.

In the last decade, memory technology has gained a pivotal position as the functionality and performance of today's computing system are increasingly dependent on the characteristics of the memory subsystem. Thus, the semiconductor industry is definitely in need of a new memory technology that combines the best features of current memories in a fabrication technology compatible with the CMOS process flow, scaled beyond the present limits of RAM and Flash technologies, while being nonvolatile. Recently, "memristor"-based memory has emerged as a promising NVM candidate [6]. "memristor"-based memory technologies have experienced an increasing interest as next generation non-volatile memory devices [7].

7.2 Memory Systems: Current Landscape

7.2.1 *Embedded memory*

Embedded memories are integrated with bulk logic CMOS technology and are vital components in System-on-a-Chip (SoC) since these devices can provide a high speed and wide bus capabilities. Embedded SRAM is based on the same process used in logic designs and thus is widely used as primary cache or level-one cache. Over the past few years, the prospects for embedding DRAM and Flash memory on-chip have received a great deal of attention. Developing a single fabrication process to maximize the performance of both memory and logic circuits has been an ongoing struggle. Due to the complexity of DRAM process technology, they are the least-used embedded memory devices. Embedded Flash (eFlash) is the current nonvolatile option for embedded memory. Two key application areas are driving advances in embedded Flash technologies:

automotive microcontrollers and smart cards. However, eFlash will become cost prohibitive at 28 nm in many applications. Five key players having more than 80% of the total business (TSMC, GF, UMC, SMIC, and Samsung) dominate the embedded memory market [8].

Stand-alone memories refer to devices that reside outside of the logic core. Flash and DRAM market is the largest, pushed by the growing need for memory in mobile devices and enterprise storage. Flash and DRAM stand-alone memories scaling is slowing down in terms of technological node. However, the 3D approach alternative allows increasing bit density for Flash [9]. The stand-alone memory market is very concentrated with five major Integrated Device Manufacturers (IDMs) having 95 % of the total business (Samsung, Micron, Sk Hynix, Toshiba, WD Sandisk) [8]. NAND and DRAM scalability is supposed to reach its limit in 2020. In this context, emerging memory technologies must "catch a running train" with huge annual R&D investments (several $B) [8]. Right now, big memory makers consider emerging NVM more of a "nice have" than a "must have".

Table 7.1 propose a comparison between stand-alone and embedded memory versus density, technology node, price, mainstream technologies and key players. NOR and NAND flash architectures are introduced. The NOR Flash architecture provides enough address lines to map the entire memory range. This gives the advantage of random access and short read times, which makes it ideal for embedded code execution. Disadvantages include larger cell size, resulting in a higher cost per bit. NAND Flash, in contrast, does not allow random access and has a much smaller cell size. As a result, NAND Flash memories are available in much higher densities compared to NOR Flash and are intensively used for data storage [9].

Table 7.1. Stand-alone vs embedded memory businesses

	Stand Alone	***Embedded***
Density	Gb – few 100Gb	Mb – few 100 Mb
Technology node	< 20nm	28 – 90 nm
Price	< $ 1/Gb	> $ 10/Gb
Main stream	Flash NAND - DRAM	Flash NOR - SRAM
Key Players	IDM's : Samsung, Micron, Sk Hynix, Toshiba, WD Sandisk	Foundries : TSMC, GF, UMC, SMIC, Samsung

7.2.2 *Alternatives to current memory landscape*

For almost 40 years, computer memory architectures have been essentially the same: volatile, high-speed memory technologies like SRAM and DRAM used for cache and main memory; magnetic disks for high-end data storage; and persistent, low speed Flash memory for storage with low capacity and low energy consumption. In the quest for innovative non-volatile memory solutions, different technologies have emerged in research over the last 15 years, such as:

- Phase-Change Random Access Memory (PCRAM or PCM);
- Spin-Transfer-Torque Magnetic Random Access Memory (STT-MRAM);
- Resistive Random Access Memory (RRAM).

Mainstream memory technologies are essentially based on the charge storage mechanism: SRAM stores charges at the storage nodes of the cross-coupled inverters, DRAM stores charges at the cell capacitor, and Flash stores the charges in a floating gate transistor. All these charge-based memories face challenges in scaling down to the 10-nm node and beyond. The easy loss of the stored charges at nanoscale results in the degradation of performance, reliability, and noise margin [10]. In contrast, emerging memory technologies store information using physical mechanisms relying on resistance change. Thanks to their scalability and performances such as fast switching speed, high retention, good endurance, and great compatibility with CMOS technology, emerging memories are believed to become a good choice for embedded applications and storage class memory. The rapid growth of these emerging technologies is synchronized with the "end" of Moore's law and promises a beyond CMOS era with significant improvements in energy-efficiency, reliability, performance and manufacturability. Moreover, these devices enable very different computing paradigms, such as neuromorphic computing, non-Boolean computing, and in-memory computing [4], thus making these alternative technology platforms an interesting playground for circuit, and application-developers.

Another major advantage of emerging memories is their ability to be easily integrated with existing logic technologies. Indeed, emerging memories offer the possibility to be integrated in the Back-End-Of-Line (BEOL), on top of the CMOS subsystem [11]. However, while interest in emerging embedded memory is strong, few, if any, SoC companies would risk critical products to an unproven technology.

7.2.3 *Memory hierarchy*

The functionality and performance of today's computing system are increasingly dependent on the characteristics of the memory subsystem. An ideal memory system would be fast, inexpensive, persistent and dense. Until now, all known memory technologies address only some of these characteristics. SRAM is very fast, but expensive, has low density and is not persistent. Cache memory is usually built out of SRAM on a single chip together with the processor. Because SRAM is inherently faster than DRAM, cache memory leverages on SRAM to provide data exchange speed comparable to the processor speed as on-chip memory eliminates lengthy delays caused by traveling to and from a separate chip. DRAM is currently 10 to 100 times slower than processors, has superior density, and is less expensive and is not persistent as well. Disks are cheap, highly dense and persistent, but very slow. Flash memory is between DRAM and disk; it is a persistent solid-state memory with higher densities than DRAM, but its write latency is much higher than the latter.

Fortunately, it is possible to design a memory system that incorporates all these different technologies in a single memory hierarchy. Such a hierarchy approaches the performance of the fastest component, the cost of the cheapest component and the energy consumption of the most power-efficient component. The idea is to quickly access the most commonly used data while still having the capacity to store large amounts of data. Moving up the memory hierarchy presented in Figure 4.1 toward the cache and registers, the memory write/read latency decreases. Moving down the memory hierarchy toward the storage, the memory capacity increases.

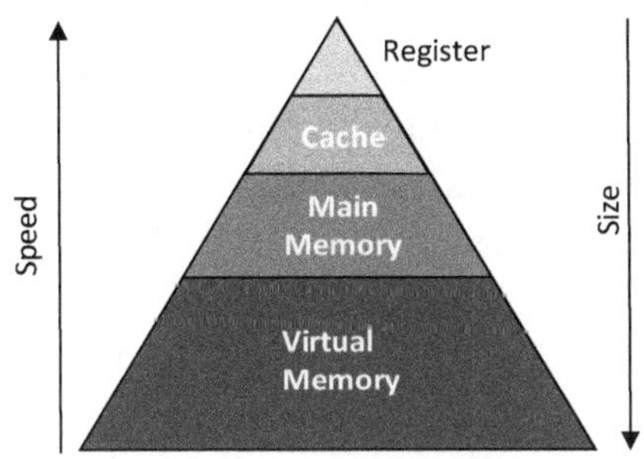

Fig. 7.1. Concept of memory hierarchy

Each level of the hierarchy presented in Figure 7.1 is described below:

Registers

Registers are typically SRAM that holds a data word, which on modern processors is typically 64 or 128 bits. Typical access time is in the clock cycle order.

Cache

Cache is usually close to the processor and is divided into levels. It holds frequently used chunks of data from the main memory. It stores the most commonly used data and instructions in a faster, but smaller memory, so on average the memory system appears fast. The cache is typically in the kilobyte to several megabyte size order.

Main Memory

Main memory is relatively fast (typical access time is in the hundreds of clock cycles) and holds most of the data and instructions that are needed by currently running programs.

Virtual memory

Virtual memory has a huge gap of latency. Solid-State Drives (SSDs), built using Flash memory technology, are an increasingly common alternative to Hard Disk Drives (HDDs). SSDs have been used by niche markets for over two decades, and they were introduced into the mainstream market in 2007. SSDs overcome some of the mechanical failures of HDDs, but they cost about ten times.

Figure 7.2 revisits the memory hierarchy considering a computer system context. The processor first seeks data in a small but fast cache that is usually located on the same chip. If the data is found (*cache hit*), it is immediately returned. If the data is not available in the cache (*cache miss*), the processor then looks in main memory. If the data is not there either, the processor fetches the data from virtual memory on the large but slow hard disk. Note that different heuristics can be used to manage cache content [12]. Processor caches are usually transparent, meaning that the same addressing space is shared between them and backing storage.

In order to create the illusion of having lots of fast memory available, it is crucial that, with a high probability, the cache contains the data that the processor is looking for. In other words, the main memory and, especially hard drive, should be accessed sporadically. This phenomenon refers to as *locality of reference*, which is the tendency of a processor to access the same set of memory locations repetitively over a short period. This can be summarized as follows:

- If you use something once, you are likely to use it again (*temporal locality*);
- If you use something once, you are likely to use its neighbor (*spatial locality*).

Finally, it is important to mention that modern computers frequently have multi-processor, multi-core and multi-thread architectures. Usually, each core has its own Level 1 instruction and data cache, and all cores in the same processor share Level 2 and 3 caches. However, this is not a rule as some architectures can explore different organizations [12].

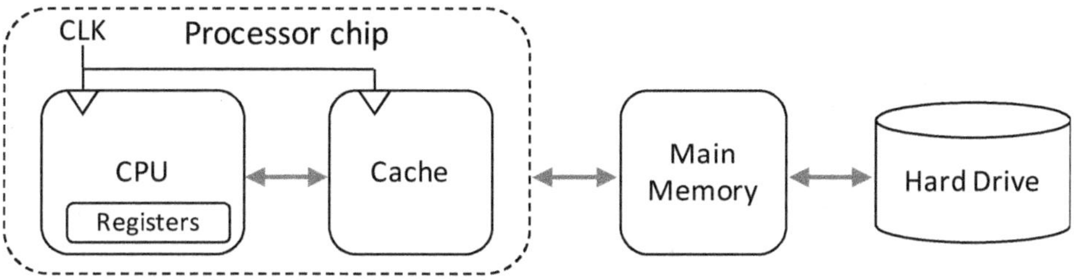

Fig. 7.2. Computer system architecture

7.3 Memory Technologies

7.3.1 *SRAM*

SRAM has long been a standard commodity-type product filling memory needs in different applications ranging from consumer electronics to high-end computers. Nowadays, it is mostly used in processors (register files and caches) and buffers. As already mentioned, SRAM is faster and easier to use than DRAM. In addition, technology scaling enabled SRAM size reduction and its dense integration on chips. On the other hand, problems related to the cost and leakage power consumption are introduced. Supply voltage control along with size reduction also made SRAM more vulnerable to soft errors and more precisely to Single Event Upsets (SEU) [13].

The conventional SRAM structure consisting of 6 transistors follows the cross coupled inverters structure presented in Figure 7.3. The 3 main operations in a SRAM are HOLD, READ and WRITE. In an array of SRAM cells, rows are called Word Lines (WLs) and columns are called Bit Lines (BLs). WLs are connected to the gates of the SRAM select transistors (M5 and M6) and BLs are connected to the drains of the select transistors. C_{BL} and $C_{BL/}$ represent BLs capacitors.

During a HOLD operation, the WL is deactivated to disconnect BL and BL/ from the SRAM core cell. As a result, data is held in the latch structure on nodes Q and Q/. WRITE and READ operations in the memory cells are accomplished through the BLs when WL is activated. During a WRITE operation, BL and BL/ are seen as input lines. If WRITE '0' operation is considered, BL and BL/ are set to '0' and V_{DD} respectively. Assuming that logic '1' is stored in the SRAM cell (i.e. Q = V_{DD} and Q/ = 0), transistors M3 and M2 are turned off, while M1 and M4 are operating in the linear region. The voltage of node Q is reduced below the threshold voltage of M1, so that M1 turns off and node voltage Q/ increases to turn off M4, resulting in a change of the stored data (i.e. Q = 0 and Q/ = V_{DD}).

During a READ operation BL and BL/ act as output lines. BLs capacitors C_{BL} and C_{BL}/ are initially pre-charged to V_{DD}. Assuming that logic '0' is stored in the cell, transistors M1 and M4 are turned off, while transistors M3 and M2 operate in the linear region (i.e. Q = 0 and Q/ = V_{DD}). After M6 and M5 are turned on, C_{BL}/ voltage will not vary significantly, since no current flows through M5. On the other hand, M3 and M6 will conduct a nonzero current and the voltage level of C_{BL} will begin to drop slightly. To complete the read operation, BL and BL/ are connected to a sense amplifier to compare the voltage difference between BL and BL/.

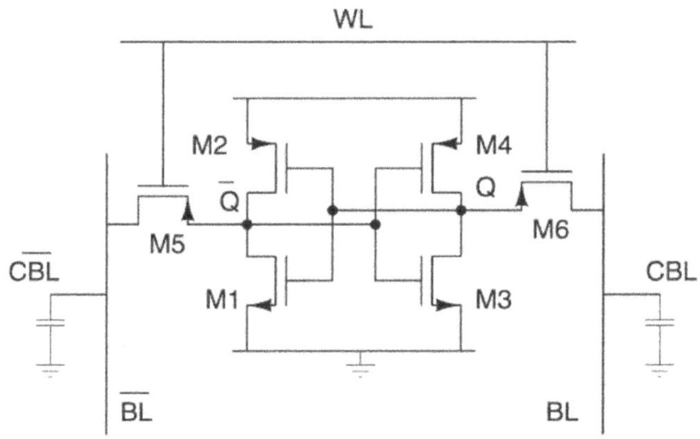

Fig. 7.3. 6-transistor CMOS SRAM cell

An SRAM memory array uses a rectangular organization to minimize the overall chip area and for ease in realization (Figure 7.4). The dimensions of each SRAM array are limited by its electrical characteristics such as capacitances and resistances of the BLs and WLs used to access memory cells at uniform delay in the array. Therefore, large size memories may be folded into multiple blocks with limited number of rows and columns. Typical SRAM memory architecture consists of a memory array of two-dimensional arrangement of cells. Memory cells are usually grouped into 8, 16 or 32 bits to form a memory word. Row decoders are needed to select one row of WLs out of a set of rows in the array according to address bits. Since the long wires of the array may have a high capacitance, row decoders are associated with logic drivers to ensure that the signal can travel the length of the row. Column decoder select a particular BL pairs out of the sets of BLs pairs in the selected row and are also associated with column drivers. This is the logic which selects which column of the SRAM array to write to and to read from. Column Decoders come along with a pre-charge circuit connected to a pair of BLs. The function of this circuit is to pull-up the BLs of a selected column to V_{DD} in order to perfectly equalize them before the read or write operation. Finally, the sense amplifier which is a differential amplifier amplifies the small difference in the BLs and translate it to full swing digital output signal. The sense amplifier is connected to the data control logic in charge of managing data in and out from the memory array.

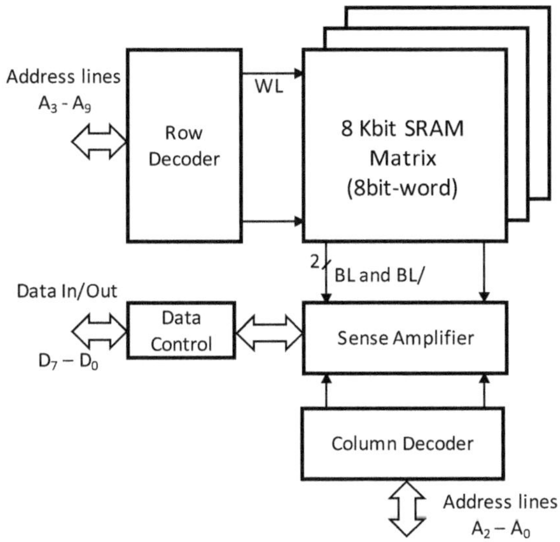

Fig. 7.4. SRAM memory array

7.3.2 *DRAM*

The DRAM cell is based on a very simple concept, the 1-Transistor 1-Capacitor (1T-1C) structure introduced by Robert Dennard (IBM) many years ago [14]. The device cell comprises an access transistor and a storage capacitor as presented in Figure 7.5a. To hold the '1' binary state, a certain amount of charge is stored in the capacitor, while the capacitor is left uncharged to represent the '0' state.

DRAM cells are grouped together in arrays, as shown in Figure 7.5b. Each cell is at a cross-point of a row and a column. Both rows and columns need to be selected through row and address decoders respectively in order to program or read the content of a particular memory word. Usually, several arrays are read at the same time to provide the contents of a memory word. Group of arrays ganged together are called banks. The READ operation is accomplished by turning ON the select transistor M1, resulting in charge sharing between the DRAM storage capacitor C_S and the bit line capacitor C_{BL}. This charge sharing causes the bit line voltage to either increase for a stored logic '1' or decrease for a stored logic '0'. Data WRITE is performed by forcing either a high or low level on the column when the cell is selected. To write a '1' into the cell, V_{DD} is applied to the bit line and a high level pulse is simultaneously applied to the word line. The applied pulse turns ON the select transistor, which charges the storage capacitor C_S to a value equals to $V_{DD} - V_{T(M1)}$. To write a '0', the bit-line voltage is set to 0 V, and the WL is pulsed to V_{DD}, which discharges the storage capacitor C_S.

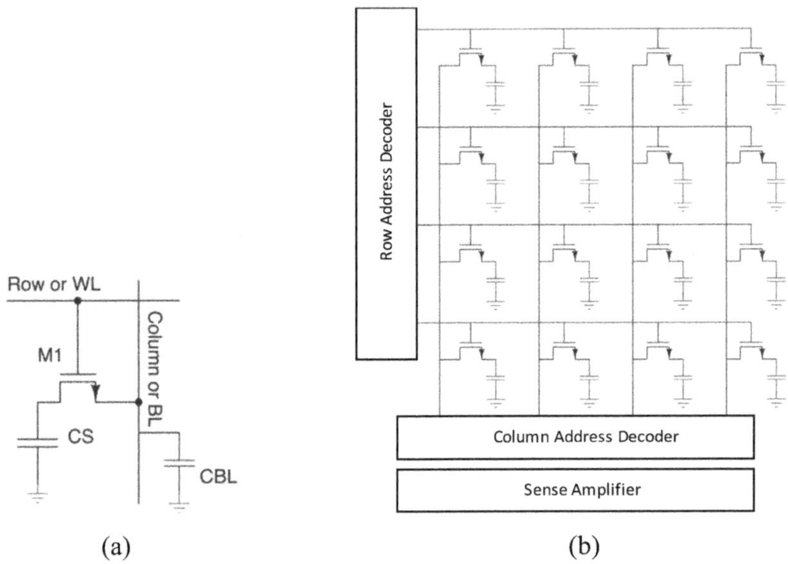

Fig. 7.5. a) DRAM cell and b) DRAM array organization

Over time, the charges stored in the cell capacitors leak away. To prevent this charge loss, each cell must be periodically refreshed so that the correct data remains stored at each bit location. The time interval between the refresh cycles is called the 'refresh time' and represents dozens of milliseconds in most current DRAM devices. Compared to SRAM, the essential characteristics of DRAM can be summarized as follows:

- DRAM cells are simple, allowing high densities at a low cost/bit;
- DRAM is slower than SRAM, yet still fast;
- Constant refreshes due to current leakage make the DRAM energy consuming.

The basic principle of operation has remained unaltered for more than three decades although the cell has evolved to complicated three-dimensional structures, including high-k materials and corrugated capacitor surfaces [15]. Since the

DRAM storage capacitor gets physically smaller with scaling, the capacitor oxide thickness must scale down sharply to maintain adequate storage capacitance. To scale the oxide, dielectric materials having high relative dielectric constant are needed. One of the most promising candidates for the next generation of embedded DRAM is the capacitor-less DRAM (1T-DRAM). Indeed, by removing the capacitor of the bitcell, all issues related to capacitor scaling are removed as well. Capacitor-less DRAM uses the floating body effect of the select transistor to store data [16].

7.3.3 *Flash*

Flash products are classified according two major application segments:

- Code storage, where the program or the operating system is stored and is executed by the microprocessor or microcontroller;
- Data storage, where data files such as image, music, and voice are recorded and read sequentially.

Flash cells and architectures can be divided in terms of access type, parallel or serial, and in terms of the used memorization mechanism [9]. Among all Flash architectures, two can be considered as industry standards: the NOR Flash, that due to its versatility is addressing both the code and data storage segments, and the NAND Flash, optimized for the data storage market. A Flash cell is basically a MOS transistor with a Floating Gate (FG), i.e., a transistor with a gate completely surrounded by dielectrics (Figure 7.6a). Compared to a classical MOS, a Flash memory cell has two gates, one above the tunnel oxide referred to as the FG and the other above the inter polysilicon dielectric layer (interpoly) referred to as the control gate (CG). Note that the FG is completely electrically isolated (Figure 7.6b). Figure 7.6c proposes a cross section of a flash memory cell, highlighting the FG and the CG [17].

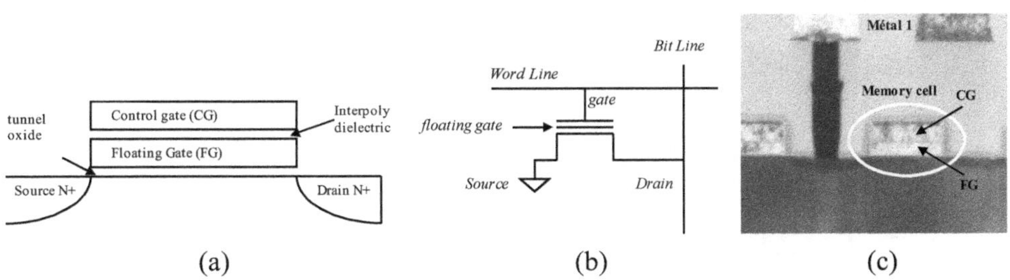

Fig. 7.6. a) Flash cell b) Flash electrical model and c) Flash cross section

Given that the floating gate is isolated, it acts as the storing electrode for the cell device. Charge injected into the FG is maintained there, allowing modulation of the 'apparent' threshold voltage (i.e. V_T seen from the control gate) of the cell. The potential of the FG determines the conductivity of the channel, and thus, the logic state of the cell. Programming or erasing a Flash cell means that charge, or electrons, are respectively added to or removed from the FG. This is achieved by applying specific bias conditions across the cell. During a PROGRAM operation, a high drain-to-source bias voltage is applied, along with a high control gate voltage (Figure 7.7a). The gate voltage inverts the channel (turns it to the ON state), while the drain bias accelerates electrons toward the drain, injecting electrons into the FG at the drain side. Programming a flash cell is performed by CHE with high kinetic energy [18]. During the ERASE operation, the drain is grounded and a high negative gate voltage is applied along with a positive bulk voltage (Figure 7.7a). In these bias conditions, charges are removed from the FG. The electrical erasure of flash memory is achieved by FN tunneling through the tunnel oxide [19]. Despite the need for relatively high writing and erasing voltages, virtually all flash chips today require only a single supply voltage and produce the high voltages using on-chip charge pumps [20]. It is also interesting to notice how CHE and FN mechanisms have been deeply investigated for MOS transistors in order to avoid their unwanted degradation effects. In Flash cells, they are exploited to become efficient program/erase mechanisms. During a READ operation, a specific gate voltage (V_{read}) is applied on the gate (Figure 7.7b). If the memory cell is programmed (i.e. high V_T), the transistor will not conduct. If the memory cell is erased (i.e. low V_T), the transistor will conduct. The resulting reading current is sensed through a sense amplifier circuit and converted to a logical value.

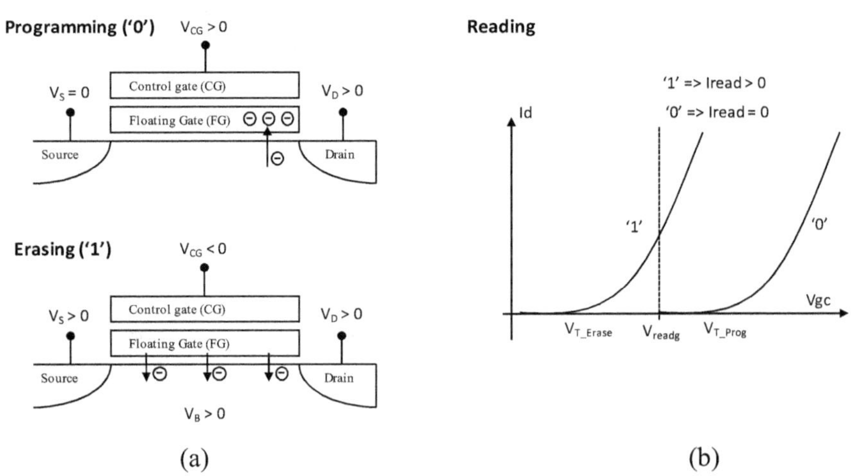

Fig. 7.7. a) Flash programming and erasing conditions; b) reading conditions

The method used to read a flash cell can cause nearby cells in the same memory block to change over time (become programmed). This is known as *read disturb*. The average number of reads is generally in the hundreds of thousands of reads between intervening erase operations. To avoid read disturb problems, the flash controller will typically count the total number of reads to a block since the last erase. When the count exceeds a target limit, the affected block is copied over to a new block, erased, and then released to the block pool [9].

In NOR-type Flash, a memory cell is made of two transistors: a select transistor and the memory cell transistor (Figure 7.8a). The bit lines run alongside the cells (connecting the columns) and the word lines connect the rows through select transistors. A cell is turned on by activating the corresponding word line. MS signal connects selected memory cells to ground (program) or to a positive voltage (erase). The key detail here is that each cell requires a WL connector and a BL connector. NAND Flash, on the other hand, links an entire cell column by running the bit line through each cell, bypassing the need for an extra bit line connector on the side (Figure 7.8b). Each column contains two selection transistors. The first one is controlled by the Select Line (SL) signal to select the addressed bit line. The second one is controlled by the Ground Select (GS) signal and connects the bit line to ground. Accessing a cell within a row implies not only to activate the word line of the selected cells, but also all the word lines of the selected bit line. As a result, NAND Flash reduces memory cell area so that a lower bit cost can be achieved.

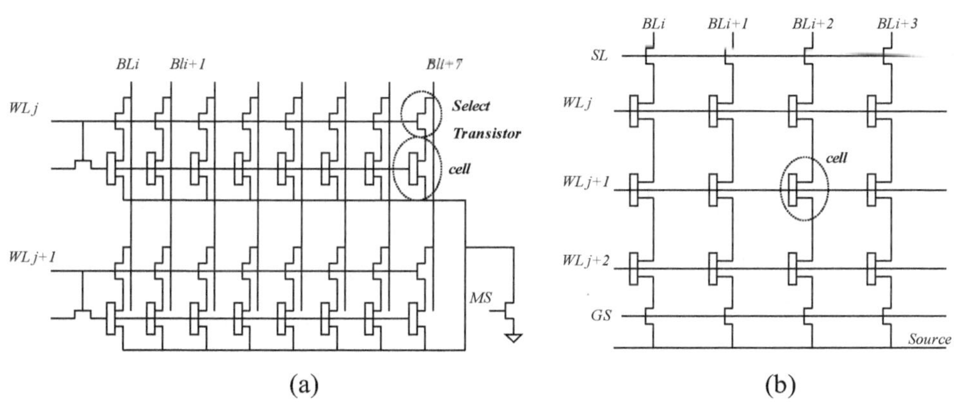

Fig. 7.8. a) NOR Flash; b) NAND Flash architectures

Figure 7.9 shows a simplified device block diagram of an embedded Flash, which includes the memory array, a row and a column decoder, an I/O buffer and a control logic associated with an address generator and control registers. Memory array peripheral circuits also include critical components such as high-voltage generators and sense amplifiers. This architecture allows easy microprocessor or microcontroller control for execution of various instructions through control logic.

The high-voltage generation unit includes a Voltage Controlled Oscillator (VCO) connected to a Charge Pump (CP). The latter is an electronic circuit that converts the supply voltage V_{DD} to a DC output voltage that is several times higher than V_{DD} (i.e., it is a DC-DC converter whose input voltage is lower than the output voltage). Unlike other traditional DC-DC converters, which employ inductors, CPs are only made of capacitors and CMOS switches, thereby allowing integration on silicon [21]. The CP output feeds a shaping circuit, which provides the signals needed during Flash memory program or erase operations. Note that the high voltage path requires high voltage transistors, whose breakdown voltages (V_{bd}) are sufficiently higher than the maximum voltages necessary for program or erase operations. The sense amplifier 'senses' the current cell under strictly-defined biasing conditions. The read operation can be achieved by comparing the cell current with a reference current usually provided by another cell normally linked with the process characteristics. In general, a differential sense amplifier compares the cell voltage after the current is converted to a reference voltage [22].

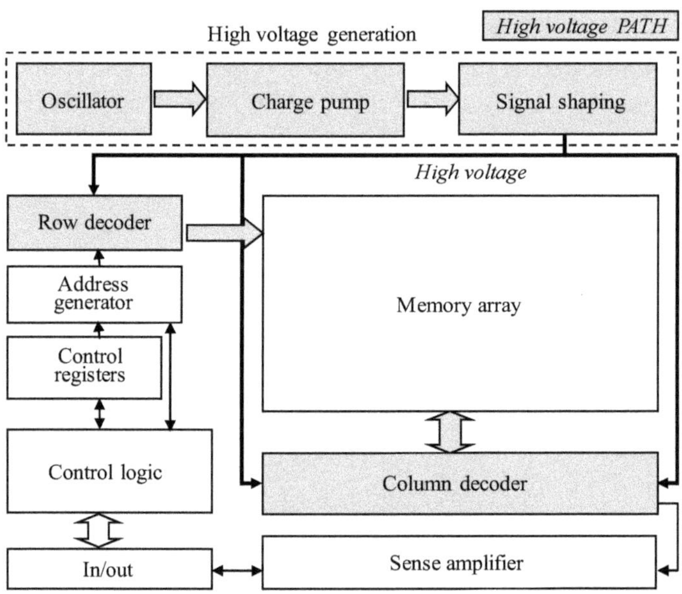

Fig. 7.9. Flash memory array

7.3.4 *Emerging memory technologies*

As already mentioned, BEOL memories are a new concept, where memory points are stacked on top of the CMOS layer. Their interest is huge as they pave the way for emerging non-volatile applications build with high performance (low power consumption, fast access) and safe circuits (thanks to non-volatility). These back-end memories (STT-RAM, PCM, and RRAM) are called memristors or 'resistive-switching elements'. The stored logic value corresponds to a resistive state (high or low) and their operating voltage is compatible with advanced CMOS technology nodes. It is worth noticing that in literature the generic term resistive RAM (also called RRAM or ReRAM) is often used to generically refer to both Oxide Random Access Memory (OxRAM) and Conductive-Bridging Random Access Memory (CBRAM).

RRAM are considered as easier to integrate than STT-RAM in advanced CMOS processes [6], because they use 'CMOS compatible' materials. However, despite high manufacturing costs, STT-RAM has SRAM-like performance and smaller footprint than the conventional 6T-SRAM. Moreover, STT-RAM technology has gained much interest in space applications due to their intrinsic immunity to radiations [23]. Compared to RRAM, PCM suffers from high current (fraction of mA) required to reset the phase change element, which limits its application in low-power devices. Even for state-of-the-art PCM at 20 nm, the write current (~100 μA [24]) is roughly ten times larger than its RRAM counterpart. Moreover, code integrity is not guaranteed during the technological integration process flow, in particular during solder reflow (at ~250 °C).

Despite being the least mature technology, RRAM is one of the most promising candidates to overcome the scaling limitation of current embedded memories such as Flash and SRAM, because of its simple configuration (i.e. MIM capacitor-like structure) that allows three-dimensional stacking of devices and multi-level capability. A major advantage of the RRAM architectures is that it allows scaling of the device to sub-nanometer dimensions. Table 7.2 presents MCUs embedded memory projections considering mainstream memories (eFlash and SRAM) and emerging memories. Convergence between RRAM and eFlash is expected to happen at the 28 nm node in 2021. SRAM substitution by MRAM and RRAM is expected after 2020 [8].

Table 7.2. Memory market scaling projections | Yole | emerging non-volatile memory report – 2018

Memory Technology	*2016*	*2017*	*2018*	*2019*	*2020*	*2021*	*2022*
SRAM (nm)	10	10	7	7	5	5	5
eFlash (nm)	28	28	28	28	20	20	20
MRAM (nm)	130	40	28	28	22	22	14
PCM (nm)	-	110	110	110	28	28	28
RRAM (nm)	**180**	**40**	**40**	**40**	**28**	**22**	**22**

In the next sections, we focus on the performances and most important properties of the main emerging nonvolatile memory technologies.

7.3.4.1 *PCM*

Phase-Change Random Access Memory (also called PCRAM, PRAM or PCM) is currently the most mature of the technologies under research. The fundamental mechanism for phase-change memory was invented in early 1960s by Stanford Robert Ovshinsky [25]. It relies on some materials, called phase-change materials that exist in two different phases with distinct properties:

- An amorphous phase, characterized by high electrical resistivity;
- A crystalline phase, characterized by low electrical resistivity.

PCM is made using a Germanium Antimony Tellurium (GST) alloy, and takes advantage of rapid heat-controlled changes in the material's physical property between amorphous and crystalline states (Figure 7.10). These states, which correspond to logic '0' and '1', are electrically differentiated by the phase-change material phase. The two phases can be repeatedly and rapidly cycled by applying heat to the material. SET and RESET state refers to low and high-resistance state respectively. To make the material amorphous, the programming region is first melted over a temperature T_{melt} and then quenched rapidly by applying a large electrical current pulse for a short time period. To crystallize the material, it is heated above its crystallization temperature T_{crys} using a moderate power pulse with a longer duration. Since the duration of this pulse varies according to the crystallization speed of the material being used, this operation tends to dictate the writing speed of PCM. Based on the large resistivity contrast between the resistances of the crystalline state and the amorphous state (typically 3-4 orders of magnitude) and the analog variation of the resistance, it is possible to employ the intermediate resistance levels to achieve a Multi Level Cell (MLC) operation. During a PCM read, a low-amplitude electrical pulse is used to sense the resistance of the cell.

Figure 7.11 presents an array of PCM cells. A 1T-1R configuration is used to build the memory array with PCM cell integrated in the BEOL on top of the select transistor. A cross-section is proposed to highlight the GST material.

PCM cells are affected by variability. Indeed, variations in the process and material parameters during device fabrication lead to variations in the critical dimension of the cells across the array. As a result, the same electrical stimuli might give rise to different thermal profiles within the active storage element and hence result in different cell resistance values [26]. Given the wide resistance distributions arising from cell variability, single pulse programming methods are not suited for MLC storage. The most common solution is to employ iterative programming schemes, where a sequence of program and verify pulses are used in a closed-loop fashion to program the cell into a particular resistance level [26].

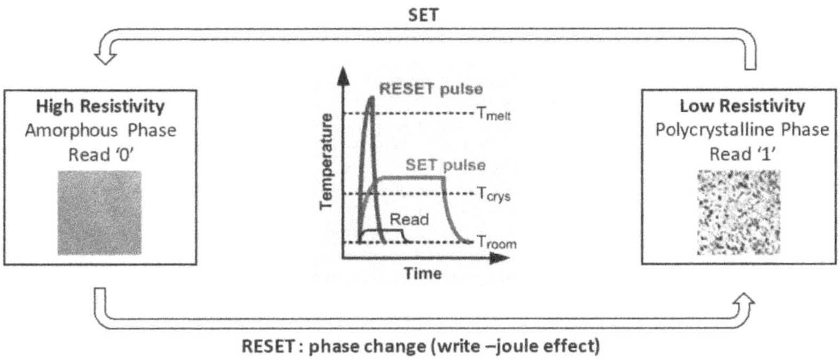

Fig. 7.10. PCM Storage mechanism based on resistance variation of a $Ge_2Sb_2Te_5$ (GST - Germanium Antimony and Tellurium) chalcogenide alloy

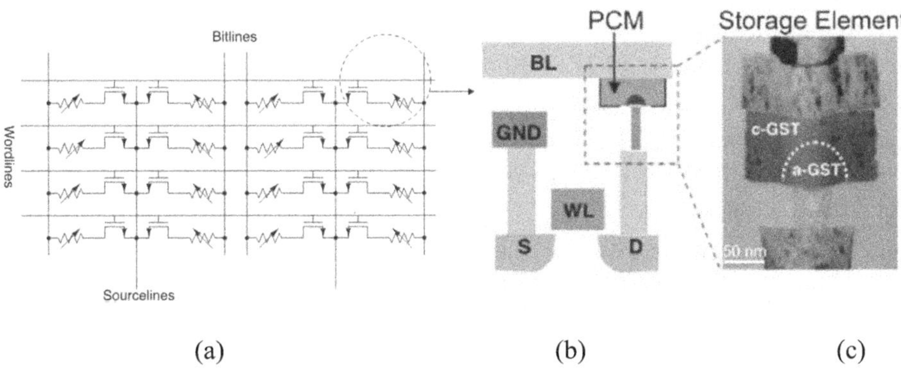

Fig. 7.11. a) An array of the PCM cells where each cell consists of b) a transistor and c) a phase-change storage element

7.3.4.2 *STT-RAM*

Magnetoresistive Random Access Memory (MRAM) is different from conventional types of memory like SRAM, DRAM, and Flash, where electric charge is used to store information. Instead of exploiting the charge of electrons, MRAM uses its spin to store data. This type of electronics is known as "spintronics." Spin-Transfer Torque (STT) is an effect in which the orientation of a magnetic layer in a Magnetic Tunnel Junction (MTJ) can be modified using a spin-polarized current. STT-RAM is projected to be the future for MRAM products as it is expected to scale down to 15 nm [27]. STT-RAM is based on the MTJ structure which is a thin-film structure comprised of many ferromagnetic and nonmagnetic layers. The resistance of the MTJ, and therefore the bit state of the memory element, switches purely with the change in polarization of these layers. Since these materials can hold their polarization virtually forever when unpowered, this technology falls into the nonvolatile memory category.

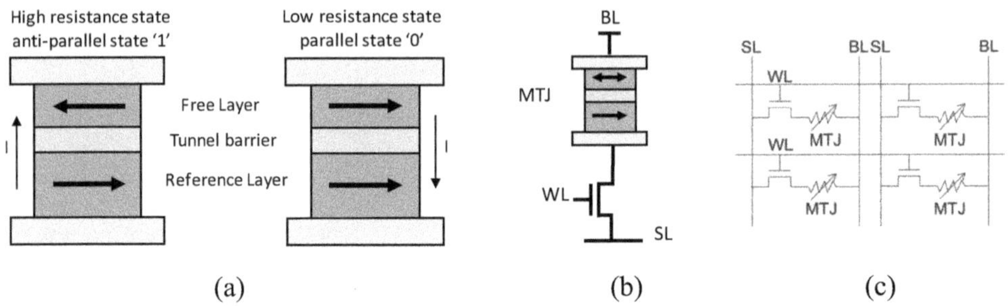

Fig. 7.12. a) MTJ states b) 1T-1R STT-RAM c) STT-RAM array

The MTJ cell has three elements: the Free Layer (FL), the Reference Layer (RL) and the tunnel barrier (Figure 7.12a). A STT-RAM memory cell is made of on transistor along with one MTJ (1T-1R configuration, see Figure 7.12b) integrated in the BEOL. Writing a '0' or writing a '1' to a cell is achieved by applying reversed direction currents. When writing '0' (MTJ in parallel state), the word line (WL) and bit line (BL) are connected to the supply voltage V_{DD}, and the source line (SL) is connected to the ground (Figure 7.12c). The NMOS transistor is either working in its saturation region for a small transistor width or in its linear region for a large transistor width. When writing '1' (MTJ in anti-parallel state), WL and SL are connected to V_{DD}, while BL is connected to the ground. The transistor is then working in its saturation region. Parallel state is associated with a low resistance R_p and the anti-parallel state is associated with a high resistance R_{ap}. The Tunneling Magneto Resistance (TMR) ratio of the MTJ, defined as $(R_{ap}-R_p)/R_p$ is typically small (<200%), thereby imposing challenges for sensing circuit designs to detect the small difference between the ON and OFF states.

There are two ways to read the cell, the 'parallel direction read' with the same direction as writing '0', and the 'anti-parallel direction read' with the same direction as writing '1'. In the parallel direction, a low voltage is applied on the BL. After activating WL, current flows from BL to SL. In the anti-parallel direction, the voltage polarity applied to BL and SL is switched and current flows in the reversed direction, from SL to BL.

STT-RAM is a very attractive nonvolatile memory, thanks to its radiation hardness, low standby power, and unlimited endurance and retention characteristics. It may lack the density to serve as the primary science data store for some missions (NAND Flash may be a better option), but applications requiring only 64 Mb per device may find this technology to be a superior option compared to NOR Flash.

7.3.4.3 *RRAM*

Resistive RAMs based on metal oxides, so-called OxRAMs are considered in this section. OxRAM is one of the most competitive candidates for future non-volatile memory applications due to its simple structure offering low manufacturing costs, fast switching speed (~1 ns), great scalability (<10 nm), low voltage operation and compatibility with current CMOS technology [28]. The OxRAM consists of two metallic electrodes that sandwich a thin dielectric layer serving as a permanent storage medium. This Metal-Insulator-Metal (MIM) structure, denoted RRAM in Figure 7.13a can be easily integrated in the BEOL in combination with advanced CMOS technologies. The MIM structure is integrated on top of Metal 4 copper layer (Cu). A TiN Bottom Electrode (BE) is first deposited. Then, a 10nm-HfO_2/10nm-Ti/TiN stack is added to form a capacitor-like structure [29]. Figure 7.13b shows the basic 1T-1R memory cell where the select transistor acts as a current limiting device. It controls the amount of current flowing through the cell according to its gate voltage bias in order to prevent memory cell damage. Figure 7.13c presents a typical 1T-1R OxRAM I-V characteristic in logarithmic scale.

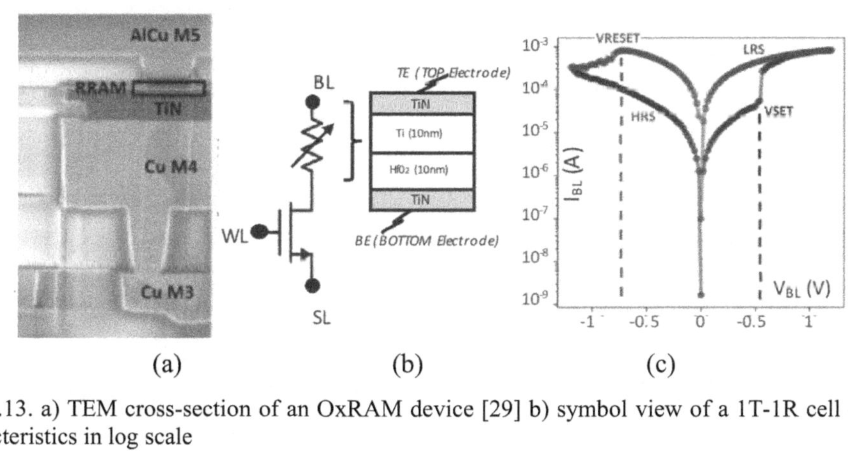

Fig. 7.13. a) TEM cross-section of an OxRAM device [29] b) symbol view of a 1T-1R cell c) I-V characteristics in log scale

Based on the I-V curve, the memory cell behavior can be seen as follows: after an initial electroforming step [30, 31] (not presented here), the memory element can be reversibly switched between a High Resistance State (HRS) and a Low Resistance State (LRS). Resistive switching in an OxRAM corresponds to an abrupt change between a HRS and a LRS. This resistance change is achieved by applying specific biases across the 1T-1R cell (i.e. V_{SET} and V_{RESET}). According to

Figure 7.13(c), V_{SET} value required to switch from HRS to LRS state is around 0.55 V, while V_{RESET} value required to switch back to the HRS state is around -0.75 V.

Common problems with OxRAMs are related to high variability in operating conditions [32] and high programming currents during FORMING, SET and RESET operations. In this context, programming algorithms and write assist circuits play a crucial role on performance and yield optimization of resistive switching memories [33]. Program-verify algorithms have been developed to ensure distinct resistance distributions. These algorithms adjust the write signal (pulse number, width or amplitude) until the cell resistance reaches the target resistance state.

The 1T-1R architecture is suitable for NOR flash-like and embedded nonvolatile memory applications where high reliability and short latency are important. However, this configuration is not suited for storage memory applications where high density is required. The simplest way to increase the memory array density is to build a passive crossbar array where only the OxRAM cells connect bit lines and word lines, resulting in the highest compactness (Figure 7.14a and Figure 7.14b). However, in the absence of the selector in the crossbar structure, a large amount of leakage current flowing through unselected cells is inevitable (*sneak-paths*) causing large amount of leakage currents [34]. The sneak path problem is one of the main issue in the construction of crossbar arrays and three-dimensional (3D) crossbar memory design. One approach to mitigate sneak currents is to use self-rectifying selectors such as diodes, integrated within a ReRAM memory cell stack. Indeed, the rectifying characteristics of the cell can suppress current conduction through unselected ReRAM cells along the sneak path. MLC offer another opportunity to improve ReRAM density [35]. Compared to the multi-layer approach, 1T-1R MLC cell relaxes the magnitude of the sneak current and voltage drop problems. This approach requires more accurate tuning of the analog resistance value of each cell [36].

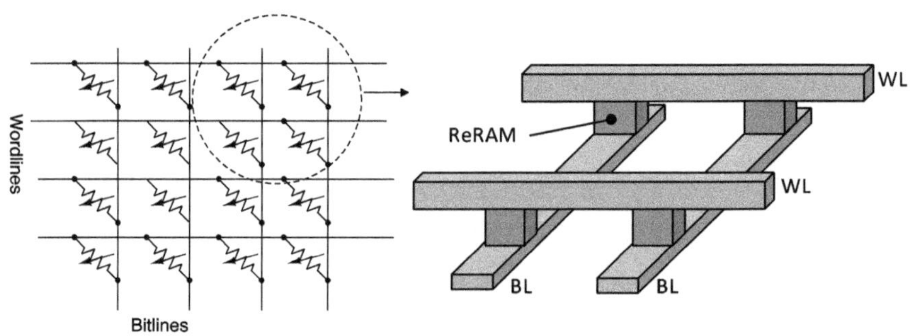

(a) (c)

Fig. 7.14. a) Crossbar memory array and b) RRAM integration in the BEOL

7.4 Conclusion

Ever more demanding applications are pushing the limits of MCU architectures due to their need for more processing power, low power consumption, and larger memory sizes. One of the most challenging demands is for larger embedded memories to hold bigger and more complex firmware. Emerging memories present a solution to these challenges. Table 7.3 compares the typical device characteristics of the emerging memory technologies and mainstream memory technologies.

Table 7.3. Memory market scaling projections | Yole | emerging non-volatile memory report – 2018

	Mainstream memories				*Emerging memory*		
Memory Technology	*SRAM*	*DRAM*	*NOR*	*NAND*	*STT-MRAM*	*PCRAM*	*RRAM*
Cell area	$>90\ F^2$	$6F^2$	$10F^2$	$<4F^2$	$6\text{-}50F^2$	$4\text{-}30F^2$	$4\text{-}12F^2$
Multibit	1	1	2	4	1	2	2
Voltage	<1V	<1V	>10V	>10V	<1.5V	<3V	<3V
Read time	1ns	10ns	50 ns	10us	<10ns	<10ns	<10ns
Write time	1ns	10ns	10u-1ms	90u-1ms	<10ns	50ns	<10ns
Retention	NA	60ms	>10y	>10y	>10y	>10y	>10y
Endurance	$>$1E16	$>$1E16	$>$1E5	$>$1E4	$>$1E15	$>$1E9	$>$1E6-12
Write E (J/bit)	~fJ	~10fJ	~100pJ	~10fJ	~0,1pJ	~10pJ	~0,1pJ

The numbers of this table are representative (not the best or the worst cases).
F: feature size of the lithography
Energy estimation is extracted at cell-level

Due to the different memorization mechanisms, the device characteristics are different among emerging NVMs. Therefore, due to their unique characteristics, different emerging NVMs have different application spaces. These application spaces are dictated by memory performances. STT-RAM is attractive as a replacement for embedded memories (e.g., SRAM or embedded DRAM). PCM and RRAM is attractive as a replacement for NOR flash for code storage and, more ambitiously, to replace NAND flash for data storage. In addition, emerging memories have the potential to revolutionize today's memory hierarchy by adding more levels in the hierarchy, e.g., creating a storage-class memory layer between the main memory and SSDs. Alternatively, hybrid systems combining CMOS logic and emerging memory are rising for hardware neuronal algorithm applications [37].

References

1. Van der Leest, V., Maes, R., Schrijen, G.J. and Tuyls, P. (2014). Hardware intrinsic security to protect value in the mobile market. *In ISSE 2014 Securing Electronic Business Processes*, pp. 188-198, https://doi.org/10.1007/978-3-658-06708-3_15

2. Aziza, H., Postel-Pellerin, J., Bazzi, H., Canet, P., Moreau, M., Della Marca, V. and Harb, A. (2020). True random number generator integration in a resistive RAM memory array using input current limitation, *IEEE Transactions on Nanotechnology*, 19, pp. 214-222, https://doi.org/10.1145/ 586110.586132
3. Meena, J. S., Sze, S. M., Chand, U. and Tseng, T. Y. (2014). Overview of emerging nonvolatile memory technologies, *Nanoscale Research Letters*, 9-526, pp. 1-33, https://doi.org/10.1186/1556-276X-9-526
4. Prall, K. (2017). Benchmarking and metrics for emerging memory, *Proc. IEEE Internat. Memory Workshop*, pp. 1-5, https://doi.org/10.1109/ IMW.2017.7939072
5. Puglia, G. O., Zorzo, A. F., De Rose, C. A., Perez, T. D. and Milojicic, D. (2019). Non-volatile memory file systems: A survey, *IEEE Access*, 7, pp. 25836-25871, https://doi.org/10.1109/ACCESS.2019.2899463
6. Vatajelu, E. I., Aziza, H. and Zambelli, C. (2014). Nonvolatile memories: Present and future challenges, *Proc. 9th International Design and Test Symposium*, pp. 61-66, https://doi.org/10.1109/IDT.2014.7038588
7. Hamdioui, S., Aziza, H. and Sirakoulis, G. C. (2014). Memristor based memories: Technology, design and test, *Proc. 9th IEEE International Conference on Design and Technology of Integrated Systems in Nanoscale Era*, pp. 1-7, https://doi.org/10.1109/DTIS.2014.6850647
8. Yole Developpement (2015). Emerging nonvolatile memory technology and market report. http://www.yole.fr/Emerging_NVM_Memory_Activity_Webts.aspx
9. Strenz, R. (2011). Embedded flash technologies and their applications: Status and outlook, *Proc. IEEE International Electron Devices Meeting*, pp. 9.4.1-9.4.4, https://doi.org/10.1109/IEDM.2011.6131521
10. Plantier, J., Aziza, H., Portal, J. M. and Reliaud, C. (2009). Investigation of EEPROM memories reliability during endurance and retention tests, *Proc. 4th International Conference on Design and Technology of Integrated Systems in Nanoscale Era*, pp. 241-246, https://doi.org/10.1109/DTIS.2009.4938063
11. Bazzi, H., Harb, A., Aziza, H., Moreau, M. and Kassem, A. (2020). RRAM-based non-volatile SRAM cell architectures for ultra-low-power applications, *Analog Integrated Circuits and Signal Processing*, 1-11, https://doi.org/ 10.1007/s10470-020-01587-z
12. Jacob, B., Wang, D. and Ng, S. (2010). *Memory systems: Cache, DRAM, disk*, (Morgan Kaufmann)
13. Castellani-Coulié, K., Aziza, H., Rahajandraibe, W., Micolau, G. and Portal, J. M. (2013). Development of a CMOS oscillator concept for particle detection and tracking, *IEEE Transactions on Nuclear Science*, 60, pp. 2450-2455, https://doi.org/10.1109/TNS.2013.2254723
14. Dennard, R. H. (1968). *U.S. Patent No. 3,387,286*, Washington, DC, U.S. Patent and Trademark Office. https://patents.google.com/patent/ US3387286A/en
15. Sunami, H., Kure, T., Hashimoto, N., Itoh, K., Toyabe, T. and Asai, S. (1983). A corrugated capacitor cell (CCC) for megabit dynamic MOS memories, *IEEE Electron Device Letters*, 4, pp. 90-91, https://doi.org/10.1109/T-ED.1984.21602

16. Pulicani, R., Goducheau, O., Degoirat, H., Aziza, H., Pérez, A. and Bergeret, E. (2010). Simulation of intrinsic bipolar transistor mechanisms for future capacitor-less eDRAM on bulk substrate, *Proc. 17th IEEE International Conference on Electronics, Circuits and Systems*, pp. 966-969, https://doi.org/10.1109/ICECS.2010.5724674
17. La Rosa, F., Niel, S., Regnier, A. and Delalleau, J. (2015). *U.S. Patent No. 9,224,482*, Washington, DC, U.S. Patent and Trademark Office, https://patents.google.com/patent/US9224482B2/en
18. Aziza, H. and Delsuc, B. (2008). Device and memory array models for flash EEPROM technology, *WSEAS Transactions on Circuits and Systems*, 7, pp. 249-258, https://dl.acm.org/citation.cfm?id=1482061
19. Aziza, H., Portal, J. M., Plantier, J., Reliaud, C., Regnier, A. and Ogier, J. L. (2012). Non volatile memory reliability evaluation based on oxide defect generation rate during stress and retention test, *Solid-State Electronics*, 78, pp. 151-155, https://doi.org/10.1016/j.sse.2012.05.069
20. Ngueya, W. S., Portal, J. M., Aziza, H., Mellier, J. and Ricard, S. (2018). An ultra-low power and high performance single ended sense amplifier for low voltage flash memories, *Journal of Low Power Electronics*, 14, pp. 157-169, https://doi.org/10.1166/jolpe.2018.1543
21. Ngueya, W. S., Mellier, J., Ricard, S., Portal, J. M. and Aziza, H. (2017). Power efficiency optimization of charge pumps in embedded low voltage NOR flash memory, *Proc. IEEE Nordic Circuits and Systems Conference: NORCHIP and International Symposium of System-on-Chip*, pp. 1-5, https://10.1109/NORCHIP.2017.8124949
22. Conte, A., Giudice, G. L., Palumbo, G. and Signorello, A. (2005). A high-performance very low-voltage current sense amplifier for nonvolatile memories, *IEEE Journal of Solid-State Circuits*, 40, pp. 507-514, https://10.1109/JSSC.2004.840985
23. Lopes, J., Di Pendina, G., Beigne, E. and Torres, L. (2016). An SEU tolerant MRAM based non-volatile asynchronous circuit design, *Proc. 16th European Conference on Radiation and Its Effects on Components and Systems*, pp. 1-4, https://10.1109/RADECS.2016.8093151
24. Kang, M. J., Park, T. J., Kwon, Y. W., Ahn, D. H., Kang, Y. S., Jeong, H., et al. (2011). PRAM cell technology and characterization in 20nm node size, *Proc. IEEE International Electron Devices Meeting*, pp. 3.1.1-3.1.4, https://10.1109/IEDM.2011.6131478
25. Ovshinsky, S. R. (1966). *U.S. Patent No. 3,271,591*, Washington, DC, U.S. Patent and Trademark Office, https://patents.google.com/patent/ US3271591A/en
26. Athmanathan, A., Stanisavljevic, M., Papandreou, N., Pozidis, H. and Eleftheriou, E. (2016). Multilevel-cell phase-change memory: A viable technology, *IEEE Journal on Emerging and Selected Topics in Circuits and Systems*, 6, pp. 87-100. https://10.1109/JETCAS.2016.2528598
27. Xu, N., Wang, J., Lu, Y., Park, H. H., Fu, B., Chen, R., et al. (2015). Physics-based compact modeling framework for state-of-the-art and emerging STT-MRAM

technology, *Proc. IEEE International Electron Devices Meeting*, pp. 28.5.1-28.5.4, https://10.1109/IEDM.2015.7409789

28. Yu, S. and Chen, P. Y. (2016). Emerging memory technologies: Recent trends and prospects, *IEEE Solid-State Circuits Magazine*, 8, pp. 43-56. https://10.1109/MSSC.2016.2546199
29. Molas, G., Sassine, G., Nail, C., Robayo, D. A., Nodin, J. F., Cagli, C., et al. (2018). Resistive memories (RRAM) variability: Challenges and solutions, *ECS Transactions*, 86, pp. 35-47, https://10.1149/08603.0035ecst
30. Hajri, B., Mansour, M. M., Chehab, A. and Aziza, H. (2017). Oxide-based RRAM models for circuit designers: A comparative analysis, *Proc. 12th International Conference on Design and Technology of Integrated Systems in Nanoscale Era*, pp. 1-6, https://10.1109/DTIS.2017.7930176
31. Hajri, B., Aziza, H., Mansour, M. M. and Chehab, A. (2019). RRAM device models: A comparative analysis with experimental validation, *IEEE Access*, 7, pp. 168963-168980, https://doi.org/10.1109/ACCESS.2019.2954753
32. Aziza, H., Hajri, B., Mansour, M., Chehab, A. and Perez, A. (2018). A lightweight write-assist scheme for reduced RRAM variability and power, *Microelectronics Reliability*, 88, pp. 6-10, https://doi.org/10.1016/j.microrel. 2018.07.065
33. Aziza, H., Bocquet, M., Moreau, M. and Portal, J. M. (2015). A built-in self-test structure (BIST) for resistive RAMs characterization: Application to bipolar OxRRAM. *Solid-State Electronics*, 103, pp. 73-78, https://doi.org/ 10.1016/j.sse.2014.09.005
34. Vourkas, I. and Sirakoulis, G. C. (2016). Memristor-based nanoelectronic computing circuits and architectures, *Cham: Springer International Publishing*, 19, https://doi.org/10.1166/jolpe.2017.1519
35. Aziza, H., Ayari, H., Onkaraiah, S., Moreau, M., Portal, J. M. and Bocquet, M. (2016). Multilevel operation in oxide based resistive RAM with SET voltage modulation, *Proc. International Conference on Design and Technology of Integrated Systems in Nanoscale Era*, pp. 1-5, https://doi.org/2016. 10.1109/DTIS.2016.7483892
36. Aziza, H., Perez, A. and Portal, J. M. (2018). Resistive RAMs as analog trimming elements, *Solid-State Electronics*, 142, pp. 52-55, https://doi.org/10. 1016/j.sse.2018.02.005
37. Aziza, H., Moreau, M., Perez, A., Virazel, A., & Girard, P. (2019). A capacitor-less CMOS neuron circuit for neuromemristive networks, *Proc. 17th IEEE International New Circuits and Systems Conference*, pp. 1-4, https://doi.org/10.1109/NEWCAS44328.2019.8961278

Part 3

Implementation and Integration

https://doi.org/10.1142/9789811210723_0008

Chapter 8

Implementation Methodologies

Zoran Stamenković
IHP - Leibniz-Institut für innovative Mikroelektronik
Frankfurt (Oder), Germany

Gildas Leger
Consejo Superior de Investigaciones Científicas,
Universidad de Sevilla
Instituto de Microelectrónica de Sevilla
Seville, Spain

8.1 Introduction

In this chapter, we describe methodologies used to implement a WLAN transceiver based on MIMO communication and signal combining in the analog front-end. The system functionalities are implemented into the separated subsystems: MAC processor, digital baseband processor, and analog front-end circuits [1]. Figure 8.1 shows the implemented transceiver.

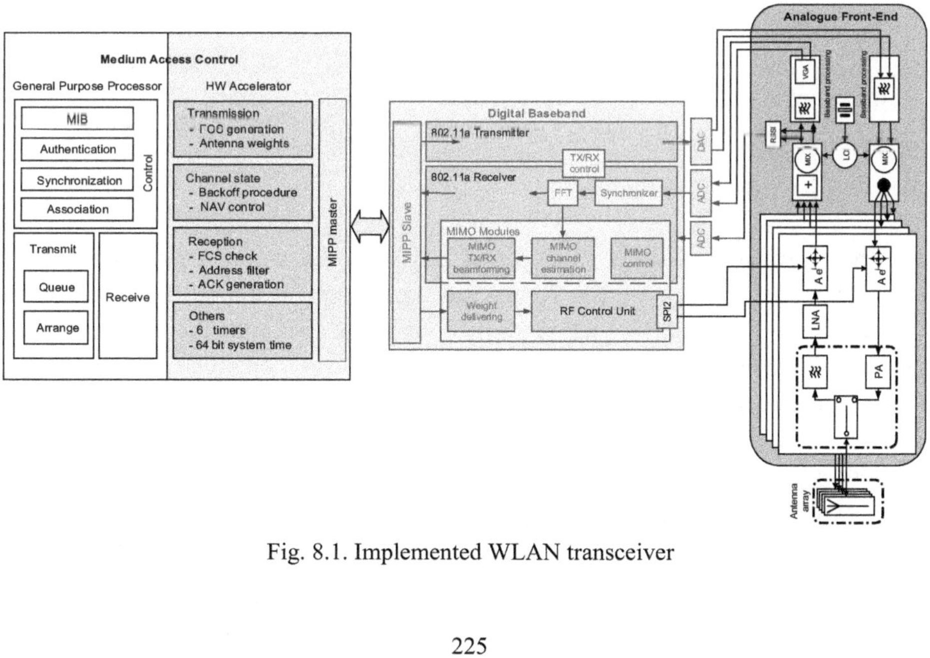

Fig. 8.1. Implemented WLAN transceiver

Multiple antennas increase the transmission reliability through spatial diversity. Redesigns have mostly been done in the physical medium-dependent (PMD) layer. They demand for changes in the physical layer convergence (PLC) and medium access control (MAC) protocols to exploit benefits of the new RF front-end [2-4]. The PLCP pursues mapping of MAC protocol data units in PMD layer compliant frame formats. This task is common for all communication schemes defined by the IEEE Standard 802.11. Moreover, the spatial diversity has to be exploited, possible impairments in the RF spatial processing have to be compensated, and the MIMO channel has to be estimated. These tasks are not part of the IEEE802.11a scheme, which specifies the single-input single-output (SISO) communication.

There are several differences between the RF-MIMO approach and the full multiplexing MIMO approach. In RF-MIMO, the same weight is used for all sub-carriers in OFDM transmissions, whereas it is possible to weight each sub-carrier independently from the others in the full MIMO transmission scheme.

Implementation of the complex integrated system as a WLAN OFDM RF-MIMO transceiver need both methodologies:

1. Semi-custom, which limits the optimization freedom but reduces the implementation time and effort. Different hardware description languages can be used. Verified hardware (technology independent and technology dependent) cores (usually general-purpose processors, embedded memory blocks, buses, controllers, and peripherals) can be integrated together with custom hardware components. Digital parts (MAC and baseband processors) of the transceiver have been implemented using this methodology.
2. Full-custom, which requires all the components to be designed, implemented, and verified at the transistor (schematic and layout) level. This implementation methodology is used to optimize the size (silicon area), speed, and energy consumption of transceiver analog and RF circuits.

8.2 Semi-Custom Methodology

Once the hardware/software partitioning has been done and the register-transfer level (RTL) description of the digital part has been prepared (in the form of generated and/or manually written HDL code), one can start the implementation phase. The semi-custom implementation methodology depends on many factors: size of the design, time constraints, available tools, and customer demands regarding cost, performance, and energy consumption. Depending on the chosen methodology, a digital design can be implemented as:

1. Application Specific Integrated Circuit (ASIC) based on a technology-dependent library of predefined standard cells, and

2. Field Programmable Gate Array (FPGA) that is reprogrammable (but less optimized) and prototyping oriented.

8.2.1 *ASIC implementation flow*

An ASIC implementation flow starts with hardware/software partitioning of the design. The manually written HDL RTL models describing new components are introduced. Most of other components are configurable processors and interfaces, which can be easily combined with new components to implement a system-on-chip [5-7]. Memory blocks and standard cells are the necessary pre-defined building modules. The simulation models, as well as area, speed, and energy consumption estimates of these building modules are generated to support post-synthesis and post-layout system simulations.

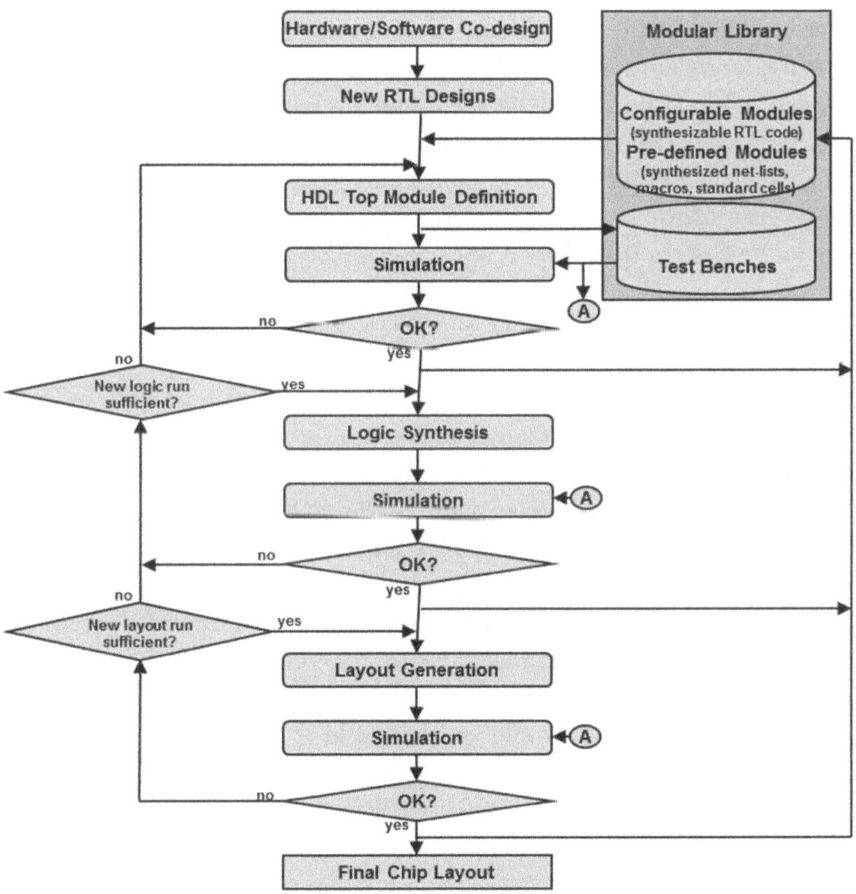

Fig. 8.2. Overview of the ASIC implementation flow

Figure 8.2 shows all main steps of the implementation flow including both logical and physical stage. The computer tools called design compilers can synthesize a gate-level representation of the design [8]. Behavioral and net-list (without and with timing information) RTL simulations are usually performed by the event-driven simulators [9, 10]. The layout tools can do the floor planning, placement, clock tree generation, routing (including the delay optimization), and verification of layout [11].

HDL modeling

A system level model has to capture the intent of the design and is the definition of what the design should do. It allows designers to understand the implications of their choices and decisions. The model should describe system behavior and performance, define a system architecture and a framework for implementation, as well as verify all the requirements regarding performance, energy consumption, cost, maintenance effort, etc. Using the developed model, the designer can consider (and compare) different design styles and do hardware/software system partitioning. Afterwards, a synthesizable hardware model (i.e. HDL top module) has to be prepared. The system model can be of a mixed HDL style (VHDL and Verilog languages, as well as behavioral and gate-level descriptions) [12, 13]. Functional verification relies on pre-defined and custom HDL testbenches written to simulate the system behavior. It consumes much time but allows tracing of all the internal signals and efficient debugging.

Simulation

A testbench can prove correctness of the RTL model against the specification. The aim of a good testbench (or testbenches) is to:

- Prove that the design functionality is as described in the specification reporting any discrepancies that must be fixed;
- Fully exercise every line of RTL code;
- Allow re-simulation of the post-synthesis and post-layout net-lists.

Figure 8.3 shows the structure of a VHDL simulation testbench prepared to verify a WLAN digital baseband. It includes the library, entity, architecture, component, configuration, constant, signal, port, and process definitions. The aim is to verify the design functionality and timing prescribed in the specification, as well as to prove the identical functional behavior of the pre-synthesis, post-synthesis, and post-layout implementations. The skill and experience of the verification engineer is the most important factor in this process.

```
-- TESTBENCH FOR DIGITAL BASEBAND

LIBRARY ieee;
USE    ieee.std_logic_1164.ALL;
USE   ieee.std_logic_unsigned.ALL;
USE   ieee.std_logic_arith.ALL;
USE    ieee.math_real.ALL;

ENTITY BB_TOP_TB IS
END BB_TOP_TB ;

ARCHITECTURE testbench OF BB_TOP_TB IS

 CONSTANT period : TIME := 12.5 ns;          -- clock period
 CONSTANT max_size : NATURAL := 1200;

 COMPONENT BB_TOP
  PORT (clk       : IN  STD_LOGIC; -- 20 MHz clock
      mipp_din   : IN  STD_LOGIC_VECTOR(15 DOWNTO 0);-- 16 bit data input (from bidir port)
      mipp_dout  : OUT STD_LOGIC_VECTOR(15 DOWNTO 0);-- 16 bit data output (to bidir port)
      mipp_reset : IN  STD_LOGIC;             -- reset from host, low active
      mipp_dstb  : IN  STD_LOGIC;             -- data strobe from host, low active
      mipp_cstb  : IN  STD_LOGIC;             -- control strobe from host, low active
      mipp_write : IN  STD_LOGIC;             -- write signal from host, low active
      mipp_wait  : OUT STD_LOGIC;             -- not ready signal to host, high active
      mipp_cca   : OUT STD_LOGIC;             -- MIPP CCA indicator
      mipp_rxd   : OUT STD_LOGIC);            -- MIPP Rx data available
 END COMPONENT;

 COMPONENT simprt
 GENERIC (design : STRING;                    -- design name = file name of data file
          file_ext  : STRING;                 -- extension for protocol data file
          comment   : STRING;                 -- arbitrary comment line
          sig_names : STRING;                 -- signal names (separated by spaces)
          sig_types : STRING;                 -- IN / OUT / BIDIR for every signal
          sig_bits  : STRING);                -- bus width of every signal

 PORT (errors    : OUT INTEGER;  -- number of compare errors
          write_prt : IN  BOOLEAN;            -- write to protocol file ?
          comp_prt  : IN  BOOLEAN;            -- compare with protocol file ?
          relaxed_X : IN  BOOLEAN;            -- allow any value for 'X' in compare ?
          clock     : IN  STD_LOGIC;          -- clock for sampling data
          vector    : IN  STD_LOGIC_VECTOR;-- signals to protocol / compare
          out_enb   : IN  BIT_VECTOR);        -- 1: bit is output / 0: bit is input
 END COMPONENT;

 SIGNAL write_prt : BOOLEAN := FALSE;         -- generate protocol file ?
 SIGNAL comp_prt  : BOOLEAN := TRUE;          -- compare with protocol file ?
 SIGNAL relaxed_X : BOOLEAN := TRUE;          -- allow any value for 'X' in compare ?

 SIGNAL errors  : INTEGER;                    -- signals for/from protocol utility
 SIGNAL vector  : STD_LOGIC_VECTOR (22 DOWNTO 0);
 SIGNAL out_enb : BIT_VECTOR (vector'RANGE);

 SIGNAL clk, mipp_oe : STD_LOGIC;  -- mipp interface signals
 SIGNAL mipp_data, mipp_din, mipp_dout : STD_LOGIC_VECTOR(15 DOWNTO 0);
```

```
 SIGNAL mipp_reset, mipp_dstb, mipp_cstb, mipp_write : STD_LOGIC;
 SIGNAL mipp_wait, mipp_cca, mipp_rxd  : STD_LOGIC;
 SIGNAL data_error : STD_LOGIC := '0';

BEGIN

 PROCESS                 -- clock generator
 BEGIN
  clk <= '1';
  WAIT FOR period/2;
  clk <= '0';
  WAIT FOR period/2;
 END PROCESS;

 PROCESS                              -- test pattern mipp interface
   PROCEDURE MIPP_ctrl (CONSTANT wait_time : TIME;
            CONSTANT data : STD_LOGIC_VECTOR(15 DOWNTO 0)) IS
             BEGIN
               IF (mipp_wait/='0') THEN
                 WAIT UNTIL mipp_wait = '0';
                 WAIT FOR 0.2*period;
               END IF;
               IF (data = "ZZZZZZZZZZZZZZZZ") THEN
                 mipp_write <= '1';
               ELSE
                 mipp_write <= '0';
               END IF;
               mipp_data <= data;
               WAIT FOR wait_time;
               mipp_cstb <= '0';
               WAIT UNTIL mipp_wait = '1';
               WAIT FOR 0.2*period;
               WAIT FOR wait_time;
               mipp_cstb <= '1';
               mipp_data <= "XXXXXXXXXXXXXXXX";
               WAIT UNTIL clk'EVENT AND clk='1';
               WAIT FOR 0.2*period;
            END MIPP_ctrl;
 ...

   -- mipp testbench
   VARIABLE random : STD_LOGIC_VECTOR(15 DOWNTO 0);
   VARIABLE frame_length  : INTEGER;
   VARIABLE set_weigths   : BOOLEAN;

   BEGIN
             mipp_cstb  <= '1';
             mipp_dstb  <= '1';
             mipp_write <= '0';
             mipp_reset <= '1';
             WAIT FOR 1.2 * period;
             mipp_reset <= '0';
             WAIT FOR 2 * period;
             mipp_reset <= '1';
             WAIT FOR 2 * period;
             random := X"1BF5";
```

```
            frame_length := 3;
            set_weigths := false;
            MIPP_ctrl(2*period, "0000000000000001");  -- get chip ID
            WAIT FOR 2 * period;
            MIPP_ctrl(2*period, "ZZZZZZZZZZZZZZZZ");  -- get chip ID
            WAIT FOR 2 * period;
            ...
            ASSERT NOT (data_error = '0')
                    REPORT "Test completed (no errors during data transfers)"  SEVERITY FAILURE;
            ASSERT NOT (data_error = '1')
                    REPORT "!! Test completed with error in data transfers!!"  SEVERITY FAILURE;

  END PROCESS;

          mipp_oe   <= mipp_write AND NOT (mipp_dstb AND mipp_cstb);
          mipp_data <= mipp_dout WHEN mipp_oe='1' ELSE (OTHERS => 'Z');
          mipp_din  <= mipp_data;                      -- imitate bidirectional bus
          vector <= mipp_data & mipp_reset & mipp_write & mipp_dstb &
                mipp_cstb & mipp_wait & mipp_cca & mipp_rxd;
          out_enb(22 DOWNTO 7) <= (OTHERS => '1') WHEN mipp_oe='1' ELSE
                              (OTHERS => '0'); -- mipp_data, bidir
          out_enb(6 DOWNTO  0) <= "1111000"; -- '1'= OUTPUT

dut:  BB_TOP          PORT MAP (clk => clk, mipp_reset => mipp_reset,
                        mipp_din => mipp_din, mipp_dout => mipp_dout,
                        mipp_dstb => mipp_dstb, mipp_cstb => mipp_cstb,
                        mipp_write => mipp_write, mipp_wait => mipp_wait,
                        mipp_cca => mipp_cca, mipp_rxd => mipp_rxd);

pt:   simprt          GENERIC MAP (
                        design => "BB_TOP", file_ext => "spr", comment => "",
                        sig_names => "mipp_data mipp_reset mipp_write mipp_dstb " &
                                "mipp_cstb mipp_wait mipp_cca mipp_rxd",
                        sig_types => "BIDIR OUT OUT OUT OUT IN IN IN",
                        sig_bits  => "16 1 1 1 1 1 1 1")
                      PORT MAP (
                        errors => errors, write_prt => write_prt,
                        comp_prt => comp_prt, relaxed_X => relaxed_X,
                        clock => clk, vector => vector, out_enb => out_enb);

END ARCHITECTURE testbench;

CONFIGURATION pre_syn OF BB_TOP_tb IS
 FOR testbench
          FOR dut : BB_TOP USE ENTITY work.BB_TOP (BB_TOP);
           FOR BB_TOP
            FOR u0       : mipp_slave USE ENTITY work.mipp_slave (behav);  END FOR;
            FOR tx_fifo  : fifo2 USE ENTITY work.fifo2 (behav);            END FOR;
            FOR tx_ram   : syncRAM2 USE ENTITY work.syncRAM2 (behav);      END FOR;
          ...
            FOR REG8_2   : reg8x8 USE ENTITY work.reg8x8 (behav);          END FOR;
           END FOR; END FOR;
          FOR pt  : simprt   USE ENTITY work.simprt   (program); END FOR;
 END FOR;
END pre_syn;
```

```
--CONFIGURATION post_syn OF BB_TOP_tb IS
-- FOR testbench
--          FOR dut : BB_TOP USE ENTITY work.BB_TOP (BB_TOP);
--           FOR BB_TOP
--           FOR u0      : mipp_slave USE ENTITY work.mipp_slave (verilog);   END FOR;
--           FOR u1 : fifo2 USE ENTITY work.fifo2 (verilog);   END FOR;
--           FOR u2 : syncRAM2 USE ENTITY work.syncRAM2 (behav);   END FOR;
...
--            FOR REG8_2 : reg8x8 USE ENTITY work.reg8x8 (verilog); END FOR;
--           END FOR; END FOR;
--          FOR pt  : simprt USE ENTITY work.simprt (program); END FOR;
-- END FOR;
--END post_syn;
```

Fig. 8.3. An example of the VHDL simulation testbench

Testbenches sometimes provide SRAM and PROM behavioral models, as well as the emulation of system peripherals (UART, I^2C, GPIO, SPI, EPP, etc.). To test the system based on a general-purpose processor, one can write a test program in the assembler and compile it into a bit stream ready for loading into a boot PROM. The program performs the processor initialization and tests the basic data transactions on the system ports. Testbenches can also generate cycle based test vectors for the final electrical testing of chips at the end of fabrication process.

Logic synthesis

Synthesis tools take the design to generate a gate-level equivalent implementation with specified timing constraints, i.e. to synthesize it into a target standard cell library. It is a collection of microelectronic devices implementing basic logic functions in a given technology. Characteristics of the library and its cells play a fundamental role in the process of automatic synthesis of digital circuits [14-17]. Building a standard cell library requires development of the physical layout for all the leaf cells ranging from simple gates to latches and flip-flops. Synthesis tools can then build arithmetic blocks like adders and multipliers. Each standard cell contains necessary port and connectivity information required for simulation and synthesis. Additionally, data on typical rise and fall delays, load characteristics, maximum clock frequency and other specifications are available in the form of simulation and synthesis library files, which can be used by standard simulation and synthesis tools. The last synthesizing step is generation of Verilog net-list and standard delay format (SDF) files. A net-list describes system components in terms of instance names associated with names of the standard cells. Net names are associated to groups of standard cell pin names. To shorten the synthesis time, it is necessary to prepare the synthesis scripts with design constraints (Figure 8.4). Finally, the post-synthesis system functionality has to be verified by simulation.

```
# Initialization file for Synopsys Design Compiler #
# Set cache directory
set cache_read  /tmp
set cache_write /tmp
# Define design library
sh rm -rf WORK
sh mkdir WORK
define_design_lib WORK -path WORK
# Set search paths and libraries (do not forget memories)
set search_path  "/designkits/sgb25tm1tm2/synopsys  /designkits/sgb25tm1tm2/encounter/LEF  \
                       /synopsys/sy200712sp4/libraries/syn ."
set link_library    "SESAME-LP2_SS_2.25_125.db sgb25v_io.wc.db dw_foundation.sldb \
                         GSRAMV_64_256_16mbf.typ.db GSRAMV_64_512_16bf.typ.db \
                         GSRAMV_64_512_32bf.typ.db GSRAMV_64_96_24bf.typ.db \
                         Flash_2t_32_cor4_m.typ.db ACF_RAM.db"
set target_library   "SESAME-LP2_SS_2.25_125.db sgb25v_io.wc.db"
set synthetic_library "dw_foundation.sldb"
set physical_library  "abstract1m4m5thin.lef SESAME_IOm5thin.lef"
set default_name_rules verilog
set verilogout_no_tri true
# Script file to compile with Synopsys Design Compiler #
cd ./../rtl/gtech
read -f verilog DW_ahb.v
read -f verilog DW_apb.v
read -f verilog DW_apb_gpio.v
read -f verilog DW_apb_uart.v
read -f verilog DW_memctl.v
cd ./../rtl
analyze -format verilog -lib WORK m4k.v
analyze -format verilog -lib WORK mbb.v
...
analyze -format verilog -lib WORK amba_glue.v
analyze -format verilog -lib WORK amba_sys.v
elaborate amba_sys -arch "verilog" -lib WORK -update
uniquify
create_clock -name "Clk" -period 10 -waveform {"0" "5"} {"Clk"}
set_clock_skew -ideal -uncertainty 0.5 Clk
set_dont_touch_network Clk
compile -map_effort medium
compile -map_effort high -incremental_mapping
cd ./../db
write -hierarchy -output "amba_sys.db"
cd ./../netlist
write -f verilog -hierarchy -output "amba_sys.v"
cd ./../timing
write_sdf -version 1.0 "amba_sys.sdf"
cd ./../report
report_timing > amba_sys.timing
report_area > amba_sys.area
report_power > amba_sys.power
check_design > amba_sys.check
report_timing
```

Fig. 8.4. An example of user's initialization and synthesis scripts

Physical layout

An existing standard cell library can be modified according to the requirements of a new silicon technology changing physical layout (Figure 8.5) of all the cells [18, 19]. Automation helps this process, but eventually logic designers and layout engineers create process-optimized layout that takes advantage of special layers, local interconnect, and many other features unique to a given fabrication process.

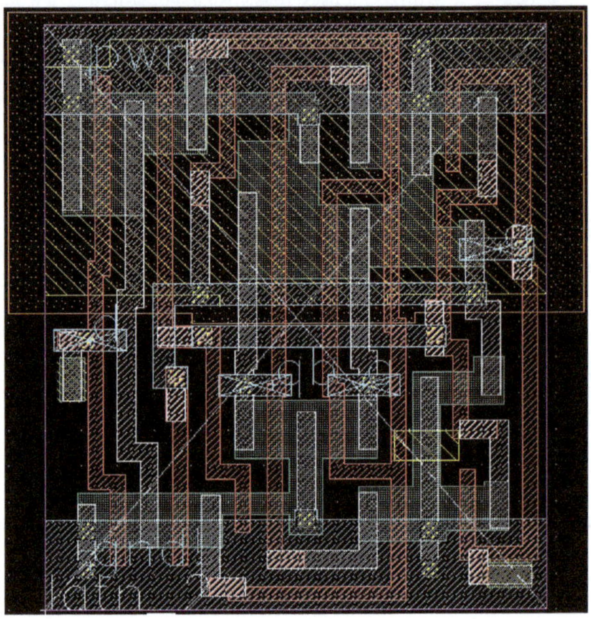

Fig. 8.5. Physical layout of a latch

Layout tools quickly produce a silicon virtual prototype of the physical design, which provides both rapid feedback on chip performance and a fully functional, physically feasible layout. With this physical prototype, the front-end designers quickly explore the impact of their implementation choices on chip performance and physical feasibility. The back-end designers can generate a floor plan and placement optimized for rapid, reliable design closure. A layout generation flow includes the floor planning with power planning, placement, clock tree generation, routing, and verification.

Floor planner uses a hierarchical and interactive approach, combining automatic functions and interactive editing to plan locations for blocks and core rows. It can predict and assess the effects of physical layout before you place and route the design. This speeds up the design process by minimizing costly iterations. The primary goal of floor planning is to create rows where the placer can place cells. Each row is assigned to a site type, limiting the number of cells that can be placed in a row. Initialization sets up the physical design based on the data you read in. It creates a core area with core rows, and sets up cell and routing grid tracks. If the standard cell model is tall and narrow, horizontal rows are created. If the standard cell model is short and wide, vertical rows are created. Initialization does not place any cells. The floor planner estimates the required layout area (die size), defines the layout domain (shape and aspect ratio), and plans and modifies the row configuration. The row configuration implies the position, spacing, orientation, and row type. One can also align rows, adjust the channel space, and modify the cell distribution within rows. Finally, it is possible to edit groups and group constraints, resize the layout area, and modify the global routing grid.

After the floor plan is ready and read in, the next step is to create power paths in the design. Then power rings and power stripes can be added. Using the plan power and connect ring commands, one can plan and modify power paths before routing them. The artwork window in Figure 8.6 shows a floor plan including I/O and power pads, standard cell rows, memory blocks, power rings, and power stripes.

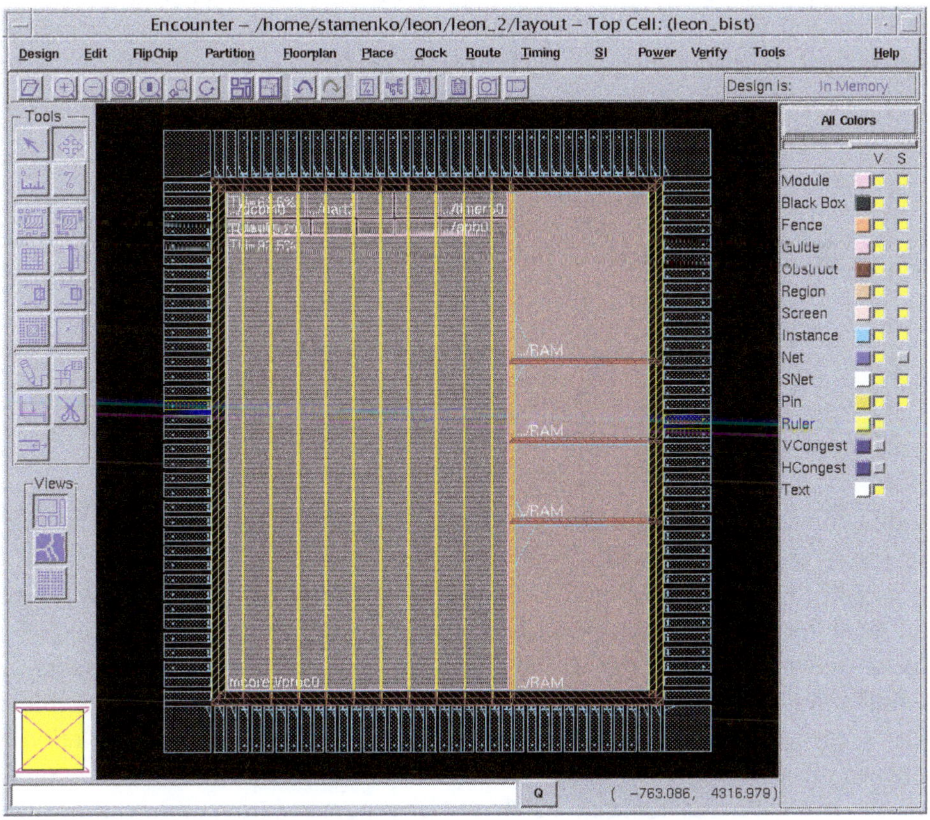

Fig. 8.6. An SOC floor plan before placement

Generally, I/O pins are placed before blocks. If there are some predetermined I/O pin placement requirements, a constraint file should be used to place I/O pins. Otherwise, I/O pins and cells can be placed at the same time. After this, we can analyze the timing at different points in the design with ideal clock, as there are no clock trees yet. Layout tools are capable to perform in-place optimization resizing gates (including flip-flops), and inserting buffers and inverters to correct timing and electrical violations. The primary goal of placement is to position components on the floorplan so that the number and length of interconnections is minimal. The placer determines legal placement locations by considering the floorplan and the site requirements of a macro cell instance. Automatic grouping and global placement is available. The placer anticipates routing (group-related and timing-related) requirements by considering the resulting positions of sets of pins that represent nets. It supports a net-list-driven placement.

Clock delay varies with position on the integrated circuit chip. It is necessary to deliver the clock to memory elements with acceptable skew and sharpness of clock edges. Clock distribution (skew and edges) is controlled using an automatic clock tree generator. The generator constructs an optimized clock tree, minimizes the skew and insertion delay, and selects the optimal size (driving strength) of buffers and inverters (Figure 8.7). It can perform a hierarchical clock tree synthesis. The clock source pin and buffer/inverter names must be defined in a constraint file.

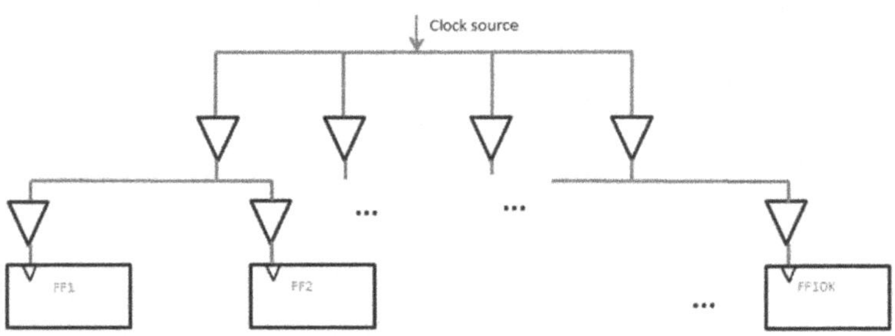

Fig. 8.7. An example of the clock tree structure

At this point, a fully placed design is ready for routing. In general, the routing is done in three stages: the routing of power nets, the routing of clock trees, and the routing of the remaining nets. The router incorporates the power routing, global and final routing, and extraction and physical verification. Capabilities of the power router allow creating a power grid, routing core-area power buses, and adding filler cells. The connect ring command connects existing rings, blocks, stripes, rows, and pads with I/O rings. There is a critical synergistic relationship between placement and routing for all aspects of design closure, including timing, power, cross talk, and congestion. That is why a post-routing optimization step is necessary.

The primary objective is to fully complete the required interconnect. The routed design is not always free of connectivity (unconnected nets) and geometry (shorts and design rule) violations. The final router searches and repairs these violations automatically. At this phase, parasitic capacitances and resistances from the layout are extracted and a Verilog net-list, an SDF file and a layout file of the design are generated. It ends with the simulation (including the generated SDF file) or timing analysis to verify the design performance. This operation annotates delays back to layout net-list instances and interconnections. All the RTL behavioral description, logically synthesized net-list, and physical layout net-list have to be functionally identical.

Embedded memory

A memory generator [20, 21] is a tool that can create memories (RAM or ROM blocks) in a range of sizes as needed. The customer usually wants a particular number of words (depth) and bits (width) for each memory ordered. Each of the final building blocks (physical layout) is a stand-alone, densely packed, pitch-matched array. Using complex layout generators and adopting state-of-the-art logic and circuit design technique, today's embedded memories can reach extreme density and performance. Designers can choose the memory aspect ratio according to the requirements of SOC level layout. Each memory generator is a set of various, parameterized generators that generate an array of custom pitch-matched leaf cells (Figure 8.8) and its schematic, extract a net-list used for both layout vs. schematic and functional verification, and create models for gate-level simulation and static timing analysis.

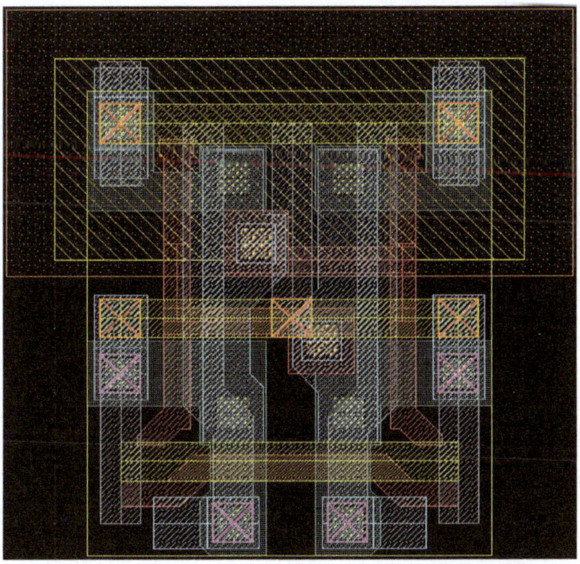

Fig. 8.8. Physical layout of a 6-transistor SRAM cell

8.2.2 *FPGA implementation flow*

Field Programmable Gate Arrays (FPGAs) are integrated circuits that are based on configurable logic blocks (CLBs) connected via programmable interconnects (Figure 8.9). I/O blocks (grouped in banks to support different I/O standards) provide communication with the external world. Most FPGAs are reconfigurable, allowing a relatively quick refinement and optimization of hardware with no additional costs. This feature distinguishes FPGAs from Application Specific Integrated Circuits (ASICs), which are custom manufactured. Although one-time programmable (OTP) FPGAs are available, the dominant types are SRAM based, which are reprogrammable. The designer can modify the HDL description of the components [22] and then use an FPGA vendor toolchain for the synthesis of the bit-stream to configure the FPGA [23].

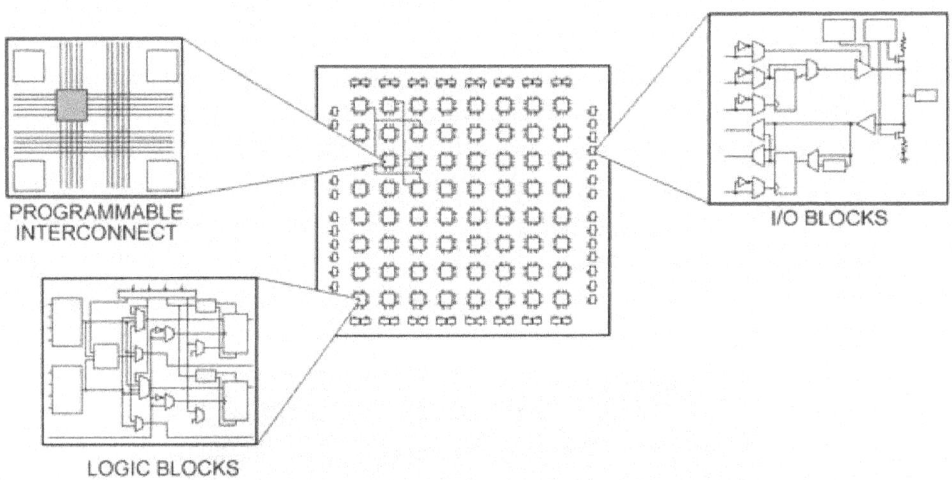

Fig. 8.9. Structure and building blocks of an FPGA

Every CLB consists of a configurable switch box with 4 or 6 inputs and logical cells (or slices). The cells contain some selection circuitry (typically, multiplexer), flip-flops, and combinational elements, depending on the technology. The switch box is highly flexible and can handle combinational logic, luck-up tables, shift registers or RAM (which is included in the CLB to implement on-chip memory). Programmable interconnect points (switch boxes outside the CLB) enable different routing combinations within a single interconnect crossing. FPGA routing is unsegmented: each wiring segment spans only one logic block before it terminates in a switch box. The routing is done through programmable interconnect points connecting with longer channels and switch boxes in the CLB.

8.3 Full-Custom Methodology

As will be discussed in Chapter 10, the performance of the RF section is very sensitive to parasitics of the layout, package, and printed circuit board (PCB). The designer must thus consider these parasitics as early as possible in the design cycle. Though many advances have been made in the field of analog and RF design automation [24, 25] including layout-aware optimization, most solutions are based on some form of design space exploration and optimization algorithms like genetic algorithms. These black-box approaches call for numerous simulation cycles, which unfortunately is too computationally expensive for high-complexity designs. Such automated tools may be used for simpler building blocks, as well as Intellectual Property (IP) blocks, but a hierarchical full-custom approach is usually preferred. Indeed, designers' expert knowledge of the sensitive nodes and good layout practices usually provide the fastest results. Such an expert knowledge is extremely difficult to encode into an algorithm.

Full-custom methodology is mainly used for the analog front-end. Figure 8.10 shows the diagram of a traditional top-down full-custom design flow. The system-level (i.e. the WLAN transceiver) is optimized to select an adequate architecture and get block-level performance targets. The optimization metrics are the system level specifications, that is to say: the performance specifications of the RF transceiver. Then, the different blocks (mixers, ADC, DAC, LNA, PA, filters, etc.) are designed at transistors level to meet these targets. The transistor-level simulations are carried out using the preferred EDA tool and the design KIT provided by the chosen foundry. Once transistor-level versions of the building blocks become available, the task of top-level assembly can begin, where system-level simulations are carried out to verify that the non-idealities introduced by the building blocks do not alter performance. This is a fundamental step since the high-level models of the blocks used to select and optimize the system architecture are usually not accurate enough to encode all the subtleties of transistor-level behavior. The compatibility of high-level ideal models with transistor-level simulation is put to good use here to speed-up partial simulations.

The next step involves the layout of the building blocks and the extraction of parasitics. Another verification step is carried out using transistor-level simulations to check that the post-layout blocks still meet their performance targets. System layout is then carried out and global routing parasitics are extracted. As well, package and PCB parasitics would also be considered here. Interactions are modeled and taken into account for the final top-level simulation.

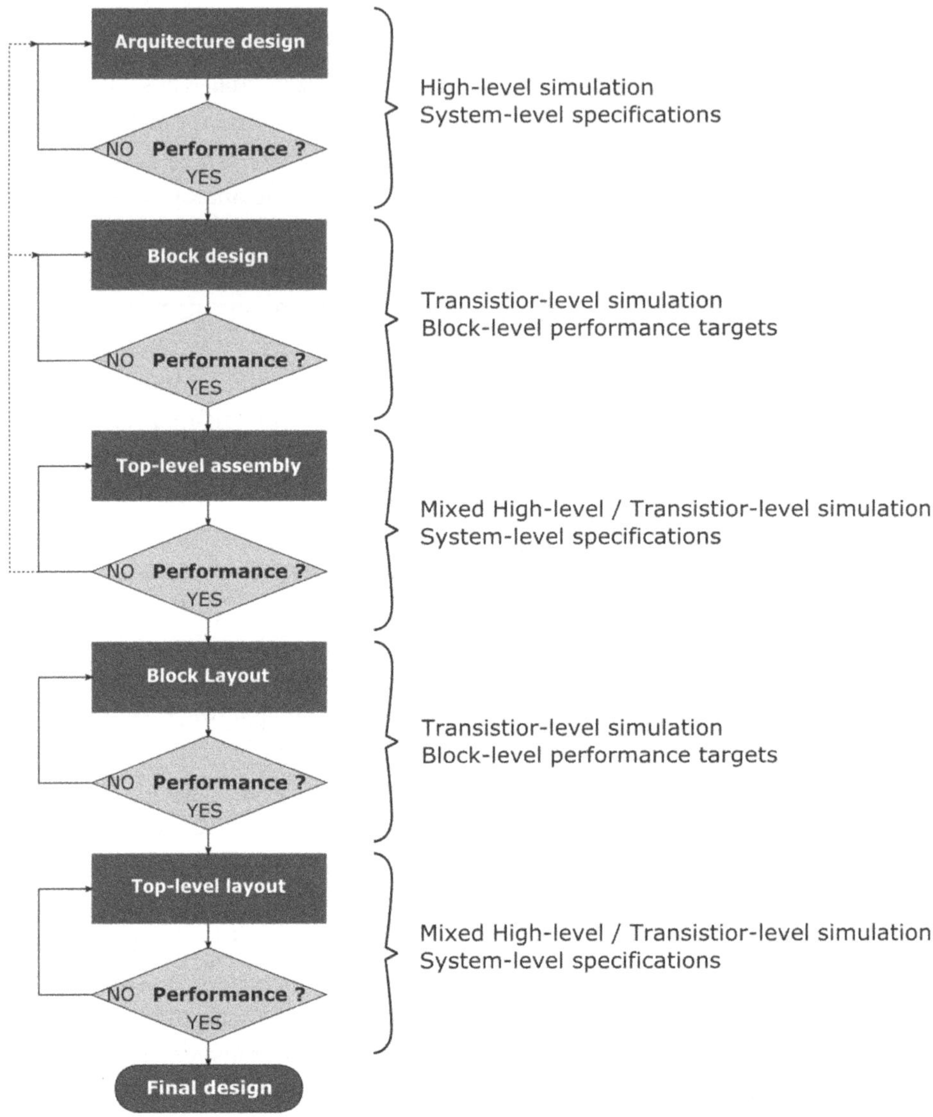

Fig. 8.10. Conventional full-custom flow diagram

The different simulation levels do not necessarily make use of the same simulation tools. System-level simulation can resort to specific high-level simulation languages like Matlab Simulink, or languages embedded in the EDA tool suite like Verilog-A or System-Verilog. Transistor-level simulations resort to SPICE variants and differential equation solvers. Parasitic extraction is performed by specific tools, but the results are mapped directly onto a netlist for transistor-level simulation. Finally, package and PCB modeling usually require some form of electromagnetic simulation, even if some compact model is generated to enable its use at transistor level.

At each step, a feedback loop maps the newly introduced non-idealities back onto the higher level to verify that performance criteria are met. However, if too many iterations are necessary to meet the targets, it may be necessary to step back to a higher level. For instance, after several intents, the block designers may realize that the performance target of their blocks are unfeasible or leave too little margin for yield considerations. In such a case, it would be necessary to alter the top-level architecture and the transceiver error budgets. This sort of feedback is represented by the dotted lines, and usually implies a strong impact on the project schedule.

Unfortunately, modern transceivers are too complex to achieve first-time right silicon with such a linear design flow. Many iterations over the different feedback loops may be necessary to meet the specifications. Hence, the top-down flow is usually mixed with a bottom-up process that feeds the models with early estimates of the parasitics. This allows for a clever selection of the architecture including correction mechanisms and fine-tuning capabilities.

References

1. Stamenkovic, Z., Tittelbach-Helmrich, K., Krstic, M., Perez, J., Via, J. and Ibanez, J. (2009). Architecture of an analog combining MIMO system compliant to IEEE802.11a, *Proc. ICT-MobileSummit*, pp. 1–8.
2. Eickhoff, R., Kraemer, R., Santamaria, I. and Gonzalez, L. (2009). Developing energy-efficient MIMO radios, *IEEE Vehicular Technology Magazine*, 4, pp. 34–41.
3. Via, J., Santamaria, I., Elvira, V. and Eickhoff, R. (2010). A general criterion for analog Tx-Rx beamforming under OFDM transmissions, *IEEE Trans. on Signal Processing*, 58, pp. 2155–2167.
4. Stamenkovic, Z., Tittelbach-Helmrich, K., Krstic, M., Ibanez, J., Elvira, V. and Santamaria, I. (2011). MAC and baseband processors for RF-MIMO WLAN, *EURASIP Journal on Wireless Communications and Networking*, 1/207, pp. 1–13.
5. Wolf, W. (2002) *Modern VLSI Design: System-On-Chip Design*, (Prentice Hall, New Jersey, USA).
6. Henkel, J. (2003). Closing the SoC design gap, *Computer*, 36, pp. 119–121.
7. Rowen, C. (2004) *Engineering the Complex SoC: Fast, Flexible Design with Configurable Processors*, (Prentice Hall, New Jersey, USA).
8. https://www.synopsys.com/implementation-and-signoff/rtl-synthesis-test.html
9. https://www.mentor.com/products/fv/modelsim/
10. https://www.cadence.com/en_US/home/tools/system-design-and-verification.html

11. https://www.cadence.com/en_US/home/tools/digital-design-and-signoff/soc-implementation-and-floorplanning.html
12. Ashenden, P. J. (2008) *The Designer's Guide to VHDL*, (Morgan Kaufmann, San Francisco, USA).
13. Ashenden, P. J. (2008) *Digital Design: An Embedded Systems Approach Using Verilog*, (Morgan Kaufmann, San Francisco, USA).
14. Hashimoto, M., Fujimori, K. and Onodera, H. (2003). Standard cell libraries with various driving strength cells for 0.13, 0.18 and 0.35 μm technologies, *Proc. 8th Asia and South Pacific Design Automation Conference*, pp. 589–590.
15. Aitken, R. and Becker, S. (2004). Cell library techniques using advanced transistor structures, *Proc. International Conference on Integrated Circuit Design and Technology*, pp. 199–204.
16. Chandrasekar, S. Varshney, G. and Visvanathan, V. (2005). A comprehensive methodology for noise characterization of ASIC cell libraries, *Proc. 6th International Symposium on Quality of Electronic Design*, pp. 530–535.
17. Naga, L. M. and Mullangi, P. (2018). Design and development of an ASIC standard cell library using 90 nm technology node, *Proc. International Conference on Computer Communication and Informatics*, pp. 1–6.
18. https://www.arm.com/products/silicon-ip-physical/logic-ip
19. http://www.dolphin-ic.com/standard-cell.html
20. https://www.synopsys.com/dw/ipdir.php?ds=dwc_sram_memory_compiler
21. https://www.dolphin-technology.com/memory-compilers
22. Wilson, P. (2016) *Design Recipes for FPGAs (Using Verilog and VHDL)*, (Elsevier, Amsterdam, Netherlands).
23. https://www.xilinx.com/support/documentation/sw_manuals/xilinx10/books/docs/qst/qst.pdf
24. Rutenbar, R. A., Gielen, G. G. E. and Roychowdhury, J. (2007). Hierarchical modeling, optimization and synthesis for system-level analog and RF designs, *Proceedings of the IEEE*, 95, pp. 640–669.
25. Fakhfakh, M., Tlelo-Cuautle, E. and Castro-Lopez, R. (2013) *Analog/RF and Mixed-Signal Circuit Systematic Design*, (Springer-Verlag, Berlin, Germany)

https://doi.org/10.1142/9789811210723_0009

Chapter 9

Semi-Custom Implementation of Digital Circuits

Zoran Stamenković[1] and Miloš Krstić[1, 2]
[1]IHP - Leibniz-Institut für innovative Mikroelektronik
Frankfurt (Oder), Germany
[2]University of Potsdam
Potsdam, Germany

9.1 Introduction

RF-MIMO operates compatibly to the WLAN 802.11a and amends some parts of this standard such as the physical and data link layer [1-4]. Figure 9.1 shows the investigated and developed communication layers. The RF-MIMO physical layer is based on an extended analog front-end, which weights multiple transmit/receive signals by means of analog circuits. It is controlled by an IEEE802.11a digital baseband, which is extended for RF-MIMO specific operation. Likewise, the MAC ensures data and control flow between the new baseband and logical link control (LLC). Finally, the compatibility to IEEE 802 is achieved by a compliant LLC.

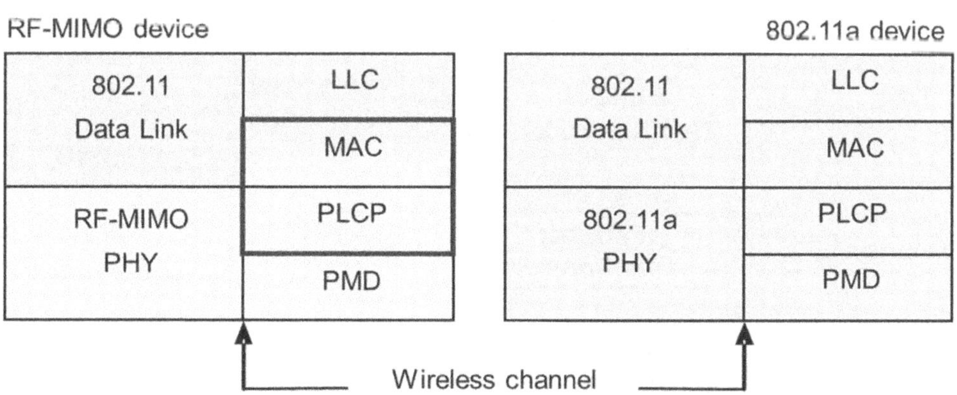

Fig. 9.1. Modified communication layers in an RF-MIMO WLAN transceiver

Adjustments of the layers are performed in such a way that a RF-MIMO device can operate with other RF-MIMO and IEEE 802.11a devices simultaneously without affecting neither the performance enhancements of the first nor the standard performance of the second. Modifications in the physical layer convergence protocol (PLCP) and MAC of the RF-MIMO device have to be provided for ensuring compatibility to the IEEE 802.11a standard. The physical medium dependent (PMD) sublayer is not affected by amendments because each receive and transmit path is designed for an IEEE 802.11a link budget.

Next two sections describe the implementation of digital parts of the proposed RF-MIMO transceiver (MAC and digital baseband processors).

9.2 MAC Processor

The MAC hardware (Figure 9.2) is based on a MIPS32 4KE processor core that performs most of the MAC functionality. The core communicates with the rest of the system through the X-bus. Two targets are connected to the X-bus: General Purpose Input/Output (GPIO) and Peripheral Bus Controller (PBC). The MIPP port implemented via GPIO provides an interface to the physical layer [5, 6]. It is a combination of a 16-bit parallel bidirectional data bus and some control and handshake signals.

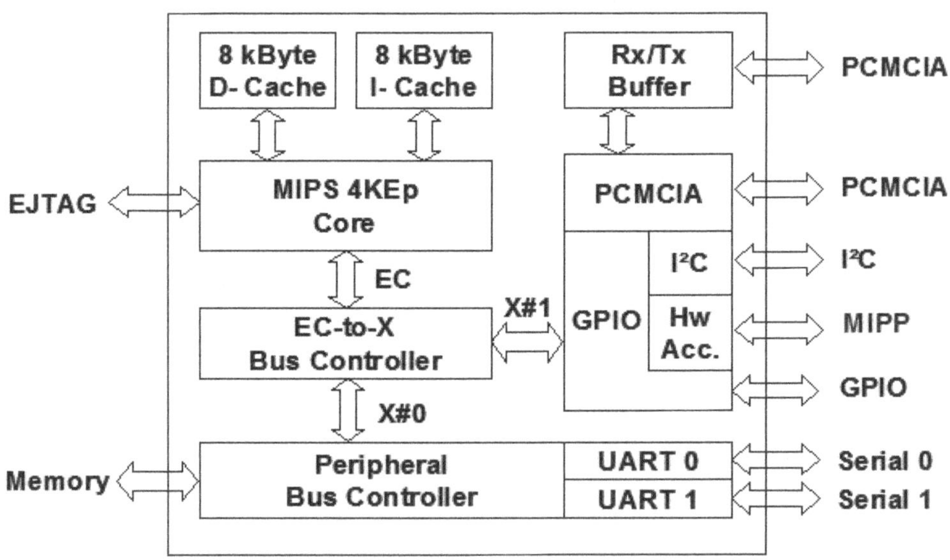

Fig. 9.2. MAC hardware components

Most of the timing-critical MAC functionality is implemented into a hardware accelerator. This unit acts as a data pipe between the processor and the IEEE

802.11 physical layer connected via the MIPP port. An interface to other external components has been provided through the I^2C bus and two serial ports. The I^2C bus controller is implemented as a part of the GPIO. Personal Computer Memory Card International Association (PCMCIA) interface is connected to the processor via GPIO too. The serial ports are controlled by two UARTs connected via PBC. The PBC also provides an interface to the external SRAM, flash and other memory-like peripherals. The compiled MAC protocol software (processor specific machine code requiring 300 kB of memory) is stored in the flash memory.

The implemented SOC includes a MIPS32 4KE core configuration with instruction and data caches. We have also provided basic EJTAG options including the Test Access Port (TAP) to enable programmability of the external flash memory containing the MAC executable code. This makes maintaining of the MAC very flexible allowing the user to change the MAC protocol simply by the software.

As mentioned before, the purpose of the hardware accelerator is to reduce the burden of the protocol processor by implementing time-critical functions of the IEEE 802.11a MAC layer [5, 6]. The functionality of the hardware accelerator can be divided into four major tasks, namely frame transmission, frame reception, channel-state monitoring, and timer functions. Figure 9.3 shows the general block diagram of the hardware accelerator structure.

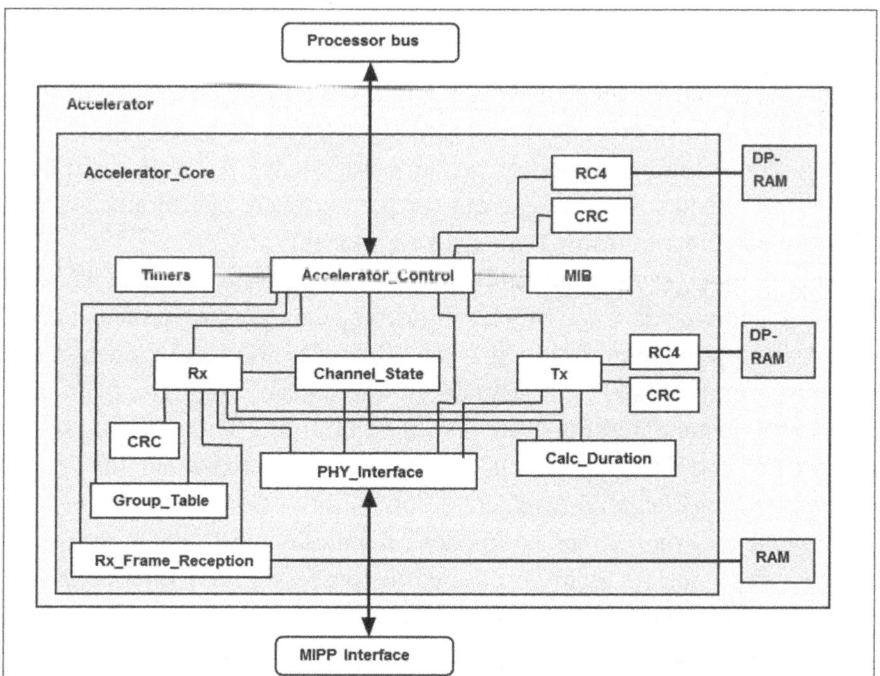

Fig. 9.3. Structure of a MAC hardware accelerator

A standard memory bus connects the hardware accelerator to the processor bus. Additionally, one interrupt line is provided. On the other side, the MIPP interface connects the accelerator with the baseband processor. The requests from the processor bus are interpreted by the *Accelerator_Control* component. It is responsible for applying correct data to the data bus in the case of read access from the processor, controlling the interrupt handling, decrypting frames, and for interpreting the requests from the processor.

Frame transmission is handled by the *Tx* component. It generates the required control signals for the *PHY_Interface* to indicate the start of a new frame, and processes the frame data received from the processor. The *PHY_Interface* component implements the MIPP protocol for communication with the baseband processor. Furthermore, it forwards control signals, like the state of the wireless medium and the presence of received data, to other components.

Channel_State component keeps track of the state of the wireless medium and implements the 802.11 logical carrier-sense mechanism. Based on this state information, it controls the back-off procedure that is performed before a new frame can be sent.

Frame reception is controlled by the *Rx* component. Frame data is fetched from the *PHY_Interface* when there is an indication of available data. The data is analyzed to determine the frame type and to decide if the station is an intended receiver of that frame. Additionally, the frame data is passed to the CRC algorithm and stored in a buffer to be retrieved later by the processor. After error-free reception, an acknowledgement frame might have to be sent back to the frame source, and a notification to the processor is generated.

Timers control the operation of certain components. All station attributes and timing characteristics that are relevant for the hardware accelerator are stored in the *MIB* (management information base) component.

A couple of algorithms are implemented as separate components: the *CRC* component implements the CRC-32 algorithm, the *RC4* component implements the RC4 encryption algorithm, and the *Calc_Duration* component calculates duration of a frame based on its length and the transmission rate. *Calc_Duration* component is instantiated only once, so that an arbiter controls the access to it, whereas *CRC* and *RC4* algorithms are separately implemented for *Tx*, *Rx*, and *Accelerator_Control* components.

Finally, the *Group_Table* component manages a list of up to 8 multicast addresses. The table is configured by the processor and queried by the *Rx* component to find out if an address identifies a multicast group the station belongs to. The *Rx_Frame_Reception* component controls a buffer for one received frame. It receives frame data from the *Rx* component and delivers this data to the processor on later request. The frame is stored in RAM, so that two addresses – the write pointer and the read pointer – must be updated.

As a MIMO extension, the transmitter contains a table of antenna weight coefficients for distinct connections. It transfers the respective weight coefficient to the PHY layer before sending a frame. When a frame exchange sequence is finished, it sets some configurable default weight coefficients which should be good enough to receive a short RTS frame from any station. From the source address contained in the RTS frame, the optimal weight coefficients for that connection can be deduced and set in the PHY layer before receiving the (possibly long) frame itself.

The receiver comprises a CRC checker, a frame address filter, and the generation of acknowledgements and CTS frames. The control component, as a broker between MAC and PHY, sets and reads the PHY parameters, controls the timers for handshake of the MIPP interface, and stores the received data from PHY after any set/write command from MAC.

The arbiter controls the MIPP handshake and the access to bi-directional data bus. A special priority mechanism has been developed to prevent undesired delays in the data flow and raise the data reliability. The priority mechanism is implemented as a state machine driven by signals responsible for reset, sending the frame data, sending and receiving the control data, and receiving the frame data. Transmitted data have the highest priority. Then the control data come. After writing to the MIPP interface, the arbiter automatically will read one word from PHY. This atomic set of instructions prevents from unexpected data loss. Reading of the frame data from PHY has the lowest priority. Of course, when the reset occurs the state machine will stop for given number of clock cycles and go to idle state.

The RTL model of the MAC processor is a mixture of Verilog (MIPS 4KE core, buses, and cache memories) and VHDL (peripherals, hardware accelerator, and RAMs) components. The MIPS 4KE is a licensed soft core [7]. The VHDL code of the hardware accelerator, as well as, the simulation models of memories have been manually written. As the size of HDL code of the complete MAC processor is huge, we have selected the manually prepared VHDL model of CRC component shown in Figure 9.4 as a representative example.

```
--**********************************************************************
--          Written by Klaus Tittelbach-Helmrich
--          32 bit polynomial (802.11 FCS):
--                      x^32 + x^26 + x^23 + x^22 + x^16 + x^12 + x^11 +
--                      x^10 + x^8 + x^7+ x^5 + x^4 + x^2 + x^1 + x^0
--          Processes 8 data bits in parallel. Calculates inverted CRC,
--          i.e. saves the final inversion of CRC bits. Operates in
```

```
--          reverse direction, i.e. CRC bits have to be transmitted
--          with LSB first (like the data bytes themselves).
--*****************************************************************

LIBRARY  ieee;
USE      ieee.std_logic_1164.ALL;

ENTITY Crc32a IS
 PORT (clk  : IN  STD_LOGIC;                    -- clock signal
       init : IN  STD_LOGIC;                    -- synchr. CRC clear, high active
       dvld : IN  STD_LOGIC;                    -- write strobe, high active
       din  : IN  STD_LOGIC_VECTOR( 7 DOWNTO 0);-- 8 bit write data word
       crc32 : OUT STD_LOGIC_VECTOR(31 DOWNTO 0));-- 32 bit CRC output
END Crc32a;


ARCHITECTURE rtl OF Crc32a IS

 SIGNAL crc : STD_LOGIC_VECTOR(31 DOWNTO 0);  -- register for 32 bit CRC

BEGIN

 PROCESS (clk)
 BEGIN
   IF clk'EVENT AND clk='1' THEN
    IF init='1' THEN
     crc <= (OTHERS => '0');
    ELSIF dvld='1' THEN
     crc(31) <= NOT (crc( 1) XOR crc( 7) XOR din( 1) XOR din( 7));
     crc(30) <= NOT (crc( 0) XOR crc( 1) XOR crc( 6) XOR crc( 7)
            XOR din( 0) XOR din( 1) XOR din( 6) XOR din( 7));
     crc(29) <=     crc( 0) XOR crc( 1) XOR crc( 5) XOR crc( 6) XOR crc( 7)
            XOR din( 0) XOR din( 1) XOR din( 5) XOR din( 6) XOR din( 7);
     crc(28) <= NOT (crc( 0) XOR crc( 4) XOR crc( 5) XOR crc( 6)
            XOR din( 0) XOR din( 4) XOR din( 5) XOR din( 6));
     crc(27) <=     crc( 1) XOR crc( 3) XOR crc( 4) XOR crc( 5) XOR crc( 7)
            XOR din( 1) XOR din( 3) XOR din( 4) XOR din( 5) XOR din( 7);
     crc(26) <=     crc( 0) XOR crc( 1) XOR crc( 2) XOR crc( 3)
            XOR crc( 4) XOR crc( 6) XOR crc( 7)
            XOR din( 0) XOR din( 1) XOR din( 2) XOR din( 3)
            XOR din( 4) XOR din( 6) XOR din( 7);
     crc(25) <= NOT (crc( 0) XOR crc( 1) XOR crc( 2) XOR crc( 3)
            XOR crc( 5) XOR crc( 6)
            XOR din( 0) XOR din( 1) XOR din( 2) XOR din( 3)
            XOR din( 5) XOR din( 6));
     crc(24) <=     crc( 0) XOR crc( 2) XOR crc( 4) XOR crc( 5) XOR crc( 7)
            XOR din( 0) XOR din( 2) XOR din( 4) XOR din( 5) XOR din( 7);
     crc(23) <=     crc( 3) XOR crc( 4) XOR crc( 6) XOR crc( 7) XOR crc(31)
            XOR din( 3) XOR din( 4) XOR din( 6) XOR din( 7);
     crc(22) <=     crc( 2) XOR crc( 3) XOR crc( 5) XOR crc( 6) XOR crc(30)
            XOR din( 2) XOR din( 3) XOR din( 5) XOR din( 6);
     crc(21) <=     crc( 2) XOR crc( 4) XOR crc( 5) XOR crc( 7) XOR crc(29)
            XOR din( 2) XOR din( 4) XOR din( 5) XOR din( 7);
     crc(20) <=     crc( 3) XOR crc( 4) XOR crc( 6) XOR crc( 7) XOR crc(28)
            XOR din( 3) XOR din( 4) XOR din( 6) XOR din( 7);
     crc(19) <=     crc( 1) XOR crc( 2) XOR crc( 3) XOR crc( 5)
```

```
            XOR crc( 6) XOR crc( 7) XOR crc(27)
            XOR din( 1) XOR din( 2) XOR din( 3) XOR din( 5)
            XOR din( 6) XOR din( 7);
      crc(18) <=    crc( 0) XOR crc( 1) XOR crc( 2) XOR crc( 4)
            XOR crc( 5) XOR crc( 6) XOR crc(26)
            XOR din( 0) XOR din( 1) XOR din( 2) XOR din( 4)
            XOR din( 5) XOR din( 6);
      crc(17) <= NOT (crc( 0) XOR crc( 1) XOR crc( 3) XOR crc( 4)
            XOR crc( 5) XOR crc(25)
            XOR din( 0) XOR din( 1) XOR din( 3) XOR din( 4) XOR din( 5));
      crc(16) <=    crc( 0) XOR crc( 2) XOR crc( 3) XOR crc( 4) XOR crc(24)
            XOR din( 0) XOR din( 2) XOR din( 3) XOR din( 4);
      crc(15) <= NOT (crc( 2) XOR crc( 3) XOR crc( 7) XOR crc(23)
            XOR din( 2) XOR din( 3) XOR din( 7));
      crc(14) <= NOT (crc( 1) XOR crc( 2) XOR crc( 6) XOR crc(22)
            XOR din( 1) XOR din( 2) XOR din( 6));
      crc(13) <= NOT (crc( 0) XOR crc( 1) XOR crc( 5) XOR crc(21)
            XOR din( 0) XOR din( 1) XOR din( 5));
      crc(12) <=    crc( 0) XOR crc( 4) XOR crc(20)
            XOR din( 0) XOR din( 4);
      crc(11) <= NOT (crc( 3) XOR crc(19) XOR din( 3));
      crc(10) <= NOT (crc( 2) XOR crc(18) XOR din( 2));
      crc( 9) <= NOT (crc( 7) XOR crc(17) XOR din( 7));
      crc( 8) <= NOT (crc( 1) XOR crc( 6) XOR crc( 7) XOR crc(16)
            XOR din( 1) XOR din( 6) XOR din( 7));
      crc( 7) <= NOT (crc( 0) XOR crc( 5) XOR crc( 6) XOR crc(15)
            XOR din( 0) XOR din( 5) XOR din( 6));
      crc( 6) <=    crc( 4) XOR crc( 5) XOR crc(14)
            XOR din( 4) XOR din( 5);
      crc( 5) <=    crc( 1) XOR crc( 3) XOR crc( 4) XOR crc( 7) XOR crc(13)
            XOR din( 1) XOR din( 3) XOR din( 4) XOR din( 7);
      crc( 4) <=    crc( 0) XOR crc( 2) XOR crc( 3) XOR crc( 6) XOR crc(12)
            XOR din( 0) XOR din( 2) XOR din( 3) XOR din( 6);
      crc( 3) <= NOT (crc( 1) XOR crc( 2) XOR crc( 5) XOR crc(11)
            XOR din( 1) XOR din( 2) XOR din( 5));
      crc( 2) <= NOT (crc( 0) XOR crc( 1) XOR crc( 4) XOR crc(10)
            XOR din( 0) XOR din( 1) XOR din( 4));
      crc( 1) <=    crc( 0) XOR crc( 3) XOR crc( 9)
            XOR din( 0) XOR din( 3);
      crc( 0) <= NOT (crc( 2) XOR crc( 8) XOR din( 2));
    END IF;
  END IF;
 END PROCESS;

 crc32 <= crc;

END rtl;
```

Fig. 9.4. VHDL code of a CRC component

A VHDL testbench has been written to simulate the system behavior. It allows tracing of all the internal signals and efficient debugging. The VHDL simulation testbench provides external SRAM and PROM behavioral models as well as the emulation of the behavior on the system's peripherals. To test the system, a test program has been written in the MIPS assembler, compiled into a bit stream, and loaded into the boot PROM. The program performs processor initialization and tests all of the basic data transactions on the system ports.

The system-on-chip has been fabricated in a 0.25 μm CMOS technology [8]. The CPU, hardware accelerator, and memory blocks take the most of 40 mm^2 of the chip area. Due to intensive calculations, these two components consume most energy. The total power consumption at operating frequency of 80 MHz is 1.2 W. The hardware accelerator uses about 3 kB of memory (5x512 byte single-port RAM and 2x256 byte dual-port RAM). Figure 9.5 shows the SOC photo.

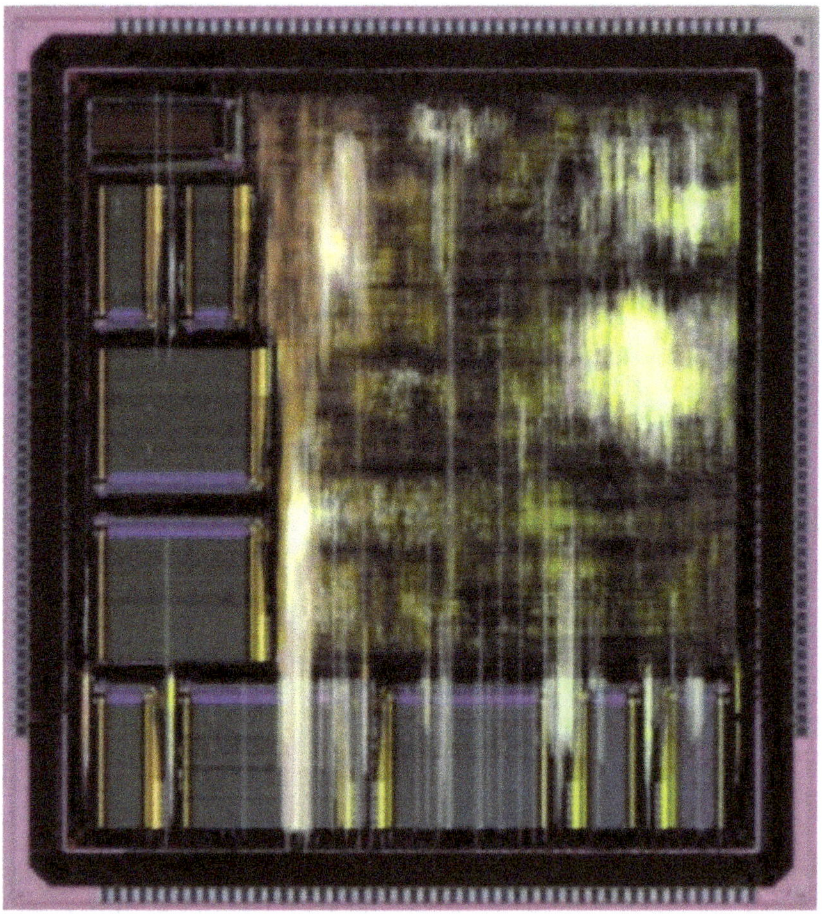

Fig. 9.5. Photo of the WLAN MAC processor SOC

9.3 Digital Baseband Processor

A detailed block diagram of the digital baseband processor is shown in Figure 9.6. It is composed of two main parts: an IEEE 802.11a standard digital baseband processor [9] and new baseband modules required by the RF-MIMO front-end architecture [10]. The new functionalities are grouped into two main modules: channel estimator and RF weights (or beamforming) block. These new modules will be active only when a RF-MIMO training frame is detected by the Tx/Rx control block, which transfers the signal field data to the control block in order to start the procedure (channel estimation and beamforming).

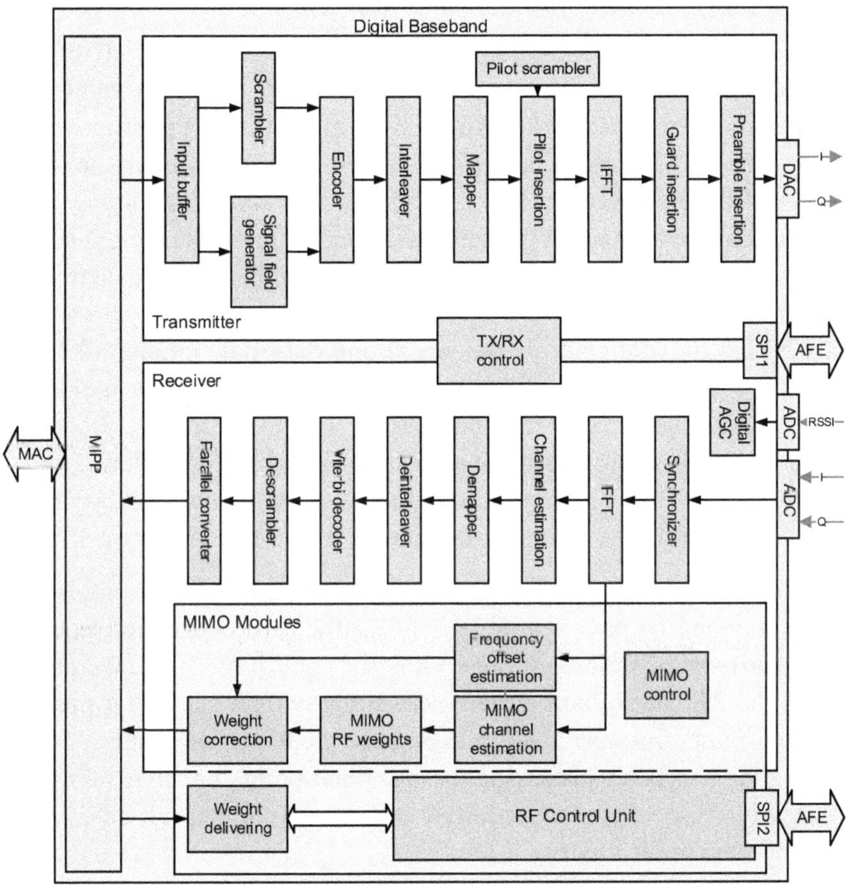

Fig. 9.6. RF-MIMO digital baseband processor

More precisely, the digital baseband processor integrates the following modules:

- *Channel estimation*: This module estimates the n_T x n_R MIMO channel based on the n_T x n_R training OFDM symbols of the received training frame. It works in the frequency domain taking the FFT signal provided by the IEEE 802.11a processor as input using a least squares estimation method.

- *RF weights*: It takes the estimated MIMO channel as input and computes the optimal TX/RX beamforming weights using the Max-SNR algorithm. It is the most important block in terms of complexity.
- *Frequency offset estimation*: Due to the residual frequency error at the output of the conventional IEEE 802.11a synchronizer, it might be necessary to include a frequency offset estimator working in parallel with the channel estimation and RF weights modules. To estimate the frequency offset, it is necessary to transmit an additional training symbol, resulting in a training frame of $n_T x n_R + 1$ training symbols.
- *Weight correction*: This module multiplies the weights by a unitary (e.g. rotation) matrix in order to compensate the effects of the residual frequency offset and specific Tx/Rx beamformers used in the training phase.
- *Weight delivery*: It transfers the calculated optimal weights to the MAC processor (the weight updating). In addition, it allows applying (from the baseband) the predefined set of weights during training (the weight setting) and transferring (from MAC) the optimal or default weights during data transmission or reception (the weight uploading).
- *Control*: This module controls the signal and data flow among all RF-MIMO blocks. It receives from the TX/RX control block information included in the training frame signal field (the number of TX/RX antennas, the number of training symbols), as well as some activation and synchronization signals.
- *RF control unit*: This is a control interface between the baseband processor and analogue front-end. It is an integrated part of the baseband processor.

All the RF-MIMO blocks are activated only when a training frame is received. Therefore they can be powered down while either processing conventional data frames or transmitting training frames. Only the control block, the weight delivery block, and the RF control unit remain active at any time because it must transfer and set the weights from the MAC processor to the RF control unit.

The digital baseband processor uses a fixed-point model to allow making decisions with respect to the required precision (number of bits, number of iterations applied in the algorithms, etc.).

Channel estimator

The channel estimator uses the n_T x n_R training OFDM symbols included in a training frame. Each training symbol is affected by a specific pair of Tx and Rx beamformers. A conventional least squares algorithm is used to estimate the n_T x n_R equivalent SISO channels at the 52 active sub-carriers.

Some design decisions have been made in order to simplify implementation of the channel estimator. First, the identity matrix has been selected for the TX and RX beamforming matrices used during the training stage. Second, RF-MIMO training symbols will be identical to the IEEE 802.11a long training symbols composed of 52 sub-carriers modulated by Binary Phase Shift Keying (BPSK) values.

As Figure 9.7 shows, the channel estimator works in the frequency domain (after FFT) and includes an optional post filtering procedure to smooth the resulting frequency responses. From an implementation point of view, the LS estimator requires very few hardware resources (just sign inverters and control logic), but the post filtering process could be expensive in terms of memory and multiply-accumulate units (while providing marginal BER improvement). For this reason, only a LS version of the channel estimator block has been initially designed.

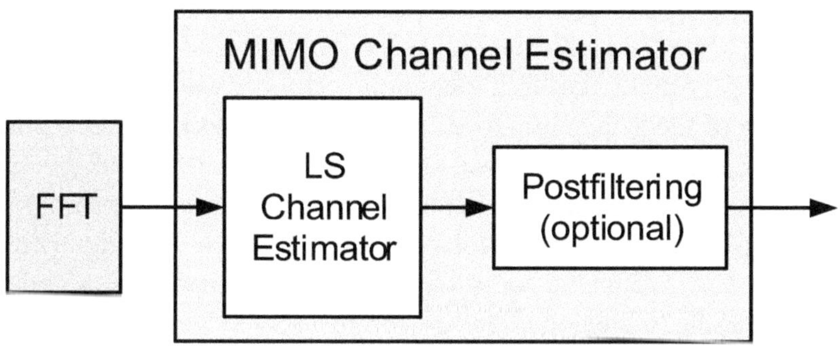

Fig. 9.7. Implemented channel estimator

Beamforming weights

The Max-SNR beamforming algorithm has been chosen because other criteria proposed in [3]3 use the Max-SNR solution as a starting point. Furthermore, our choice simplifies the implementation of this block without significant deterioration of the system performance. The proposed algorithm reduces to the maximization of energy of an equivalent SISO channel or, in other words, to the maximization of the received signal-to-noise ratio

$$\underset{w_T, w_R}{\arg\max} = \sum_{k=1}^{N_c} \left| w_R^H H_k w_T \right|^2, \text{ s.t. } \quad \|w_T\|^2 = \|w_R\|^2 = 1,$$

where the $n_T \; x \; n_R$ matrix H_k is the MIMO channel response at the k-th subcarrier, and w_T, w_R are the transmit and receive analog beamformers, respectively. These are complex vectors containing the RF weights to be applied by the analog front-end.

The input signals of the RF weights block come from the channel estimator whose outputs are the 52 sub-carrier samples for each one of the 16 (considering a link with four antennas at the transmitter and receiver sides) equivalent SISO channels. Notice also that all operations are carried out with complex numbers.

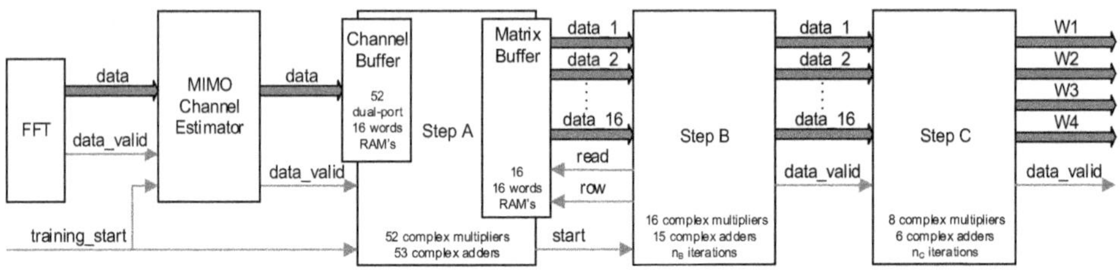

Fig. 9.8. Max-SNR beamforming weights calculation

Figure 9.8 presents implementation steps of the Max-SNR beamforming algorithm. Step A is creation of the 52 column vectors $\boldsymbol{x_k}$ where the i-th element of $\boldsymbol{x_k}$ is the sample of the k-th sub-carrier for the i-th equivalent SISO channel. The size of $\boldsymbol{x_k}$ is n_T x n_R (16 in this case). It also creates the 52 rank-one matrices $\boldsymbol{X_k}=\boldsymbol{x_k}\boldsymbol{x_k}^H$ of 16 x 16 dimension and adds these 52 matrices in a sum $\boldsymbol{Y}$. Step B calculates the z dominant eigenvector of the sum matrix. The common way to calculate this dominant eigenvector is to perform the singular value decomposition (SVD). However, the implementation of a complete SVD is not needed as it would use too many resources. The alternative solution is the power method which was finally implemented. This method is probably the simplest one for finding the largest eigenvector of a matrix. From the z vector of 16x1 dimension obtained by Step B, we construct the $\boldsymbol{Z}$ matrix of 4 x 4 dimension resized by columns. Step C calculates the SVD maximum eigenvector of $\boldsymbol{Z}$ in order to extract the first row of the $\boldsymbol{U}$ matrix. Again, it is not necessary to perform the complete SVD. A beamforming weight coefficient can be calculated as the dominant eigenvector of the product $\boldsymbol{ZZ}^H$ where $\boldsymbol{Z}^H$ is the Hermitian of matrix $\boldsymbol{Z}$. Thus Step C can be split in two substeps: the first one is a matrix multiplication and the second is a 4 x 4 power method. The resultant vector of this last power method is the $\boldsymbol{w_R}$ beamforming weight under the Max-SNR criterion.

The first task of the weight delivery block consists of transferring the calculated optimal weights to the MAC after a training frame has been received. This is so-called weight updating and it is a straightforward procedure (Figure 9.9). The beamforming weights are provided directly by the RF weights block (or by the weight correction block if finally needed).

The next task is to transfer the optimal or default weights from the MAC to the RF control unit (RFCU) during transmission or reception of data frames. This procedure, called weight uploading, has been easily implemented by allowing a direct connection between the MAC processor and RFCU as shown in Figure 9.10. Finally, the last task is to apply the predefined set of weights during transmission or reception of a training frame: this procedure is denoted as the weight setting.

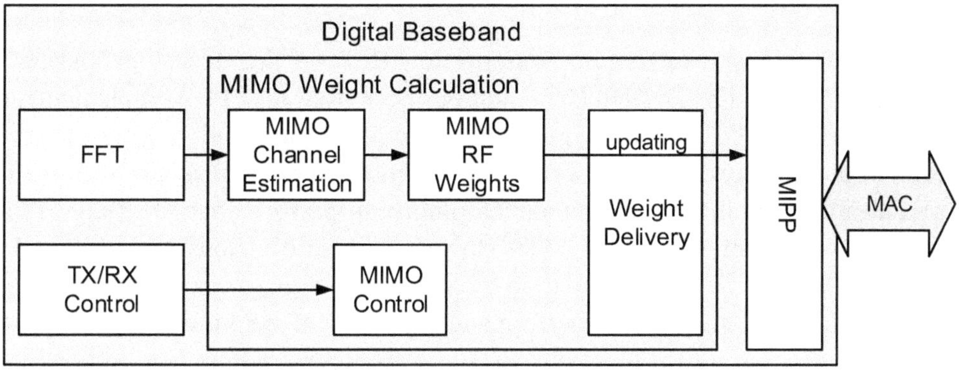

Fig. 9.9. Implementation of the weight updating

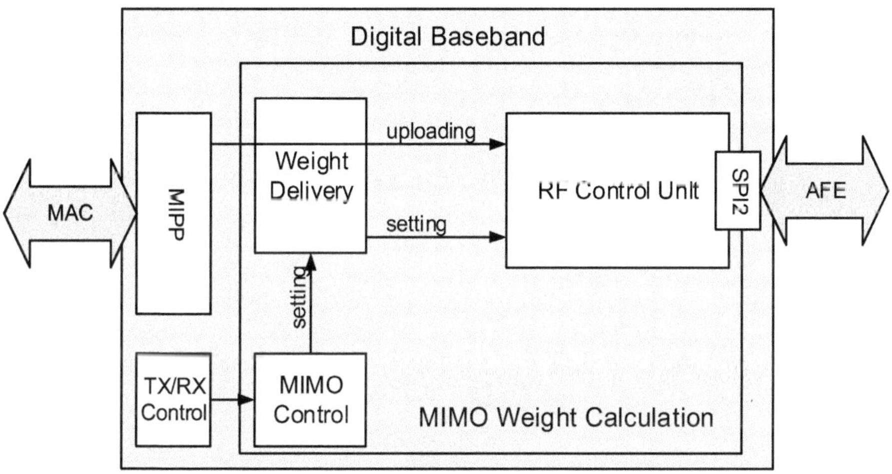

Fig. 9.10. Implementation of the weight delivery

Frequency offset estimator

Any residual frequency offset that occurs after the synchronizer stage of the conventional IEEE802.11a receiver distorts the weight calculations during training. Therefore, it could be necessary to estimate and compensate that residual frequency offset by transmitting two training symbols using the same pair of TX and RX beamformers. Under the assumption that the residual frequency offset is lower than the sub-carrier spacing, the following expression describes a maximum likelihood frequency offset estimator [3]:

$$\hat{\Delta f}_{ML} = \frac{1}{2\pi\Delta t}\,\text{angle}\left(\sum_{k=1}^{Nc} s_1[k]s_2^*[k]\right),$$

where N_c is the number of active subcarriers; s_1 and s_2 are the OFDM training symbols used for frequency estimation; and Δt means the time between symbols s_1 and s_2.

Figure 9.11 shows the implementation flow of the described baseband processor. The Xilinx System Generator has been used for FPGA implementation of the RF-MIMO baseband modules (channel estimation, frequency offset estimation and beamforming weights calculation). This tool is a plug-in to the Matlab's Simulink that enables developing high-performance DSP algorithms without writing an RTL HDL code. However, some decisions have been taken regarding the RF weights calculation block to reach a good compromise between FPGA utilization and system performance: number of iterations for each power

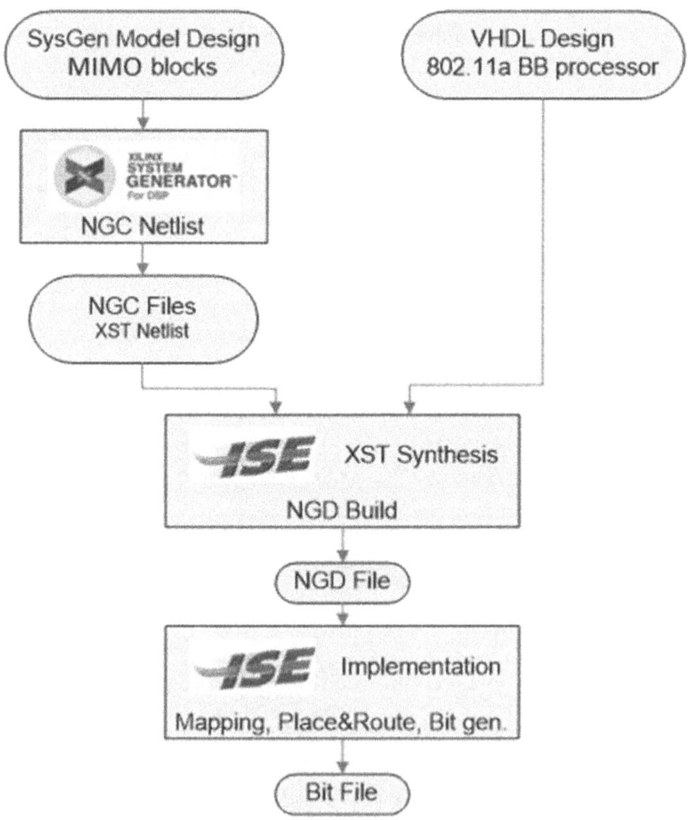

Fig. 9.11. FPGA implementation of the digital baseband processor

method is limited to 5 and interfaces between the blocks shown in Figure 9.8 are 8 bits wide.

A VHDL model of the IEEE 802.11a standard baseband processor has been manually prepared and previously verified. As the size of VHDL code of the complete baseband processor is huge, we have selected the manually prepared VHDL model of the receiver's deinterleaving component shown in Figure 9.12 as a representative example.

```
library ieee;
use ieee.std_logic_1164.all;
use ieee.std_logic_arith.all;

package deinterleaver_v1_pack is
  type interleaver_info_vector is array(0 to 7) of integer;
  constant START_ADDR : interleaver_info_vector := (0, 24, 60, 108, 180, 276, 420, 612);
  constant NUM_BITS : interleaver_info_vector := (24, 36, 48, 72, 96, 144, 192, 216);
  constant NUM_ENTRIES : integer := 48;
  constant ENTRY_ADDR_WIDTH : integer := 6;
  constant ENTRY_SEL_WIDTH : integer := 3;
  constant INPUT_PORTS : integer := 1;
  constant METRIC_WIDTH_INTERLEAVER : integer := 5;--4;
  constant BITS_PER_SC : integer := 6;
  constant BITS_PER_PORT : integer := BITS_PER_SC*METRIC_WIDTH_INTERLEAVER;
  constant INPUT_WIDTH_INTERLEAVER : integer := BITS_PER_PORT*INPUT_PORTS;
end deinterleaver_v1_pack;

library ieee;
use ieee.std_logic_1164.all;
use ieee.std_logic_arith.all;
library work;
use work.deinterleaver_v1_pack.all;

package deinterleaver_v1_comp_pack is
component muxer_24_120bit is
        generic ( m_width : integer := 4 );
        port (clk, nres : in std_logic;
        din : in std_logic_vector(29 downto 0);
        dout : out std_logic_vector(m_width-1 downto 0);
        sel : in std_logic_vector(2 downto 0) );
end component;
component mram_deinterleaver
        port (addra: IN std_logic_VECTOR(8 downto 0);
        addrb: IN std_logic_VECTOR(8 downto 0);
        clka: IN std_logic;
        clkb: IN std_logic;
        dina: IN std_logic_VECTOR(29 downto 0);
        doutb: OUT std_logic_VECTOR(29 downto 0);
        wea: IN std_logic_vector(0 downto 0));
end component;
component deinterleaver_rom
        port (addra: IN std_logic_VECTOR(9 downto 0);
        clka: IN std_logic;
        douta: OUT std_logic_VECTOR(17 downto 0));
```

```
end component;
component deinterleaver_v1
          port (clk : in std_logic;
        nres : in std_logic;
        new_input_block : in std_logic;
        termination_block : in std_logic;
        datamode : in unsigned(2 downto 0);
        new_data_in : in std_logic;
        data_in : in std_logic_vector(INPUT_WIDTH_INTERLEAVER-1 downto 0);
        new_data_out : out std_logic;
        data_out_1 : out std_logic_vector(METRIC_WIDTH_INTERLEAVER-1 downto 0);
        data_out_2 : out std_logic_vector(METRIC_WIDTH_INTERLEAVER-1 downto 0);
        new_output_block : out std_logic;
        terminate : out std_logic;
        overflow : out std_logic );
end component;
end package;

library ieee;
use ieee.std_logic_1164.all;
use ieee.std_logic_arith.all;

entity muxer_24_120bit is
  generic ( m_width : integer := 4 );
  port (
    clk, nres : in std_logic;
    din : in std_logic_vector(29 downto 0);
    dout : out std_logic_vector(m_width-1 downto 0);
    sel : in std_logic_vector(2 downto 0) );
end muxer_24_120bit;

architecture arch of muxer_24_120bit is
begin
  process(clk, nres)
    variable index : integer;
  begin
    if nres = '0' then
      dout <= (others => '0');
    elsif rising_edge(clk) then
      index := conv_integer(unsigned(sel));
      case index is
        when 0 =>
          dout <= din(29-0*m_width downto 29-1*m_width+1);
        when 1 =>
          dout <= din(29-1*m_width downto 29-2*m_width+1);
        when 2 =>
          dout <= din(29-2*m_width downto 29-3*m_width+1);
        when 3 =>
          dout <= din(29-3*m_width downto 29-4*m_width+1);
        when 4 =>
          dout <= din(29-4*m_width downto 29-5*m_width+1);
        when 5 =>
          dout <= din(29-5*m_width downto 29-6*m_width+1);
        when others =>
          dout <= (others => '0');
      end case;
    end if;
```

```
  end process;
end arch;
library ieee;
use ieee.std_logic_1164.all;
use ieee.std_logic_arith.all;

library work;
use work.deinterleaver_v1_pack.all;
use work.deinterleaver_v1_comp_pack.all;

entity deinterleaver_v1 is
  port (clk : in std_logic;
        nres : in std_logic;
        new_input_block : in std_logic;
       ...
       data_out_1 : out std_logic_vector(METRIC_WIDTH_INTERLEAVER-1 downto 0);
       data_out_2 : out std_logic_vector(METRIC_WIDTH_INTERLEAVER-1 downto 0);
       new_output_block : out std_logic;
       terminate : out std_logic;
       overflow : out std_logic );
end deinterleaver_v1;

architecture arch of deinterleaver_v1 is

subtype int8 is integer range 0 to 7;
constant i1_l : integer := 2*ENTRY_ADDR_WIDTH+2*ENTRY_SEL_WIDTH-1;
...
constant i4_r : integer := 0*ENTRY_ADDR_WIDTH+0*ENTRY_SEL_WIDTH;

-- write side
signal entry_counter : unsigned(7 downto 0);
signal check_entry_counter : std_logic;
signal wr_basepointer : unsigned(8 downto 0);
signal init_rd : std_logic;

-- ram block signals for metrics & permutation
signal metric_wr : std_logic_vector(0 downto 0);
signal wr_addr : std_logic_vector(8 downto 0);
signal write_data : std_logic_vector(29 downto 0);
...
signal perm_rd : std_logic;
signal perm_rd_addr : std_logic_vector(9 downto 0);
signal perm_rd_data : std_logic_vector(17 downto 0);

-- propagating parameters
signal terminate_int : std_logic;
signal datamode_int : unsigned(2 downto 0);
...
signal terminate_rd : std_logic;
signal datamode_rd : unsigned(2 downto 0);
signal rd_basepointer : unsigned(8 downto 0);

-- read side
signal save_bit : std_logic;
signal bit_counter : unsigned(9 downto 0);
...
signal perm_index : unsigned(9 downto 0);
```

```
signal base_addr : unsigned(9 downto 0);
begin

process(clk, nres) -- input machine
  variable bitlength_selection : int8;
begin
 if nres = '0' then
   metric_wr <= "0";
   init_rd <= '0';
   check_entry_counter <= '0';
   ...
   num_bits_next <= (others => '0');
   datamode_next <= (others => '0');
   basepointer_next <= (others => '0');
   terminate_next <= '0';
 elsif rising_edge(clk) then
     metric_wr <= "0";
     init_rd <= '0';
     check_entry_counter <= new_data_in;
   if new_input_block = '1' then -- store new block data
       datamode_int <= datamode;
       terminate_int <= termination_block;
   end if;
   if new_input_block = '1' then -- write address logic and bit counter
       wr_basepointer <= wr_basepointer + NUM_ENTRIES;
       wr_addr <= wr_basepointer + NUM_ENTRIES;
       entry_counter <= conv_unsigned(1,8);
     elsif new_data_in = '1' then
       entry_counter <= entry_counter+1;
       wr_addr <= unsigned(wr_addr)+1;
     end if;
     ...
 end if;
end process;

...
process(clk, nres)
 variable v : int8;
begin
 if nres = '0' then
   base_addr <= (others => '0');
   perm_index <= (others => '0');
   ...
   perm_rd_addr <= (others => '0');
   rd_addr_1 <= (others => '0');
   rd_addr_2 <= (others => '0');
   sel_1 <= (others => '0');
   sel_2 <= (others => '0');
   sel_1_dly <= (others => '0');
   sel_2_dly <= (others => '0');
 elsif rising_edge(clk) then
    v := conv_integer(datamode_rd);
   base_addr <= conv_unsigned(START_ADDR(v), 10);
   if new_rd_block = '1' then
     perm_index <= (others => '0');
   elsif get_data_v(1) = '1' then
     perm_index <= perm_index+1;
```

```
    end if;
    ...
    rd_addr_1 <= rd_basepointer_c3 + conv_unsigned(unsigned(perm_rd_data(i1_l downto i1_r)),9);
    rd_addr_2 <= rd_basepointer_c3 + conv_unsigned(unsigned(perm_rd_data(i3_l downto i3_r)),9);
    sel_1 <= (others => '0');
    sel_2 <= (others => '0');
    sel_1(ENTRY_SEL_WIDTH-1 downto 0) <= perm_rd_data(i2_l downto i2_r);
    sel_2(ENTRY_SEL_WIDTH-1 downto 0) <= perm_rd_data(i4_l downto i4_r);
    sel_1_dly <= sel_1;
    sel_2_dly <= sel_2;
  end if;
end process;

hardcore_muxer1 : muxer_24_120bit
  generic map (m_width => METRIC_WIDTH_INTERLEAVER)
  port map (
     clk => clk,
     nres => nres,
     din => read_data_1,
     dout => data_out_1,
     sel => sel_1_dly );

hardcore_muxer2 : muxer_24_120bit
  generic map (m_width => METRIC_WIDTH_INTERLEAVER)
  port map (clk => clk,
    nres => nres,
    din => read_data_2,
    dout => data_out_2,
    sel => sel_2_dly );

perm_rom : deinterleaver_rom
   port map (addra => perm_rd_addr,
     clka => clk,
     douta => perm_rd_data);

metric_ram_01 : mram_deinterleaver
   port map (
    addra => wr_addr,
    addrb => rd_addr_1,
    clka => clk,
    clkb => clk,
    dina => write_data,
    doutb => read_data_1,
    wea => metric_wr);

metric_ram_02 : mram_deinterleaver
  port map (
    addra => wr_addr,
    addrb => rd_addr_2,
    clka => clk,
    clkb => clk,
    dina => write_data,
    doutb => read_data_2,
    wea => metric_wr);
end arch;
```

Fig. 9.12. VHDL code of a deinterleaving component

The conventional IEEE 802.11a baseband processor modules occupy around 45%, whereas the additional RF-MIMO baseband modules occupy 33% of the available Virtex5 LX330 FPGA slices.

References

1. Stamenkovic, Z., Tittelbach-Helmrich, K., Krstic, M., Perez, J., Via, J. and Ibanez, J. (2009). Architecture of an analog combining MIMO system compliant to IEEE802.11a, *Proc. 18th ICT-MobileSummit*, pp. 1–8.
2. Eickhoff, R., Kraemer, R., Santamaria, I. and Gonzalez, L. (2009). Developing energy-efficient MIMO radios, *IEEE Vehicular Technology Magazine*, 4, pp. 34–41.
3. Via, J., Santamaria, I., Elvira, V. and Eickhoff, R. (2010). A general criterion for analog Tx-Rx beamforming under OFDM transmissions, *IEEE Trans. on Signal Processing*, 58, pp. 2155–2167.
4. Stamenkovic, Z., Tittelbach-Helmrich, K., Krstic, M., Ibanez, J., Elvira, V. and Santamaria, I. (2011). MAC and baseband processors for RF-MIMO WLAN, *EURASIP Journal on Wireless Communications and Networking*, 1/207, pp. 1–13.
5. Stamenkovic, Z., Miletic, E., Obrknezev, M. and Tittelbach-Helmrich, K. (2009). MAC protocol implementation in RF-MIMO WLAN, *Proc. 16th IEEE International Conference on Electronics, Circuits, and Systems*, pp. 303–306.
6. Tittelbach-Helmrich, K., Miletic, E., Wcislek, P. and Stamenkovic, Z. (2010). MAC hardware platform for RF-MIMO WLAN, *Proc. 53rd IEEE Internat. Midwest Symposium on Circuits and Systems*, pp. 339–342.
7. https://www.mips.com/products/classic
8. https://www.ihp-microelectronics.com/en/services/mpw-prototyping/sigec-bicmos-technologies.html
9. Krstic, M., Maharatna, K., Troya, A., Grass, E. and Jagdhold, U. (2003). Implementation of an IEEE 802.11a compliant low-power baseband processor, *Proc. 6th International Conference on Telecommunications in Modern Satellite, Cable and Broadcasting Service*, pp. 97–100.
10. Elvira, V., Ibanez, J., Santamaria, I., Krstic, M., Tittelbach-Helmrich, K. and Stamenkovic, Z. (2010). Baseband processor for RF-MIMO WLAN, *Proc. 17th IEEE International Conference on Electronics, Circuits, and Systems*, pp. 798–801.

https://doi.org/10.1142/9789811210723_0010

Chapter 10

Full-Custom Implementation of Analog and Mixed-Signal Circuits

Gildas Leger and Antonio Gines
Consejo Superior de Investigaciones Científicas,
Universidad de Sevilla
Instituto de Microelectrónica de Sevilla
Seville, Spain

10.1 Introduction

The architecture choice and the proper sizing of the RF transceiver entails a large part of a successful design. Figure 10.1 shows an example of such an architecture. The RF-MIMO transmitter uses a direct-up-conversion architecture. The baseband provides digital in-phase and quadrature symbols that are converted into analog signals using digital-to-analog converters. These signals are up-converted to the carrier frequency and fed to all available antennas (the multiplexing gain equals one). Before transmitted to the air by the antenna, each signal is multiplied by a complex weight: I and Q signals are adjusted in amplitude and phase relatively to the other paths. This multiplication is performed by a so-called vector modulator. Afterwards the power amplifier (PA) drives the antenna.

The RF-MIMO receiver, similarly, is based on a direct-down-conversion principle. The antenna unit is constructed of a four elements array shared with the transmitter. In each receive path, a band pass filter (BPF) is integrated to attenuate adjacent radio bands and to pass the 5 GHz band signals. The low noise amplifier (LNA) is integrated together with the vector modulator into a single subsystem and can be switched between high and low gain operation modes. The vector modulator performs the complex weight multiplication of the received signals, or equivalently changes the relative amplitudes and phases of all the antenna signals. Each of these received and weighted signals is then combined coherently and afterwards down-converted in the baseband to zero-IF. Consequently, I and Q components are generated by a voltage-controlled oscillator (VCO) and the combined signals are mixed with sinusoidal and co-sinusoidal signals. These baseband signals are low pass filtered (LPF) and then digitized by an ADC for further processing by the digital baseband processor and MIMO algorithms.

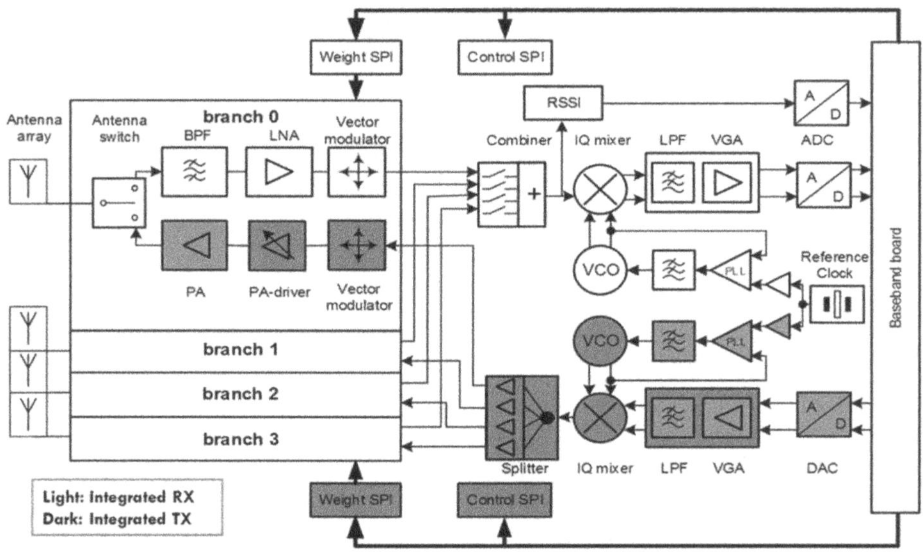

Fig. 10.1. Example of architecture for an RF transceiver with spatial diversity MIMO capabilities

This architecture is a solid example of how to implement one of the MIMO capabilities: that of enhancing Signal-to-Noise Ratio through spatial diversity. It holds the potential to fulfill the specifications of the standard. But to do so, a lot of effort has to be put in the physical design: "The devil is in the details" and a poor physical implementation can ruin a good design.

This chapter is divided in two main parts. The first will discuss the technology choice, taking into account the features of the process that are desirable as well as the valuable library blocks that can speed-up the system design. This technology selection is not usually contemplated as part of design flow but rather as an up-front decision. In the second part, we will present the full-custom design flow from a methodological perspective and give some advice on the early inclusion of these "devil's details" in the design flow: the layout parasitics, the package but also the PCB level integration to ensure that the final chip performance can be duly evaluated.

10.2 Technology Selection

There exist a wide variety of technological solutions for the implementation of a full RF transceiver font-end – understood as the complete signal chain from (and including) the antenna to the digital processing part. The particular application case will determine the adequate sweet spot in the multi-dimensional trade-off between cost, compacity and performance. Indeed, a WiFi transceiver for a given WLAN standard may have similar performance requirements for a home router and for a mobile phone, but very different compacity constraints.

From the cost perspective, in any high-volume market, it is usually better to integrate the complete circuitry in a single chip. Indeed, the so-called System-on-Chip (SoC) approach reduces the Bill of Material (BOM) and assembly complexity and the possibly high non-recurrent costs are greatly diluted by high-volume manufacturing. A fully integrated transceiver is also less susceptible to pick noise up. Yield (and test) issues can be seen as important threats for this approach but the major limitation is simply feasibility. Actually, in some cases, performance targets are unreachable for a System-on-Chip approach.

A simple example of this is the implementation of the antenna: for mm-wave regimes and beyond it may become feasible, but the sheer physical size of antennas for the current 2 to 6 GHz bands is simply not manageable. Another typical issue is the power and signal levels of the power amplifier, which prevent its integration in a low-cost CMOS process. As well, antenna switches and channel filters are seldom integrated, particularly for multi-band multi-standard chips. Notice that SoC approaches usually refer to low-cost CMOS processes. Indeed, it would be technically feasible but certainly not cost-effective to implement mixed-signal and digital circuits in a GaAs technology. As well, technologies optimized for speed like BiCMOS process are usually not suited for low-power, and the leakage of digital gates would dramatically affect the consumption of such a SoC.

If a full-SoC implementation is not feasible (which is unfortunately the case for most applications), then finding the implementation sweet-spot becomes a cost-optimization problem, constrained by the performance and compacity requirements. The specificity of this optimization problem is that the design space is not continuous: Several blocks of the system can be integrated in different semiconductor technologies and there also exist many packaging/PCB variants.

Moore's law has been dictating the evolution of semiconductor process nodes for the past decades. In general, the development of a newer and smaller technology node is welcome as a direct improvement. This, however, holds essentially for digital circuits where logic gates are designed with minimum feature size. For a digital circuit, a lower technology node means a lower area and an improvement of the power/speed trade-off. For Analogue, Mixed-Signal and RF circuits (AMS-RF), the benefits are not so straightforward. These types of circuits seldom use minimum length transistors, so the area scaling is far less direct than for digital circuits. Similarly, widely used integrated passives do not exhibit the same scaling properties as transistors. This is particularly relevant for RF devices like LNAs where a large part of the area can be consumed by integrated inductors. However, for transistors of equal dimensions, the matching in smaller technology nodes is usually better due to an improved control of the fabrication process (which is necessary to achieve the smallest features).

Smaller technology nodes do offer larger cut-off frequencies and thus speed benefits. However, the most critical drawback is the limitation of dynamic range. The conventional constant-field process scaling imposes lower power supply voltage (this voltage reduction is by the way the main cause of power consumption reduction). However, the overdrive voltage to keep transistors in saturation does not scale accordingly and the dynamic range available for analog signals is thus significantly reduced. This signal range reduction greatly affects the noise / linearity design trade-off. Reducing signal range to maintain linearity will impose a noise reduction to maintain a given SNR target, which is not always possible. For instance, the thermal noise sampled on a switched-capacitor is kT/C so reducing that noise requires increasing the capacitor, which is detrimental to the operating frequency. On the other hand, trying to maintain a large signal range make impossible the use of some linearization techniques, like the introduction of cascode transistors.

Overall, we can thus say that the benefits of technology scaling for AMS-RF circuits are mild. As an illustration, we can compare the overall performance of [1] and [2] which both implement a 4x4 MIMO WLAN transceiver. The first one is implemented in a 28nm CMOS technology while the second is in 40nm CMOS. Both areas are comparable: 12mm^2 for the former and 11.4mm^2 for the latter. The power consumption is actually 16% higher for the 28nm design in Rx mode over the 2.4GHz band, but 6% lower on the 5GHz band.

At the beginning of 2020, we were able to find only one academic contribution to WLAN transceivers in a technology below 28nm. This paper presents a polar transmitter in a 14nm tri-gate CMOS technology [3]. Notice that polar transmitters are digital friendly but do not apply to fine amplitude modulation schemes, which are mandatory for the most advanced WLAN standards. The development of AMS-RF circuits always lags behind its digital counterpart, but the 28nm node was introduced almost 10 years ago. The lack of transceivers in smaller technology nodes is thus a clear indication that process scaling does not always translate directly into performance gains.

10.2.1 *RF considerations*

Except the passive devices like channel filters and antennas, the block that poses the major issues for a fully integrated solution is undoubtedly the power amplifier. Though there exist some implementations of polar transmitters able to use "fine" amplitude modulations, like for instance the 256QAM scheme of [4], the vast majority of cases calls for class A-AB amplifiers which are particularly inefficient for high peak-to-average power ratios (PAR). The situation is even worse if significant power back-off has to be contemplated. Actually, the power efficiency is often reported at maximum output power, but this is clearly a best-case situation for a class-A amplifier.

One example of a CMOS transceiver with integrated PA can be found in [5]. The technology is 180nm CMOS but the integrated PA is biased at 3.3V and makes use of use a thin-oxide common-source input device but a thick-oxide common-gate output device (usually used for analog IOs) in a quasi-differential cascode topology in order to improve the RF performance and reliability. In that sense, the PA can be seen as an "older-node" block, since the thick-oxide transistors are not scaled at 180nm. This PA achieves around 25dBm of output power with a power efficiency of 50% and 30% in the 2.4GHz and 5GHz bands, respectively. Notice that the power efficiency is for the PA alone, not the full transceiver. This level of efficiency is achieved due to class AB operation. The downside is that pre-distortion capabilities had to be included to maintain EVM levels and in an acceptable range while complying with spectral masks. A similar solution in a 40nm CMOS can be found in [6], where the transmitter achieves a transmitted power of 23.5dBm for a 64QAM scheme on the 2.4GHz band and 17dBm for a 256QAM scheme on the 5GHz band.

The power efficiency of PAs is one of the most important metrics for battery life and long-term system reliability (higher efficiency means lower thermal dissipation and stress). Power efficiency of integrated CMOS PAs is not on par with their III-V or BiCMOS counterparts. While the issue of the total power budget can eventually be handled in single-standard transceiver, the trend towards integration of multi-standard MIMO with concurrent operation puts a significant pressure on integrated PAs. In addition, as discussed in Part 2 of this book, integrated PAs represent a significant challenge in terms of isolation. Strong RF power from the TX chain (mainly the PA) leaking through the substrate can disrupt the operation of the RX chain through LO pulling, for instance.

Besides performance, efficiency, thermal budget and robustness, yield can also be a concern. As commented above, even if the integration of CMOS PAs in SoC is feasible for low to modest output powers, it does not directly imply that it is the most cost-effective solution. Working close to the technology limits means smaller design guard-bands which immediately translates into lower yield. And for an integrated PA, a poor yield impacts the complete SoC. On the contrary, an external PA can be tested before its assembly and its yield only affects the PA cost. Finally, working close to technology limits also makes the design more challenging, which usually translates into higher design costs and longer design cycles. For competitive consumer markets, the importance of time-to-market can be paramount.

Consequently, a usual practice is to separate the core of the transceiver from the RF Front-End Module (FEM), which consists of the PA, channel filters and switches (diplexers) to the antenna(s) and eventually the LNA. This does not mean that the FEM needs necessarily to be implemented in a non-CMOS technology, but it offers the possibility to use a larger technology node with higher supply voltage for the FEM and a lower one, more digital-friendly, for the transceiver.

In modern very-deep-submicron CMOS processes, small channel length ultimately saturates the drift velocity due to high electric field (v_{scl} is the scattering-limited velocity). The relationship between the Drain current and the overdrive voltage becomes linear instead of the traditional quadratic form. Consequently, the transconductance becomes independent of the overdrive voltage, greatly limiting the design space. The transitional frequency, f_T, which is the frequency at which the current gain of a transistor is 0, can be approximated for long-length MOSFETs in saturation as,

$$f_T = \frac{g_m}{2\pi C_{gs}} \propto \frac{\mu(V_{GS} - V_t)}{2\pi L^2} \tag{10.1}$$

Where g_m is the transconductance, C_{gs} the gate to source capacitance, μ is the carrier mobility, V_{gs} is the gate to source potential, V_t the threshold voltage and L the length of the transistor. Whereas for short-length transistors it becomes,

$$f_T = \frac{g_m}{2\pi C_{gs}} \propto \frac{v_{scl}}{2\pi \mathrm{L}} \tag{10.2}$$

This simply illustrates how the benefits of technology scaling are much lower than in the past, for the most modern CMOS technology nodes.

High Electron Mobility Transistors (HEMTs) are not conceptually different from MOSFETs but take advantage of a heterostructure where the bandgap difference between two crystals is used to trap majority electrons as a 2D electron gas (2DEG) at the interface. The oxide is substituted by the wider bandgap semiconductor where the dopants are located. The physical separation between the doping impurities and the 2DEG leads to an enhanced mobility. These heterostructures can be implemented in III-V semiconductors like GaAs/AlGaAs, which also present a good intrinsic electron mobility (GaAs has an electron mobility 4 times larger than Silicon).

Heterojunction Bipolar transistors (HBT) can be considered as an improvement of Bipolar Junction Transistors (BJT) just like the HEMT can be viewed as an improvement of MOSFET. A smaller bandgap semiconductor is used for the base to increase the hole barrier from the base to the emitter, which in turn allows higher base doping levels. In Si/SiGe processes, the amount of Germanium in the base is graded to lower the base bandgap at the collector's junction. This allows to assist electron drift with a built-in voltage. In an HBT (and in a BJT), the current gain at DC is not infinite like for a MOSFET. However, the concept of transitional frequency still applies since the capacitive behavior also degrades that current gain. Making the approximation that the capacitive behavior of the HBT is dominated by the base charging, we get that,

$$f_T = \frac{1}{2\pi\tau_F} \tag{10.3}$$

where τ_F is the base charging time that depends on the base doping level.

Table 10.1 summarizes the pros and cons of different RF technologies. The power row obviously corresponds to the capacity of the technology to handle large power (and hence supply voltage), and the integration raw corresponds to the capability to integrate many different devices (for SoC approaches).

As it can be seen, the unit cost of the non-CMOS processes can be offset by the performance gains and re-usability. Indeed, as we will see later, recent advances in high-performance packaging technology have drastically reduced the impact of the extra routing parasitics introduced by "discrete" integration.

Table 10.1. Pros and cons of different RF technologies, adapted from [7]

Parameter	InP HEMT	GaAs HEMT	GaN HEMT	CMOS	InP HBT	SiGe HBT
f_T	++	+	-	=	++	+
Cost	--	-	+	++	--	+
Power	=	=	++	--	=	-
Noise	--	--	=	--	-	-
Linearity	++	++	+	=	++	++
Integration	+	+	--	++	-	++
Reliability	=	=	=	+	=	-

An example of a GaAs FEM architecture can be found in [8], where the PA manages an output power level of 100mW. This FEM implements a shunt/series T/R-switch with a voltage doubler to provide sufficient conductivity. In this way, an insertion loss of only 0.75dB and 0.8dB are achieved for the TX and RX paths, respectively. The Isolation level of this switch is 24dB. It is worth noticing that this level of performance was achieved 20 years ago. Paper [9] also presents an innovative architecture for a dual-band front-end module (FEM) for WiFi and MIMO radios, implemented in GaAS. In addition to the dual-band pHEMT PA it implements a Single-Pole 3-Throw (SP3T) switch-plexer. The TX path of the switch achieves 0.1 dB compression at 33.5 dBm power level with and insertion loss below 1dB for both bands, and an isolation above 18dB. The insertion loss of the Rx path diplexer is a bit larger (>2dB) for a band selectivity of 15 dB.

Paper [10] makes a comparison between 45nm RF SOI-CMOS process and 130nm SiGe Bipolar technology. The application target is Ka-band FEM modules, but the conclusions are very similar to those in the sub-6GHz bands. Essentially, state-of-the-art CMOS process has an edge in terms of Noise-Figure (which favors LNA design) but cannot match SiGe for high transmitted power levels due to the limits in the oxide breakdown voltage.

10.2.2 *Baseband considerations*

Independently of the technological considerations for the implementation of the RF FEM and the definition of its exact boundary (for instance, whether or not the LNA is part of the FEM), it appears clearly that the core of the analog and mixed-signal section of the transceiver, at least up to the mixer level, will most likely be implemented in CMOS.

Substrate resistivity is an important factor of substrate noise immunity. This can be particularly relevant to cope with injection pulling of oscillators due to the coupling with high frequency harmonics propagating through the substrate. Ideally, a high substrate resistivity with a more conductive epitaxial layer would be desirable. This sort of configuration is common in modern technology processes. Following the same noise immunity reasoning, fully depleted (and partially depleted, to a lower extent) SOI process can also be desirable. In addition to substrate noise isolation, SOI is also beneficial for low supply voltage since some benefits can be drawn from specific body biasing schemes. Together with SOI, another technology that reduces RF coupling is porous Silicon, though it is mostly used as a substrate for passives in higher frequency applications like mm-wave.

With respect to the most common process options for AMS-RF design, the presence of high-quality passives is necessary: Metal-Insulator-Metal (MIM) capacitors provide precise matching with a capacity between 1 and 2fF/μm^2. Metal Oxide Metal (MOM) capacitors are the second best choice for capacitors. High-Q inductors are also a desirable feature, which usually involves a thick top metal. If the process is not fully depleted SOI, double- and triple-well are also welcomed to get rid of body-effect and to provide extra shielding.

Though the development costs of Application-Specific Integrated Circuit (ASIC) design can be very high, they are generally offset by volume production. In addition, re-usability is a cornerstone of SoC approaches. If a design house uses a stable and well-characterized process for a long time, it generally put together a library of building blocks that are fully characterized, silicon-proven, and will greatly speed-up the development cycle of new systems. Such blocks can be data converters, voltage regulators, microprocessors, memory blocks, filters, mixers, etc., but also lower-level components such as current references and bandgaps. The complexity of porting such a library to a new technology node is high, but this can eventually be alleviated purchasing some of these blocks to third-party Intellectual Property (IP) companies.

10.3 AMS-RF Top-Level Implementation

Through the reading of this book, it should clearly appear that transceivers are complex systems designed by a team of domain experts: analog base-band designer, RF designer, digital designer, antenna designer, layout experts, etc. Though some system-level models do exist from early design phases, these are not accurate enough to guarantee a successful physical implementation.

Each domain has its own simulation tools and various vendors can be found in the Electronic Design Automation (EDA) market that provide detailed solutions.

For the system integration phase, special care has to be taken at the domain boundaries. The analog and RF sections may work perfectly, as well as the digital section. However, it may occur that the connection between the two lead to malfunction. For instance, the capacitive load of the DAC analog section in the transmitter chain may be larger than expected by the digital designer in charge of the decoder. Such a simple issue may lead to fail timing requirements in some corners and degrade the DAC effective resolution. More often than not, the issue arises from the fact that domain experts do not speak the same language and do not properly specify the interface requirements.

In what follows, we will describe a modern RF design flow and consider the three most important threats to system-level integration: analog and digital co-simulation, full-custom layout and PCB-level co-simulation.

10.3.1 *Modern full-custom design flow*

In Chapter 8, we have briefly described a conventional top-down full-custom design flow, from the system-level to the final layout. At each step of this flow, simulations are carried out to verify that the system and/or its building blocks fulfill their performance targets. If not, the task at hand is iterated to introduce some corrections. For instance, if the bandwidth of the LNA is too small due to the capacitive parasitics introduced by the layout, a new layout is generated to try to lower these parasitics. If after a few iterations the bandwidth is not met, the designer would probably opt to modify the design of the LNA to give it more driving capability at the cost of an increased power consumption. Such a vertical flow gives a lot of control to the designer, but these back-and-forth iterations can introduce large delays in the project schedule.

For a complex system like a modern MIMO transceiver, the design team will be large enough so that it makes sense introducing more parallelism in the different tasks. However, such an approach requires a lot of effort to be put in verification to ensure the consistency of the design decisions taken by the team members [11].

A parallel flow can be sketched as in Figure 10.2. The architecture design is initiated with educated guesses of the main performance parameters of the building blocks. Once a candidate architecture is selected and the performance targets are obtained after some high-level optimization loop, the different blocks are being designed in a traditional manner. However, the main difference with the conventional design flow is that at each design iteration of a block a refined high-level model of the block is generated that takes into account as many non-idealities as possible. Such a practice enables the system designer to refine his own performance evaluation and rapidly update the performance targets if deemed necessary. For instance, if a transistor-level LNA design has more gain than specified, it may allow the system design to relax some constrains down the receive chain.

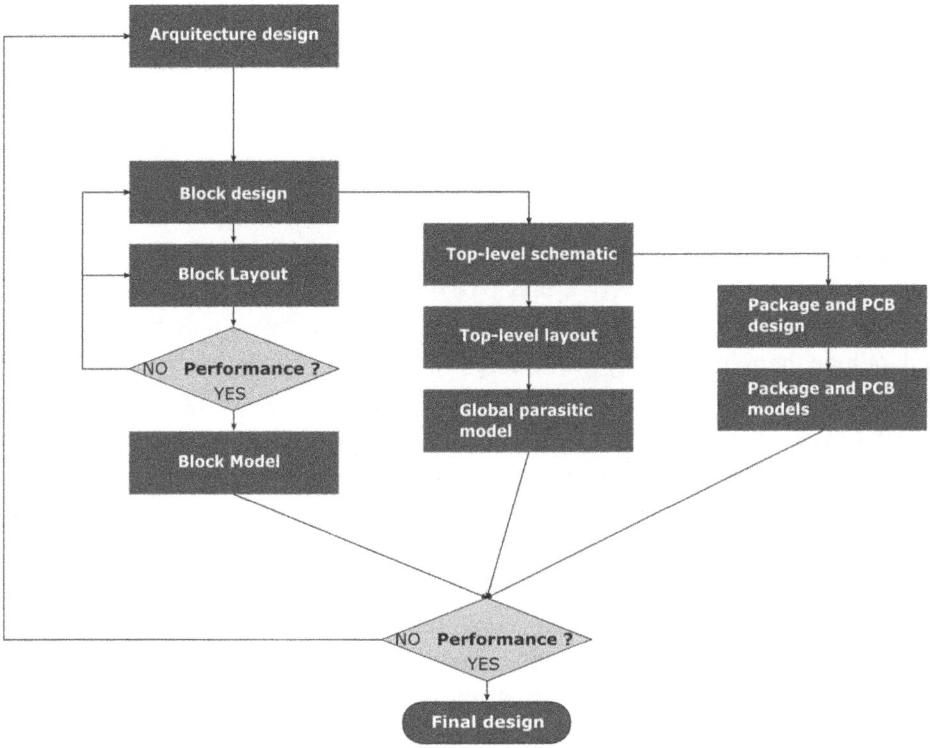

Fig. 10.2. Parallelized full-custom design flow

As soon as preliminary block designs (with tentative layouts) are available, the system assembly can begin to generate a floor-planning of the chip, allocate pins and evaluate global rooting parasitics. These early layouts allow evaluating global routing parasitics that may alter the loading conditions of the block designers. Similarly, the designers in charge of the package and PCB interaction can start to work as soon as a tentative pad ring is available. Electromagnetic simulations are used to provide a model of package and PCB parasitics.

The basic idea of such a flow is to anticipate the issues and limitations that may appear along the design cycle by feeding back detailed information as soon as possible to the system level. Model development is thus fundamental for such an approach.

Notice that Figure 10.2 actually embeds the whole process in a global feedback loop. This loop can (or should) be viewed as a global optimization loop that gets refined as the design progresses, starting from high-level simulations with almost ideal model and finishing with a mix of transistor-level simulations and system-level simulations with detailed models. The top-down bottom-up optimization method described in Chapter 6 of Part 2 would fit in such a scheme. Indeed, design re-use, block design automation, or the purchase of IP blocks can seamlessly be introduced in that flow and greatly speed-up the design cycle.

10.3.2 *Analog and digital co-simulation*

RF and analog design flow are very similar and rely on details and accurate compact models of both active and passive devices. Main EDA tool providers usually work closely with foundries in order to build these reliable compact models and include them in their Physical Design Kits (PDK) that gather basic components and gates library as well as parametrized-cells (p-cells) for full-custom design.

However, it is important to remember that any model makes some simplifications and does not exactly represent the reality of the true physical mechanisms. There is a clear trade-off between model accuracy and simulation time. Each domain has its own sweet spot in this trade-off. Analog and RF require accurate time-domain simulations to account for large-signal non-linearities, as opposed to the small-signal approximation used for simple frequency-domain analysis. These time-domain simulations rely on some form of non-linear differential equation solver for which the computational burden depends on the global signal accuracy target as well as on the scale difference between the simulation time and the smallest time step.

On the other hand, the computational burden of digital simulation is much more directly related to the sheer number of transistors. Indeed, gate models only consider two defined states for IOs and implement pre-computed in LUTs. The accuracy requirements for digital simulations rather reside in the proper definition of the transition delays and its relationship with load and corner supply voltages.

It is conceptually possible to simulate a complete transceiver using transistor-level compact models, but the simplest EVM simulations would last too much for the allowable time-to-market. Indeed, any logic transition would trigger several small time steps in the differential solver, which would have to resolve the complete circuit.

To speed-up this process, mixed-mode simulators have appeared in the market that automatically handle the translation between two simulation domains. Digital domain is simulated Verilog, VHDL or any equivalent language, whereas Analog and RF domains are simulated at transistor-level with compact models. These tools automatically recognize (from a configuration view) the simulation domain associated to each building block and handle the translation between domains by inserting virtual A/D and D/A wrappers, transparent to the user. However, some care has to be taken to correctly parametrize these "converters" to handle non-idealities like loading impedance, threshold level, hysteresis, etc.

In order to avoid any functional error, it is preferable to maintain a strict system-level hierarchy as early as possible in the projects. All the building blocks should have a unique symbol view and the different tests should always be performed from the system top-level view. In such a workflow, the tests are handled through specific configuration files that define which view to use for each block. If the same block is instantiated several times, it is also possible to define a different view for each instance, if deemed necessary. The main advantage of such an approach is to ensure that there are no mistakes in the system-level routing. Indeed, a very common mistake that can lead to long debugging time is the mismatch in bus ordering, for instance. By maintaining a single symbol per block, we force the different views or models of that block to be pin-compatible.

All the blocks should have at least a model for digital simulation and a model for analog simulation. Obviously, the level of detail and accuracy of these models will depend on its intended use, and several of them can be implemented.

Typically, analog and RF blocks should have a transistor-level view, one including extracted layout parasitics, a model of the ideal function written in an analog high-level description language like Verilog-A, Verilog-AMS or VHDL-AMS, a model of the ideal function written in a digital high-level description language, and an empty view (which is usually valid for both analog and digital simulators). The empty views are used in phases of the design centered on other unrelated parts, in order to save computational burden. For instance, the receiver chain could be set to an empty view for designing the transmitter chain, at least in the first part of the design. The high-level analog model is of much help in optimization loops of other related blocks, as commented in Chapter 6 of Part 2. For instance, the optimization of the Power Amplifier may use a high-level model of the preceding mixer.

The digital blocks typically have a functional HDL model, a post place & route digital model with the corresponding delay files, an empty view just like for analog blocks, and an analog high-level model. Large digital circuits usually implement many more functions than their analog counterparts do. Thus, it is almost impossible to develop an analog model that would implement all these functions and respond correctly to any possible input.

The analog models of the digital blocks should support only those functions that interact with the analog section in the principal modes. Other functions can be simplified or obviated. For instance, it is quite likely that a critical digital block implement scan chains for test purpose. This is the kind of functionality that would be obviated in an analog model because it likely does not interact with the analog section (though the digital designer must still make sure that no flip-flop in the scan chain is connected directly to the analog section, or that the bit flips during scan mode are not critical to its integrity). Another example: configuration registers for the analog section programmability and tuning may be loaded through an SPI interface. For the analog model, the configuration registers are mandatory, however, they could simply be implemented as a static register in the analog model because the read and write procedures can be considered as an internal process of the digital section (here again, one must make sure that the reading and writing does not cause unacceptable bit-flips in the configuration registers).

10.3.3 *Layout considerations*

Low-level

The Physical Design Kit of the semiconductor technology selected for the implementation of the transceiver contains, in addition to the basic cells and their models already commented earlier, a number of design rules that must be followed to perform the layout of these cells. Most of these design rules are usually integrated in the EDA layout tool, either as a simple Design Rule Check (DRC) process that will raise and locate some warnings if rules are infringed, or as a more interactive process embedded in the drawing of the different shapes.

Notice that these design rules do not simply mark the limits of what is technologically achievable, like the minimum width of metal lines, but also include the validity conditions of the models that have been provided. Indeed, modern deep sub-micron technologies have been pushed to the limit and many effects that could safely be obviated in larger technology must now be taken into account. For instance, the distance of an active region to the well or to another active region can influence significantly the electron mobility or the threshold voltage. Similarly, the stress induced by shallow-trench-isolation (STI) alters the electrical properties of the closest active regions. This is due to the fact that a deliberate amount of stress in actually produced in the channel of NMOS and PMOS transistors to increase the carrier mobility.

This is one of the reasons why specific p-cells are usually provided for RF transistors, which are significantly larger than their "regular" counterparts. They are not fundamentally different from the perspective of the fabrication process, but they include a number layout "guard-bands" that ensure that their operation will be as close as possible to that of the model.

Analog and RF design strongly relies on matching. Current mirrors and differential pairs are classical examples for transistors, but a ladder of identical resistors is also a common example for passives. Here again, proximity effects can be deleterious. It is recommended to use layout structures as regular as possible, with arrays of identical fingers surrounded by grounded dummy fingers as a best practice. For prototyping of first-time-right design, it is also highly recommended to implement thorough fine-tuning programmability (to control bias currents, reference voltages, etc.).

Power-hungry blocks, like Power Amplifiers, must also bear special attention for thermal effects and power handling. Electromigration is the most common one and maximum current rules will drive the dimensioning of metal lines. But some layout techniques can also be considered to favor thermal management, like for instance providing a dissipation path to the thick top metal. As detailed in Chapter 11, package selection is one of the most critical aspects for power and thermal management.

Finally, for RF design, the impact of routing parasitics can be very significant. Thus, it is necessary to generate low-level layout as early as possible in the design cycle. It is most of the time useless to spend a lot of time to optimize a block at schematic level without the inclusion of the most relevant layout parasitics.

Top-level

For system-level layout, as for the building blocks, it is desirable to perform early top-level floor-planning. A common mistake of junior designers is to floorplan for minimum area. Such a practice usually impacts significantly the layout time, since it is usual to have to modify building blocks to cope with parasitics, add some test or tuning capabilities, route some additional signals. Hence, it is better to leave sufficient spare space for busses, for decoupling networks. The last system-level layout iterations can eventually be used to save some area.

For a full-custom analog layout (or an analog-on-top system-level integration) it is also a good practice to alternate metal layers for global routing, as usual for digital circuits: even layers for vertical and odd for horizontal. This is not strictly mandatory, but it usually makes layout modifications easier. The worst-case current consumption estimates should also be taken into account to draw the different power rails. In addition to the electromigration issue that most designers are familiar with, some analog blocks with poor Power Supply Rejection Ratio may be particularly sensitive to IR drop. In large systems, the location of the power-hungry blocks and the supply routing scheme must be considered. For demanding signals, it may be necessary to stack several levels of metals to lower the overall square resistance and avoid wide rails. In such case, routing corridors across the stacks should be managed.

For signal tracks that are very sensitive to resistance, it may be preferable to avoid too many vias in series. Actually, the identification of analog signals particularly sensitive to parasitics is important for a successful design: the global floor-planning should readily try to minimize the routing of critical nets. However, a trade-off exists between the different parasitics and layout compactness: the resistance of a metal trace is proportional to its length, but inversely proportional to its width. However, the crossing of two metal lines create a capacitor whose plate are formed by the crossing section: the wider the lines, the larger the parasitic capacitor. As well, a capacitor can also form between the metal trace and the substrate.

Two adjacent metal traces running in parallel in the same layer, as for instance in a bus, can also form a significant parasitic capacitor. Special care should be taken to avoid such couplings between critical signals and assure meeting design for manufacturability (DfM) constrains. The first qualitative action that can be taken is to properly order the signals in the bus. DC signals can usually be made almost insensitive to capacitive coupling with dynamic aggressors by including a small amount of decoupling capacitance.

For sensitive dynamic signals, particularly RF ones, including decoupling capacitance is usually not an option to meet the required bandwidth requirements. In such cases, these sensitive dynamic signals should not be placed close to potential aggressors (like a clock signal, for instance). It is a better bet to surround them by quiet DC signals. Actually, for particularly sensitive signals, shielding can be used: this practice consists in surrounding (totally or partially) the sensitive track by grounded metal lines, as seen in Figure 10.4.

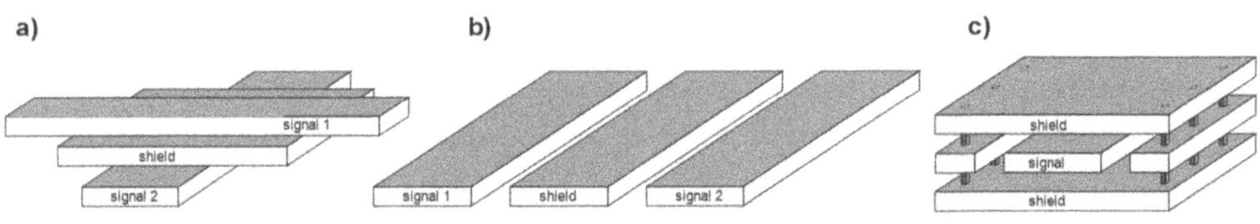

Fig. 10.4. Sensitive track shielding: a) crossing; b) bus; c) full isolation

Notice that such ground needs to be as quiet as possible and should probably go to a specific pin. As well, care must be taken in the design of the decoupling network such that the shielding signal does not pick up and propagate unwanted noise to the critical nets it was supposed to protect. The other obvious action that can be taken to reduce capacitive coupling is to increase the space between tracks. This option is quickly limited by compactness requirements, but it is one of the reasons why the initial floorplan should let sufficient room for routing.

10.3.4 *Co-simulation with package and PCB models*

We have already spoken of the analog/digital simulation boundary. However, and particularly for RF circuits, there is another simulation boundary that must be handled at system level. As operating frequency increases, it is mandatory to account for the effects of non-ideal packages and their interactions with the PCB.

Generally speaking, the sensitivity of digital signals is much lower than that of their analog counterparts. A careful design of signal traces, driving buffers and pull-up or pull-down resistors is usually sufficient to keep sufficient eye opening. If high frequencies have to be reached, differential signaling, like for instance LVDS (Low-Voltage Differential Signaling), should provide adequate signal integrity. For high-resolution and/or high frequency analog signals, keeping an adequate noise and distortion performance is not an easy task.

Three scenarios can be contemplated:

- A model of the package is made available by the manufacturer. In such a case, the package is simply another building block to consider in the outer (system-level) optimization loop. Unfortunately, even in this best case, some parts of the process still need some assessment. The leads and the pads may be fully and accurately modeled but the bonding wires are not included. The design of the pad ring and the choice of the supply domains are closely related to the floor-planning. The designer must thus properly select the pin ordering to control which signals will be adjacent. In the same way as for bus design, sensitive signals should be placed between quiet signals. The designer can sometimes play with the die position in the cavity to assign shorter wires to critical signals. Depending on the pad and package size, it is sometimes possible to perform multiple bonding. A zero-order approximation would lead to divide the inductance by the number of parallel wires. Unfortunately, this is not exactly the case due to mutual inductance coupling.
- A model of the package is not available, but the package can be measured using test circuits. Many RF simulators accept the direct integration of an S-parameter matrix, which alleviate the burden of producing a compact model.
- If the physical dimensions of the package are known, a full-wave 3D Electro-Magnetic (EM) simulator can be used to extract the package parasitics. Unfortunately, such an approach is error-prone if the dimensions and different constituting materials are not precisely known. It is also extremely computationally demanding.

EM simulators actually resolve the Maxwell's equations provided some boundary conditions:

$$\begin{aligned}
\nabla \cdot \mathbf{D} &= \rho_f \\
\nabla \cdot \mathbf{B} &= 0 \\
\nabla \times \mathbf{E} &= -\frac{\partial \mathbf{B}}{\partial t} \\
\nabla \times \mathbf{H} &= \mathbf{J}_f + \frac{\partial \mathbf{D}}{\partial t}
\end{aligned} \tag{10.4}$$

ρ_f is the free charge density and $\mathbf{J}_f$ is the current density of these free charges.

The magnetizing field (**H**) and the displacement field (**D**) are related to the magnetic (**B**) and electric (**E**) fields in linear materials through,

$$\begin{aligned}
\mathbf{D} &= \varepsilon \mathbf{E} \\
\mathbf{H} &= \frac{1}{\mu} \mathbf{B}
\end{aligned} \tag{10.5}$$

The permittivity ε and the permeability μ are scalars in isotropic materials but can be tensors for anisotropic ones.

To solve this system of differential equations, a number of different methods exits, all with pros and cons. Among the most important ones we have:

- The Method of Moments in the frequency domain is possibly the most famous one. It is the preferred method for 2.5D problems where the fields are defined in 3D but the sources that originate them are confined in 2D planes. Layered structures like PCB or System-in-Package (SiP) substrates are typical application examples. MoM solves problems stated in the form of electric (or magnetic) field integral equation. In the MoM, it is the boundary surfaces (most of the time the conductive strips) that are discretized by the expansion functions. An example of commercial EDA tool based on this technique is Momentum/ADS from Keysight EEsof EDA
- The Finite Element Method is the preferred option for full-wave 3D applications in the frequency domain. The field is discretized over the full volume using (most of the time) tetrahedral shapes. This operation is done by a tessellation algorithm that anticipates where finer granularity will be necessary. Over a unit element, the field is a linear combination of the field values at the edges. The determination of these values then involves

finding the combination of minimum energy. An example of 3D FEM-based solver is Ansys HFSS.

- The Finite Difference Time-Domain method treats, as its name indicates, Maxwell's equations in the time domain converting the time derivatives in time differences with a small time-step. The fields are thus discretized both in space and time. CST Studio Suite from Dassault Systèmes is a tool that relies on this approach.

In SiP solutions, not only the interconnects have to be considered. Indeed, some of the passives can be integrated directly on the RF substrate. The advantage of this approach is the reduced fill-factor, but also that such passives exhibit a good quality factor, enabling efficient filtering solutions. An example of high compactness can be found in [12], where a channel band-pass filter built with patch resonators is integrated on the backside of a BGA package. For the passives integrated in the substrate, and particularly for inductors, it is necessary to resort to EM simulators. As we will see in next chapter, compact SiP designs may have some substrate embedded inductors located very close or even underneath a die. This is another situation where an EM simulator would be used to estimate he potential crosstalks.

The accuracy of embedded passives is not perfect, and the filters implemented in the RF substrate, as well as the dies, must thus be tested prior to assembly to guarantee an adequate total yield. Indeed, since the defects in all the assembled parts are independent (at least in first order), the total yield without testing is the product of the individual yields. Testing the parts independently is a process often called known-good-die integration. RF substrates failing specifications are normally discarded because corrective actions involve laser trimming and are usually too expensive in low-cost high-volume markets.

In [13], however, a solution is proposed to break the dependency between the parts. Some extra tuning capabilities are implemented within the ASIC in order to cope with the possible variations of the passives integrated within the RF substrate. A calibration procedure then identifies the best tuning knobs. In that way, the effective yield of RF substrates is increased, and the total integration cost is improved.

Any circuit or SiP is deemed to be mounted on a PCBs. Even if the PCB design may not be the chip manufacturer's responsibility, some guidelines are usually provided. For complex systems, it is also usual to find reference / evaluation board. For that purpose, it is thus necessary to have at least a first order approximation of the external circuitry: passives, regulators, buffers, etc. If these models are available early in the system design cycle, corrective actions can be taken: the complete PCB can be included in the so-called Package/System co-design. As commented above, extra tuning capabilities can be introduced, not only on-chip but also on-board (or on-substrate in the case of SiPs). To give an example, if the target Tx insertion loss of the diplexer switch is deemed impossible to reach with available components, there is no other path than designing the PA for higher gain.

To conclude, we can thus say that the design of Analog Front-End for modern WLAN is a quite holistic process. Therefore, the traditional sequential design flow (where a few iterations between steps were necessary to take a few corrective actions) cannot compete with a parallel one (where educated guess and first order models of the full system, and eventually its application environment, have to be available as early as possible). Such a design flow is much more complex to handle, as it requires a good coordination of a design team where everybody has to be conscious of the potential impact of his design decisions on the other domains.

References

1. Kawai, S. *et al.* (2018). An 802.11ax 4x4 High-Efficiency WLAN AP Transceiver SoC Supporting 1024-QAM With Frequency-Dependent IQ Calibration and Integrated Interference Analyzer, IEEE J. Solid-State Circuits, 53(12), pp. 3688–3699.
2. Chen, T.-M. *et al.* (2017). An 802.11ac dual-band reconfigurable transceiver supporting up to four VHT80 spatial streams with 116fs rms-jitter frequency synthesizer and integrated LNA/PA delivering 256QAM 19dBm per stream achieving 1.733Gb/s PHY rate, Proc. *IEEE International Solid-State Circuits Conference, ISSCC*, San Francisco, CA, USA, pp. 126–127.
3. Madoglio, P. *et al.* (2017). A 2.4GHz WLAN digital polar transmitter with synthesized digital-to-time converter in 14nm trigate/FinFET technology for IoT and wearable applications, Proc. *IEEE International Solid-State Circuits Conference, ISSCC*, San Francisco, CA, USA, pp. 226–227.
4. Zhu, Q. *et al.* (2017). A Digital Polar Transmitter With DC–DC Converter Supporting 256-QAM WLAN and 40-MHz LTE-A Carrier Aggregation, IEEE J. Solid-State Circuits, 52(5), pp. 1196–1209.
5. Degani, O. *et al.* (2008). A 1×2 MIMO Multi-Band CMOS Transceiver with an Integrated Front-End in 90nm CMOS for 802.11a/g/n WLAN Applications, Proc. *IEEE International Solid-State Circuits Conference, ISSCC*, San Francisco, CA, USA, pp. 356–619.
6. Yan, S. T. *et al.* (2017). An 802.11a/b/g/n/ac WLAN Transceiver for 2x2 MIMO and Simultaneous Dual-Band Operation With +29 dBm Psat Integrated Power Amplifiers, IEEE Journal of Solid-State Circuits, 52(7), pp. 1798–1813.
7. Roodaki, P. M. *et al.* (2011). A survey of millimeter-wave technologies, Proc. *International Conference on Electrical and Control Engineering*, pp. 5726–5728.
8. Yamamoto, K. *et al.* (1999). A 2.2-V operation, 2.4-GHz single-chip GaAs MMIC transceiver for wireless applications, IEEE J. Solid-State Circuits, 34(4), pp. 502–512.

9. Huang, C.-W. P. *et al.* (2009). Innovative architecture for dual-band WLAN and MIMO frontend module based on a single pole, three throw switch-plexer, Proc. *IEEE Radio Frequency Integrated Circuits Symposium, RFIC*, Boston, MA, USA, pp. 281–284.
10. Li, C. *et al.* (2018). Ka Band FEM Design Comparison with 45nm RFSOI CMOS and High Performance SiGe BiCMOS, Proc. *14th IEEE International Conference on Solid-State and Integrated Circuit Technology, ICSICT*, Qingdao, pp. 1–4.
11. Chang, H. and Kundert, K. (2007). Verification of Complex Analog and RF IC Designs, Proceedings of the IEEE, 95(3), pp. 622–639.
12. Zhang, Y. P. *et al.* (2006). A study of dual-mode bandpass filter integrated in BGA package for single-chip RF transceivers, IEEE Transactions on Advanced Packaging, 29(2), pp. 354–358.
13. Goyal, A. *et al.* (2010). RF substrates yield improvement using package-chip co-design and on-chip calibration, Proc. *IEEE Electrical Design of Advanced Package Systems Symposium*, pp. 1–4.

https://doi.org/10.1142/9789811210723_0011

Chapter 11

Integration

Gildas Leger and **Antonio Gines**
Consejo Superior de Investigaciones Científicas,
Universidad de Sevilla
Instituto de Microelectrónica de Sevilla
Seville, Spain

Zoran Stamenković
IHP - Leibniz-Institut für innovative Mikroelektronik
Frankfurt (Oder), Germany

11.1 Introduction

WLAN transceivers are complex systems and the circuits necessary to comply with the last MIMO standard cannot (at least for current technologies) be fully integrated on a unique die of semiconductor from the MAC layer to the RF antenna array. Up to now, it is cheaper and wiser to split the different functions into several chips that can then be connected at several levels. How many discrete chips should be considered? In how many different technologies? These questions are a matter of system optimization and are very application dependent. For instance, a WLAN transceiver for a baseband station does not have the same compactness requirements as a transceiver in a mobile phone.

In the first section of this chapter, we present different packaging technologies, from the ones adapted for stand-alone chips to the ultimate Silicon-in-Package solutions that can combine and stack several dies in different technologies onto thin and densely connected substrates with integrated passives. The second section briefly discusses antenna technology and the last one provides an example of a practical realization of a MIMO WLAN transceiver.

11.2 Package Technology

An integrated circuit has to be connected to the outer world in order to be used. The first level of such connection is realized through a "package". Everybody is familiar with the small black plastic boxes with metallic legs: the dual-in-line package that is used for through-hole PCB mounting. In such a package, the pads on the periphery of the IC are connected to the package leads with metal (typically, gold/aluminum) wire bonds.

However, packaging technology has evolved significantly to meet ever more stringent miniaturization requirements, to the point that there is sometimes not any black box that we could identify as a package. That is the reason why we put the quotes in the first place.

In the next few pages, we will consider two main scenarios: the System-on-Chip and the System-In-Package. Notice, however, that the boundary between the two approaches in not always so clear, nor is the commercial nomenclature that can be found throughout the Internet. The interested reader can refer to [1] for details on state-of-the-art package modeling.

11.2.1 *System-on-Chip (SoC)*

This approach relies on the power of the technology scaling process known as Moore's law. As the minimum feature size decreases, the number of transistors that can be squeezed into a given silicon dice increases. We have thus moved from complex systems implemented from many discrete ICs in their individual packages soldered on a Printed Circuit Board (PCB), to complex systems implemented with fewer and fewer ICs. This implies fewer components to assemble. Reduced board complexity usually implies lower cost (including test cost) and better performance (due to reduced parasitics).

We have already commented above that it becomes feasible, from a performance viewpoint, to integrate a full transceiver with its associated processing logic and memory on a single dice. This, however, may not be (yet) the most cost-effective solution.

The benefits of the SoC approach include low cost, a reduced power dissipation, the best compactness and, in general, short design cycles for application-specific ICs (ASICs). This last point, though, may eventually be challenged if the application target is too close to the technology limits.

11.2.1.1 *Packaging*

As commented above, older packages with long leads are not very suitable for RF application, where signal connections should be as small as possible to minimize insertion loss. Hence, the baseline packaging option is QFN (quad-flat no-lead, illustrated in Figure 11.1): the die is attached to a metal plate (the exposed pad) that is used for thermal dissipation and eventually as a ground plane. Bonding wires connect the IC padring to the corresponding package pins in the periphery of the package, which are prepared for surface mounting.

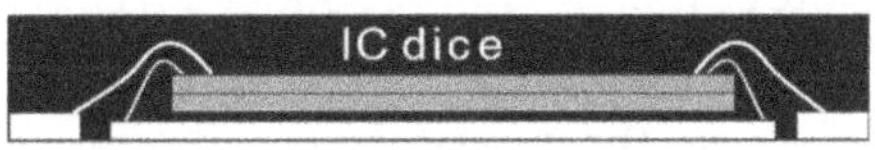

Fig. 11.1. QFN package illustration

However, the best performing package solution for a bare die is flip-chip. In that solution, the pads need not to be located in the periphery of the chip. After the conventional semiconductor processing steps, solder balls (or Cu micro-pillars) are deposited on the pads and the chip can be attached to the PCB with the active side facing the board, as shown in Figure 11.2, justifying the name.

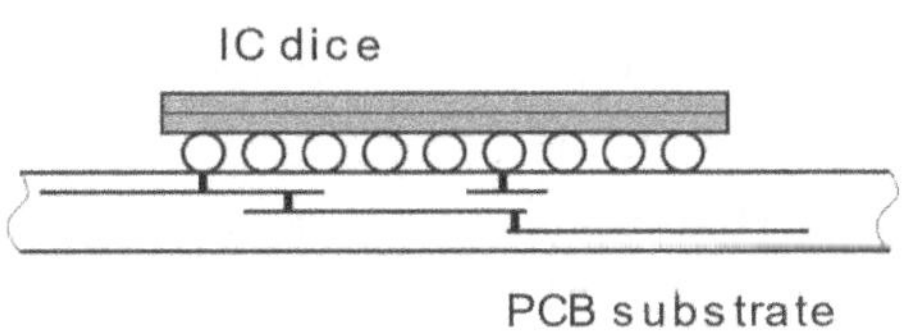

Fig. 11.2. Flip-chip solution

Parasitics related to bonding are much reduced for a flip-chip approach with respect to conventional wire bonding techniques. However, the fact that the die is placed face-to-face with the PCB gives rise to potential coupling issues in RF circuits. On-chip inductors are normally implemented in the thick top-layer metal of the IC and they are thus close to the PCB metallization. This kind of coupling has to be taken into account since it may de-tune an RF design.

Wafer-level chip-scale package, also known as Fan-In WLP, implements a redistribution layer using thin-film technology, prior to the deposition of the solder balls. This technique allows for a regular and dense grid of contacts, which may not be optimum for the pad location on the die.

Embedded Wafer Level Ball Grid Array Packaging (or Fan-Out WLP), is a very similar technique. The semiconductor wafer is diced, and an artificial wafer is reconstructed by separating the dies and molding them into a structural compound. This new "wafer" is processed with the same thin-film technology to redistribute the contacts throughout the extended area. The extra area in the perimeter around the die can also be used to contact the top level of the package which can be metalized.

11.2.1.2 *Horizontal extension*

SoC generally refers to the monolithic integration of a complete system in a unique semiconductor technology process. However, it can also be understood as any device with chip-scale integration and interconnects. This is a first place where the SoC / SiP boundary becomes fuzzy.

2.5D integration refers to the possibility of connecting different dies on a silicon (usually silicon, we should say) interposer. Such an interposer can be processed using the same back-end lithographic process as conventional wafers. It thus offers several layers of metallization with chip-scale feature size, eventually on both faces. What makes it different from a conventional back-end processing are the Through-Silicon Vias (TSVs) that are implemented to connect the two faces of the interposer. The diameter of TSV is usually in the range of a few microns and their depth is that of the interposer thickness, typically in the range of 50μm. So one side of the interposer is used to connect the different dies in a flip-chip manner (usually with micro-bumps of around 10μm, much smaller than those that would be used for PCB assembly), and the other face is used to contact with the PCB trough larger C4 bumps (micro-balls for assembly on laminate substrate are typically one order of magnitude larger than micro-bumps). This type of integration allows heterogeneous technology combination but is quite close to monolithic integration in terms of performance.

11.2.1.3 *Vertical extension*

While 2.5D integration places the dies side-by-side on a unique interposer, 3D integration, as the name indicates, considers a vertical stacking of the dies. This allows enormous compactness gains but comes at the cost of significant challenges in terms of thermal management and coupling. The fundamental difference with 2.5 D integration is that the TSVs are not implemented on the passive interposer but must be included in the active dies. This generates a number of complications for the chip design, and specific rules have to be contemplated to handle the stress-related modification of electrical characteristics. Notice that the active dies can be directly stacked (this is the case for some memory stacks) or also use an intermediate interposer as a redistribution layer (this would be the case to stack memory on top of a microprocessor).

In [2], it is shown that Through Silicon Vias (TSVs) are suitable for RF signal routing. More concretely, the copper vias in that particular paper, which have an aspect ratio of 20 for a length of 250μm, can be modeled as depicted in Figure 11.3. The via resistance is estimated to be around R_V=15mΩ, the inductance L_V=53pH and the Capacitance C_V=2.4pF. The resistance R_{sv} depends on the resistivity of the substrate. Notice that these dimensions are fairly large with respect to what is achievable today. The performance is thus expected to improve significantly as integration capabilities progress.

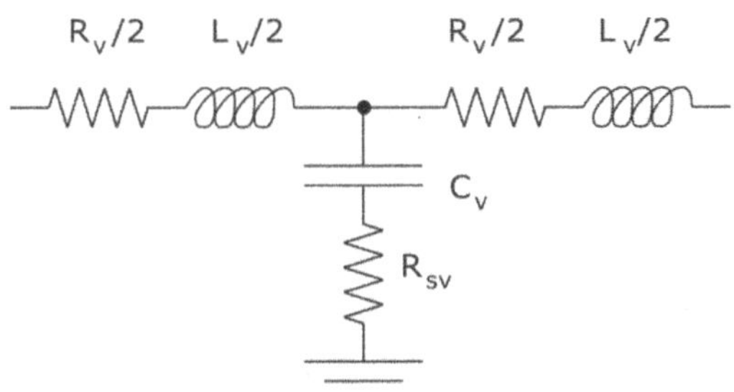

Fig. 11.3. Through-Silicon-Via lumped-element model

11.2.2 *System-In-Package (SiP)*

In Chapter 10, we have discussed in the "technology selection" section that it may be a better bet to separate the RF Front-End from the rest of the transceiver. Similarly, it is probably still better to implement the digital logic and memory in an optimized CMOS process rather than in a BiCMOS process together with the transceiver RF core.

Fortunately, integrating system parts on different technology does not necessarily imply several packages with their high parasitics. The progresses in packaging technology have allowed gathering several dices in a single package; hence the name: *System-in-Package* (SIP).

In a similar manner to what has been commented for 2.5D and 3D integration, the different chip modules are usually placed on a discrete substrate resembling a miniature PCB, in a 2D fashion. 3D packaging is also an option known as *Package-on-Package* (PoP). In this case, the modules are staked together and connected to the bottom substrate in a variety of ways, as seen in Figure 11.4: **a)** The oldest way (but still in use) to stack several dies involves wire-bonding in a pyramidal fashion. **b)** Through-mold vias in Fan-Out WLP can also be used to contact the top of the package with solder bumps at the bottom. In that way, another package can be soldered on top of it. **c)** A larger package on top of a smaller one can also be connected with larger soldering balls. **d)** Similarly, a larger die can serve as a motherboard for a smaller one if they are connected face-to-face. Several such levels can be used playing with the levels of the interconnects.

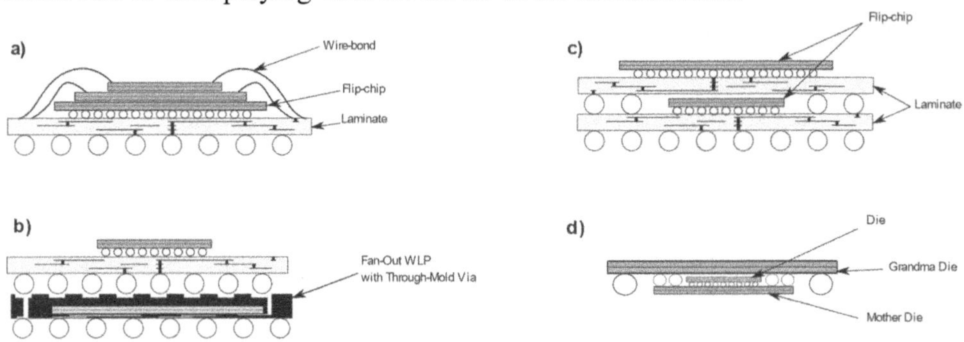

Fig. 11.4. Different chip stacking solutions for Package on Package approaches

The market driving forces towards more integration have pushed the concept of SiP very far. It is now possible to integrate several such substrates with their respective components on a "mother" substrate and package everything in a single bundle. SiP offers the possibility to also include passives integrated in dedicated substrates (or eventually embedded in the base substrate), leading to high quality factors, or even antennas if their dimensions allow it. Thermal management is of utmost importance in this type of package and heatsinks can also be included. Apart from the suitability of the reduction in interconnection length between components, which inherently reduces the associated parasitics, waveguides can also be implemented on specialized substrates that allow for very low insertion loss, even in high-frequency bands.

The SiP approach opens a wealth of integration possibilities for virtually any system in a very compact manner. It is thus a solution of choice for mobile applications. But this comes at a cost: such level of integration requires a great control of the yield of individual parts and of system test. In addition, it calls for high co-simulation capabilities to perform the system-level verification.

Substrates: There exist a wide variety of substrates for the integration of dies, some are more suitable for thermal dissipation, others for electromagnetic isolation, etc. Among the relevant substrate properties for SiP are:

- Permittivity: low-κ (κ is used in place of ε_r in material science) materials are desirable for substrates fabrics since they will minimize capacitance between metal planes for a given thickness.
- Dielectric breakdown voltage is relevant for high-power applications.

- Conductivity of metallization is also of importance to minimize parasitic resistance. In RF, and particularly considering the trend towards higher frequency, the skin depth phenomenon must be taken into account. Indeed, at higher frequencies the current flows essentially at the surface of the conductor. Table 11.1 shows the skin depth and conductivity for some common metals.
- Thermal conductivity is usually poor for substrates since dielectrics are usually much less thermally conductive than metals (the primary path of heat conductivity are usually free electrons). The thermal management of SiP thus usually involves the metallization layers, with the inclusion of heatsinks.
- Thermal expansion coefficient is also of importance to properly handle the stress induced by differential dilatation between materials. By the way, this is an important advantage of silicon interposers.
- Other mechanical properties like flexibility and thickness influence reliability.
- Finally, other specifications relate to moisture absorption, flammability, etc.

Table 11.1. Conductivity and skin depth of metals commonly used in packages

Material	**Conductivity** S/m	**Skin depth @10GHz** μm
Silver	6.3×10^7	0.634
Copper	5.8×10^7	0.661
Gold	4.1×10^7	0.786
Aluminum	3.5×10^7	0.851

Apart from the silicon substrates, which have been previously discussed, the most common substrates are Laminates and Ceramic:

- Laminate: Laminate substrates are possibly the most widely used substrates for RF. They consist in a number of layers (from 2 to 25) that are separately patterned and etched. The core layers are separated by a fabric that provides insulation and mechanical strength, known as *prepreg* or glue layers. After layer alignment, holes are drilled and plated to connect the different routing layers. One of the most used laminate substrate, FR4 is built with an epoxy resin deposited upon of structure of glass fibers. It has a dielectric constant of 3.65, a moderate thermal conductivity of 0.4W/m.K and a coefficient of thermal expansion of 230ppm/K.

- Ceramic: Low-Temperature Co-Fired Ceramic (LTCC) allows for high conductivity metallization, using metal with lower fusion temperature, while High-Temperature Co-Fired Ceramic (HTCC) allows for better dielectric properties but require the use of metals with higher melting temperature like Tungsten or Molybdenum that have lower conductivity. LTCC also enables the integration of passives on the substrate and are widely used in RF. With respect to laminates, ceramic substrates usually exhibit good thermal properties, particularly thermal expansion coefficient. For instance, one of the widely used oxides to build ceramic substrates is alumina (Al_2O_3), which has a dielectric constant of 9.5, significantly worse than the FR4 laminate, but a thermal conductivity of 22.3 W/m.K and an expansion coefficient of only 8 ppm/K.

11.3 Antennas

The application field of WLAN is evolving toward emerging advanced use in the home and business environment, such as multimedia communications or video streaming from multiple sources, with an increasing demand in bandwidth and data rate. To support this trend, more complex antenna schemes have been developed in each standard generation, from the former antenna diversity to the current MIMO technology. The design requirements for the antenna systems are especially challenging for mobile phones and other portable devices (tablets, smart clocks, etc.) [3], since the simultaneous operability between WLAN and other communications systems (LTE, Bluetooth, GPS, etc.) imposes extra restrictions in the spatial location of antennas to reduce coupling and interferences [4]. Moreover, for a given spectrum band, the antennas for different standards are not often shared to allow concurrent reception/transmission — this is the typical situation for Bluetooth and WLAN at 2.4GHz ISM band. Due to its significant impact on the performance, fill factor and cost of the complete system, the antenna selection is one of the main design parameter in a RF system. In consequence, a good understanding of the metrics and different antenna types is mandatory for an optimum selection [5–7].

11.3.1 *Antenna metrics*

Gain and efficiency: the power gain (G) is one of the most important parameters in the antenna design. It is usually characterized by a radiation efficiency parameter (η), which defines the ratio between measured radiation power and the theoretical power for a reference antenna, expressed in percentage (%) or decibels (dB).

Radiation patterns: the radiation pattern defines the intensity of the electromagnetic fields emitted by an antenna. It is usually characterized, as illustrated in the example of Figure 11.5, by a graph or polar pattern showing the power gain in the symmetry planes of the antenna. Based on this pattern, antennas can be classified as: 1) directional with a dominant direction of propagation perpendicular to the plane of polarization of the electric field, 2) sectorial with a sector-shaped radiation pattern, and 3) omnidirectional without a dominant direction in the plane of propagation. In the last case, the reference antenna for the power gain evaluation is an ideal isotropic antenna, denoted in decibels as dBi.

Impedance matching and return loss: an antenna is considered as a transducer capable of converting electromagnetic waves into electrical signals and vice versa. Therefore, it is a two-port network characterized by the scattering s-parameters.

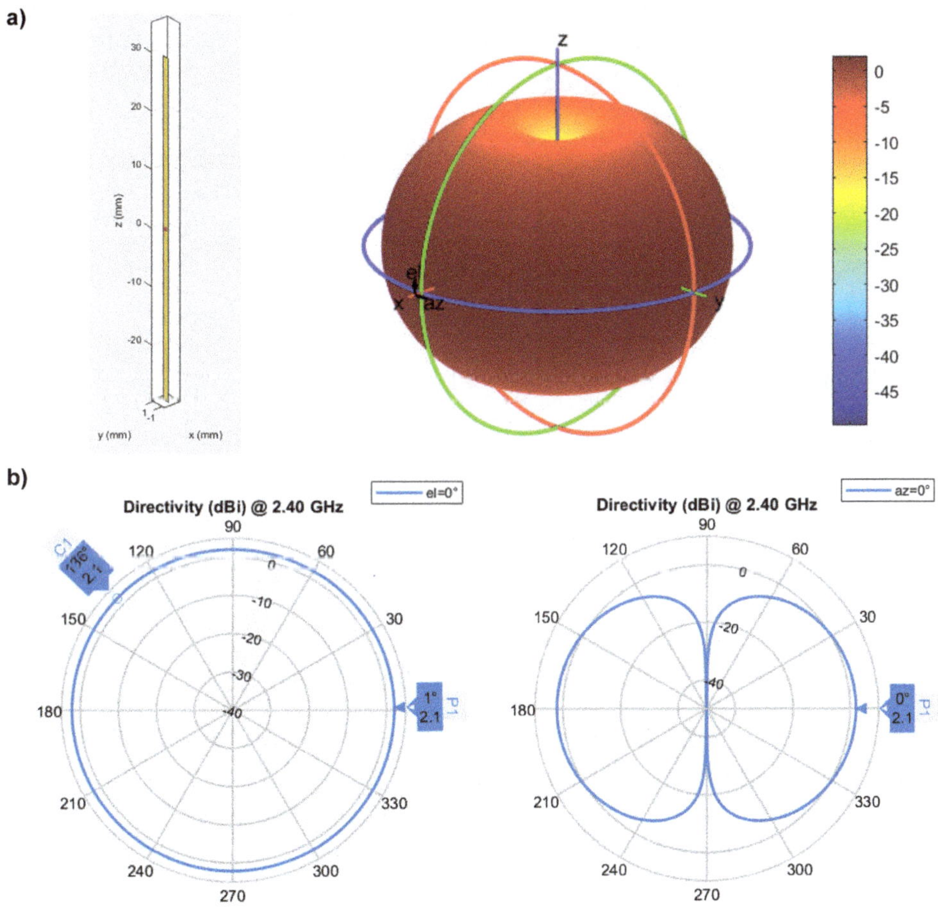

Fig. 11.5. Radiation pattern of a 2.4GHz dipole antenna: a) 3-D view, b) polar views at the horizontal plane and vertical plane with zero azimuth-angle (az = 0ºC)

To assure a maximum power transfer, special attention should be paid to the impedance matching at the input and output ports. Impedance matching is typically characterized by the Return Loss (*RL*), defined as the loss of power in the signal returned/reflected by the impedance discontinuity. In the bibliography, a negative sign is often introduced in its definition, when expressed in decibels, to be coincident with reflection coefficient (Γ) of a two-port network,

$$RL\big|_{dB} = -10log_{10}\frac{P_i}{P_r} = 10log_{10}\frac{P_r}{P_i} = 20log_{10}\Gamma \tag{11.1}$$

where P_i and P_r are the powers of the incident and reflected signals, respectively. Assuming that the impedance is matched at the interface between the antenna and the RF transceiver front-end (output port in the network), the reflection coefficient (Γ) is given by the s_{11} coefficient. One of the objectives in the antenna design is the minimization of this parameter within band of interest, as shown in Figure 11.6 for the dipole case study of Figure 11.5.

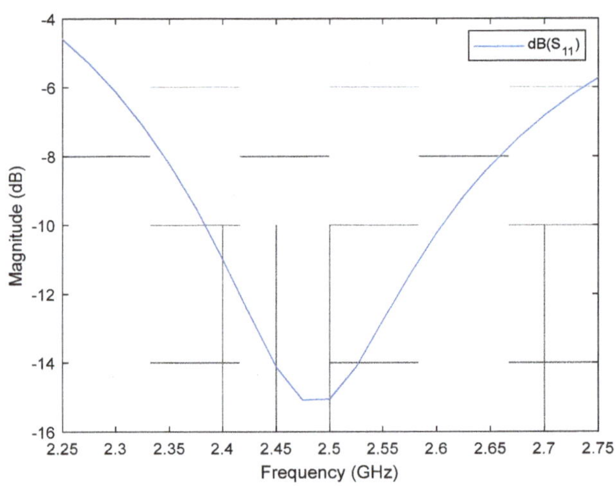

Fig. 11.6. Magnitude of the 2.4GHz dipole antenna case study

In practice, the matching at the output port is not perfectly done, leading to more realistic values of the reflection coefficient (Γ) in the order of -8 to -11dB, which can be theoretically evaluated according to the s-parameters, as well as the characteristic impedance of the antenna (Z_o) and load (Z_L), in the form,

$$\Gamma = s_{11} + \frac{s_{12}s_{21}\Gamma_L}{1 - s_{22}\Gamma_L} \quad \text{with} \quad \Gamma_L = \frac{Z_L - Z_o}{Z_L + Z_o} \tag{11.2}$$

According to the IEEE 802.11 standard, if the port is exposed the output impedance for antenna matching should be 50Ω. If the port is not accessible, there is not *a priori* any restriction, but the 50Ω has also become a standard *de facto* to assure maximum compatibility between vendors and test equipment.

Bandwidth: the frequency selectivity of the antenna is also key parameter in the RF system. For WLAN access-point (routers, wifi repeaters, etc.), independent antennas for the 2.4GHz and 5GHz bands are often considered to maximize performance and provide simultaneous band coverage.

For tight constrains in term of fill factor and cost (as common in portable devices), dual band antennas are usually considered. The design of these antennas is optimized to deal simultaneously with both the 2.4GHz and 5GHz spectrum regions, trading-off compactness and performance. In the case of a broadband design approach, the covering of the two bands of interest is achieved at expense of higher sensitivity to interferences and extra noise due to the wider bandwidth. Moreover, as the WLAN spectral content lays relatively close to the antenna band boundaries, the insertion losses due to the limited frequency selectivity become more relevant, and therefore, the achievable radiation efficiency could be degraded with respect to narrow band solutions. To overcome this drawback, the dual-band design approach can be considered. Dual-band, and more general multi-band antennas [3], have several sections, each of them acting as resonator at different frequencies to improve selectivity.

11.3.2 *Selection of antenna type*

Once we have introduced the main specification of the antenna, let us continue with some guidelines that help selecting the optimum antenna in RF system with emphasis in general indoor applications and portable devices. The information in this section for outdoor point-to-point links and fixed-point sector applications is complemented in [4–7].

Whip antenna: the use of an external whip antenna with a dedicated RF connector is indicated for high-performance applications (with power gain up to 10dBi) in which the fill factor and cost is assumable. These stand-alone solutions often show an omnidirectional radiation pattern with well-controlled bandwidth and impedance.

Single-band antennas are often based on mono-pole or dipoles topologies with one or two arms (the length of each arm being quarter of the target wavelength, i.e. $\lambda/4$), respectively. In the mono-pole case, one of the arms is replaced by a ground plan, making that the symmetry in the vertical plane of the radiation pattern is significantly affected. Although in the horizontal plane both antennas show a similar behavior and performance, dipoles tend to be more common, since due to symmetry they simplify antenna orientation for optimum reception/transmission. Multi-band whip antennas are also available in both rigid and flexible format based on arrays of dipoles [4].

PCB antenna: the direct impression of the antenna on the printed circuit board (PCB) is an interesting alternative to reduce the fill-factor when the extra PCB area occupation is not a concern. As we will show in a practical case study (section 11.4), this solution can be implemented with relative low-cost at the PCB-level, if enough clearance and distance around the antenna are available to minimize coupling and interferences. In this approach, the dielectric is a typically FR4 (Flame Retardant 4) substrate, that is a classic glass reinforced epoxy laminate with a dielectric constant permittivity around 4.3 and 0.017 dissipation factor.

The design and verification of the antenna require a relatively high RF qualification for first integration success, otherwise multiple iterations could be required for fine tuning of the final design. This process is supported by specific CAD tools for electromagnetic applications, as discussed in Chapter 10, such as Ansys HFSS or Microwave Studio CST, based on Finite Element Method (FEM) or Finite Integration Technique (FIT), respectively.

Ceramic antenna: ceramic antennas, also known in the bibliography as chip-antenna or antenna in package (AiP), offer a good balance between performance and cost when the fill-factor constrain is dominant. Hence, it has been widely adopted for highly integrated WLAN applications, for moderate power gain efficiencies in the order of 3dBi.

In the field of WLAN antennas, several commercial solutions are available for single and multi-band applications with compact sizes well below ten millimeters. Although for maximum power efficiency transmission/reception, extra elements are necessary to tune the antenna impedance due to process variability, the total size is not negligible when compared to the PCB approach.

Antenna miniaturization is supported by the development of more advanced substrates which: a) minimize the dielectric loss, and b) enhance the dielectric constant permittivity. As an example, the RT/duroid 6010.2LM laminates from Rogers Corporation ™ are quite interesting ceramic-PTFE composites designed for electronic and microwave circuit applications requiring a high dielectric constant. They provide high dielectric constants (10.2) and low loss (dissipation factor: 0.0023) to enable circuit size reduction in RF applications.

11.4 RF-MIMO Transceiver Example

The transceiver components and subsystems have been implemented in hardware (except the MAC processor, which is a hardware/software co-design). It integrates the following hardware:

- An antenna array made up of 4 multiband micro-strip-fed planar F-shaped monopoles,
- An analog front-end that uses vector modulators as an analog weighting circuitry,
- An OFDM baseband processor that supports the new analog front-end, and
- An enhanced MAC processor for MIMO specific tasks and compatibility to 802.11a.

The antenna array consists of four multiband radiating elements [8], [9] that provide dual band performance (2.4 GHz and 5 GHz bands operations). A laptop cover of 14.1" in size determines the antenna form factor. The multiband antenna is based on a micro-strip-fed planar F-shaped monopole, printed on a FR4 substrate of thickness h=0.5 mm. The overall size of the antenna is 35.9 mm x 11.17 mm. Two slots have been cut in the upper part of the antenna to obtain the desired resonance electric lengths without exceeding in size. In particular, the lower branch of the F-shaped antenna controls the resonance at 5.6 GHz, while the upper part determines the antenna resonance at 2.4 GHz. Both branches are designed to generate a resonant path for currents, which results to be less than a quarter of the free-space wavelength. It is important to observe that there is the need to obtain two $\lambda/4$ resonant paths at the two operating frequencies without exceeding in the size. Therefore, a curl path was designed in the upper part of the antenna, as shown in Figure 11.7. This solution provides a control of the resonance at the lower frequency. Consequently, this antenna can be used for multi-standards such as 802.11a/b/g/n. The antenna array has been designed using four multiband radiating elements. In order to keep low correlation between antennas, each radiator is placed in orthogonal position between them. In this way, polarization diversity can be achieved [10].

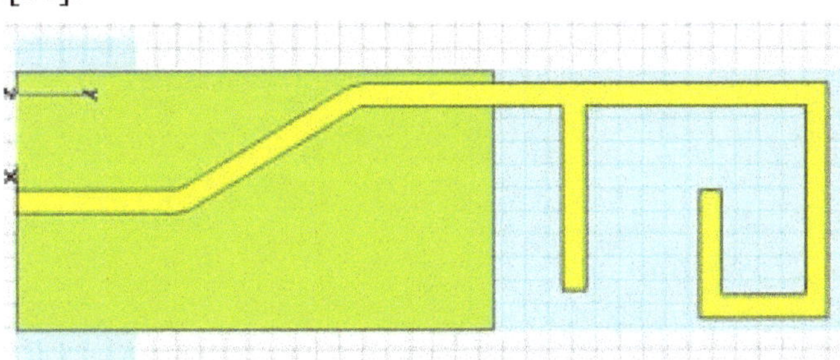

Fig. 11.7. Multiband antenna

11.4.1 *Analog front-end*

The analog front-end is designed to support the transmitted and received signal weighting in RF domain [11-13]. The transceiver is fabricated in the 250 nm SiGe BiCMOS IHP technology [14]. The photos of the transmitter and receiver RF integrated circuit chips are shown in Figure 11.8.

The incoming standard-conform signals from the antennas have a signal power of each -82 to -30 dBm and are filtered by a bandpass filter (BPF) and amplified by a low noise amplifier (LNA) before they are eventually led into the weighting module, namely a vector modulator. This module alters phase and amplitude of the incoming RF signals in a way so that, after adding them up, the summing signal has an improved signal quality leading to higher data rates and/or higher link quality. Because the receiver was intended to be fully backwards compatible to 802.11a and to enable a fair comparison with existing WLAN receivers, it is possible to switch off complete branches and let the front-end operate as a single-input single-output (SISO) device. Therefore, the following down-conversion path has to handle an approximately 12 dB increased effective dynamic range. Thus, a passive down-conversion mixer driven by a phase locked loop (PLL), which runs an 11 GHz voltage-controlled oscillator (VCO), is employed. The last analog stages incorporate variable gain and low pass filtering features to set up optimal signal conditions for the receive analog-to-digital converter (ADC). The design of the transmit path (Tx) makes use of components from the receive path (Rx) such as parts from the baseband variable gain amplifier (VGA) and filter. Furthermore, an identical weighting module is used, so that in total two vector modulators are employed per antenna branch, one for Tx and one for Rx. The transmit power control required by the authorities is realized by a variable buffer amplifier before the power amplifier (PA). The active up-conversion mixer driven by the PLL feeds its output signal to the individual Tx branches by means of splitting amplifiers. The PA and antenna switch are off-chip components and form the last part of the Tx branch.

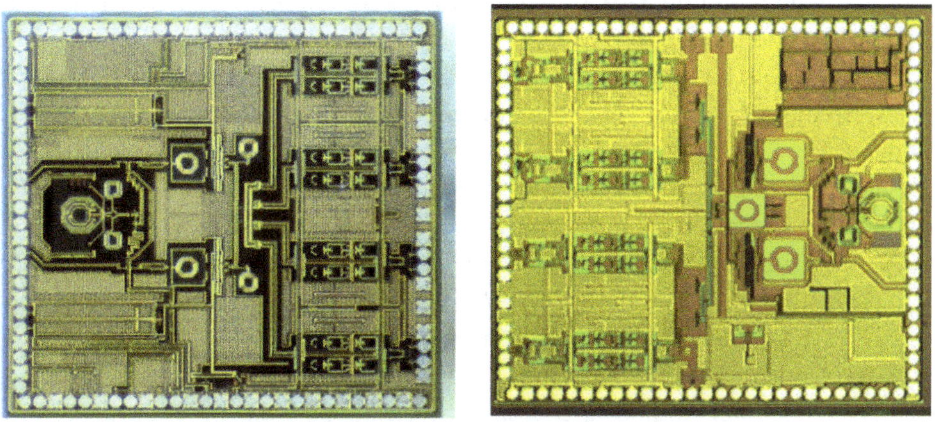

Fig. 11.8. Analog front-end transmitter (left) and receiver (right) chips

11.4.2 *Baseband processor*

The structure of the baseband processor is based on two main blocks: the compatible 802.11a parts that implement the IEEE 802.11a standard and the specific modules for MIMO channel estimation and beamformer selection [15]. The conventional 802.11a processor includes the standard functionalities: mapper, channel encoder, pilot insertion, guard interval, FFT/IFFT blocks, Viterbi decoder, etc. The new functionalities are grouped into five main blocks: channel estimator, beamforming, weight delivering, RF-MIMO control and RF control unit. Most of these modules are active only when a MIMO communication is detected by the Tx/Rx control block.

The channel estimation block estimates the n_Txn_R MIMO channel based on the $n_T n_R$ OFDM training symbols of the received training frame, where n_T and n_R stands for the number of antennas at transmitter and receiver side, respectively. It works in the frequency domain taking the FFT signal provided by the 802.11a processor as inputs and uses a least squares method. The Tx/Rx beamforming uses the estimated MIMO channel as input, computes the optimal Tx-Rx weights, and transfers them to MAC. Channel estimation and beamformer blocks are started only when a training frame is received, keeping the overall baseband processor energy consumption very close to the one of the conventional 802.11a processor.
The weight delivery block allows applying (from the baseband processor) the predefined set of weights during training and transferring the optimal or default weights during data transmission or reception. The control block controls the signal and data flow among all MIMO blocks. It receives from the Tx/Rx control block the information included in the training frame signal field (e.g., number of

Tx/Rx antennas, number of training symbols), as well as some activation and synchronization signals.

Finally, the RF control unit controls the specific MIMO functionalities of the analog front-end through an SPI and compensates the impairments of the vector modulator.

The baseband printed-circuit board shown in Figure 11.9 provides the hardware platform for the baseband processor and includes the following components:

- Xilinx Virtex-5 LX330 FPGA for implementation of digital components of the baseband processor including the RF control unit and all the interfaces.
- AD 9218 dual analog-to-digital converter, 80 Msamples/s, 2 x 10-bit, sampling Rx I/Q signal.
- AD 9763 dual digital-to-analog converter, 80 Msamples/s, 2 x 10-bit, generating Tx I/Q signal.
- AD 9200 analog-to-digital converter, 20 Msamples/s, 10-bit, sampling the RSSI signal.
- AD 7302 dual digital-to-analog converter, 20 Msamples/s, 2 x 8-bit, for future extensions.
- MAC – PHY interface (16-bit parallel interface to MAC board),
- BB – AFE interface (analog I/Q signals for Tx and Rx, RSSI input, two SPI interfaces, etc.),
- Debug outputs (3 x 16-bit) that can be connected to a logic analyzer for debugging purposes,
- VGA monitor output that can be connected to the standard computer monitor for debugging purposes,
- Components for future extension (user flash, RS232 interface, temperature sensor, etc.), and
- Power supply, clock generator, and flash storing the FPGA configuration.

Fig. 11.9. Baseband board

11.4.3 *MAC processor*

The MAC hardware platform [16] is based on two major digital chips located on a printed-circuit board in CardBus form factor (Figure 11.10):

- The MAC processor is an ASIC fabricated in IHP's 250 nm CMOS technology [14]. It provides a MIPS 4KEp 32-bit microprocessor core and the CardBus (or PCMCIA) interface to a host PC.
- A Xilinx Virtex-5 LX50 FPGA implements the MAC hardware accelerator and the MAC–PHY interface (MIPP).

The board contains many other components such as 4 Mbyte of processor flash memory and 1 MByte of RAM, clock generator, power supply for FPGA, configuration flash for FPGA, debugging and programming interfaces (UART / RS232 for MAC firmware download, logic analyzer inter-face), status LEDs, etc. The MAC protocol is fully compatible with the IEEE 802.11a standard (5 GHz band). Although there are some special MIMO features such as training frames and request-to-send (RTS) frames being transmitted before multicast frames, these are compatible with the standard and do not disturb normal WLAN traffic. The required MAC features for coexistence with the IEEE 802.11n standard, as well as the features of the IEEE 802.11h standard such as the dynamic frequency selection (DFS) and the transmit power control (TPC) are included in the required extent.

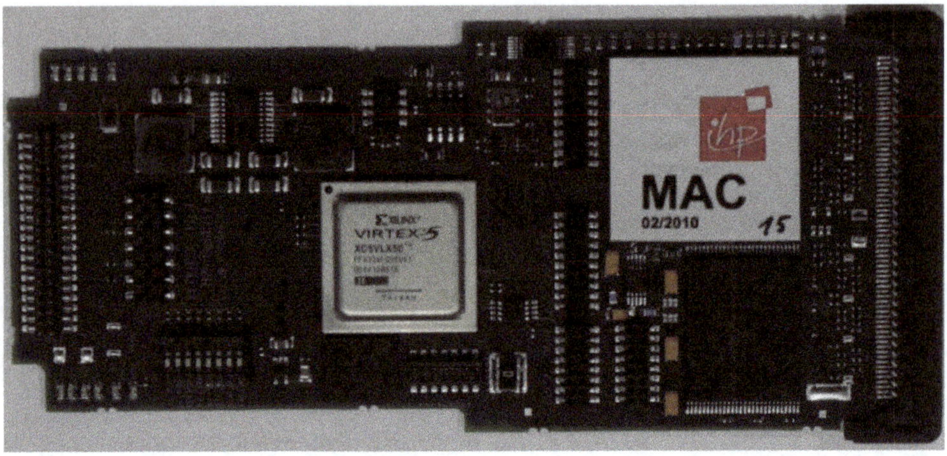

Fig. 11.10. MAC board in CardBus form factor

11.4.4 *System integration and test*

The implemented components and subsystems have been integrated in a system prototype [17]. The subsystems and the system as a whole have been simulated and tested. The system performance parameters have been measured too.

First, the performance of the analogue front-end has been evaluated in standalone operation. This allows measuring major performance characteristics without the influence of other subsystems. Second, a setup composed of the antenna array, the multi-IC analogue front-end board, the baseband board and the MAC emulator has been tested. Using this setup, the authors have evaluated the critical operations involving the baseband and the analogue front-end, namely the weight setting during training and the weight uploading during data transmission. The correct operation of the vector modulators and combiner has been verified. Moreover, several indoor and outdoor experiments to evaluate the physical layer performance of the RF-MIMO with respect to single-antenna terminals have been conducted.

On the other hand, a setup assembling the baseband processor, the MAC processor and the WLAN driver software has been used to debug and verify the interactions between mentioned subsystems under realistic conditions. Finally, the stability and robustness of the complete RF-MIMO WLAN transceiver prototype has been demonstrated.

Analog front-end

A control FPGA and USB interface on the analog front-end board allows the standalone verification of the design. Using a control program stored in FPGA, the major performance characteristics have been measured without the influence of other subsystems (baseband and MAC).

RF receiver characteristics have been observed for the signal path from the antenna input connector to the input of the commercial RFIC including an adapter cable, which is required for accessing the RF signal at this point. The measured characteristics are shown in Table 11.2.

Table 11.2. RF receiver characteristics

Parameter	Value
Gain	15.5 dB
Noise figure	7.4 dB
Input referred 1 dB compression	-34 dBm
Input referred 3rd order intersect point	-23.8 dBm
Power consumption	180 mW

RF transmitter characteristics for the signal path from the output of the MAX2828 to the antenna connector (the power amplifier, antenna selector switch and Rx/Tx switch) are shown in Table 11.3. Again, an adapter cable is incorporated for feeding signals to the power amplifier input.

Table 11.3. RF transmitter characteristics

Parameter	Value
Gain	26 dB
Input reflection	<-10 dB
Max. antenna branch leakage	-8 dB

Analog front-end and digital baseband

The setup depicted in Figure 11.11 has been used to test the correct operation of the vector modulators under the baseband processor control. Moreover, extensive indoor and outdoor experiments have been conducted in order to perform a low level characterization of the system, using as figures of merit the frame error rate (FER), the RSSI level after combining (power of the received signal after combining), and the SNR gain. These values can easily be translated into coverage improvement or increased rate in comparison to SISO system.

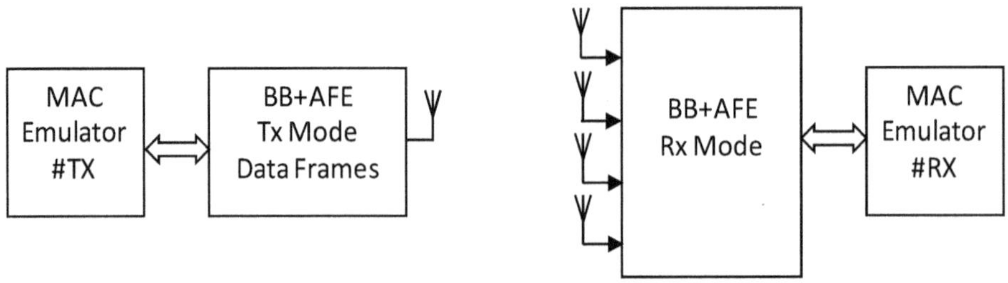

Fig. 11.11. Physical layer test setup

The experiments were performed using one transmitting antenna, four receiving antennas, central frequency of 5.26 GHz, and several frame formats (10 or 20 OFDM symbols length and 9, 18, 36, or 54 Mbps). Several distances between transmitter and receiver have been considered ranging from 8 to 20 meters in NLOS indoor scenarios and from 15 to 50 meters in LOS outdoor scenarios.

The procedure consisted of a SISO evaluation stage in which 100 data frames are transmitted and received using default weights, which means that the same weights were applied the 4 receiver antennas, showing the behavior of an equivalent SISO system. In the RF-MIMO evaluation stage, a training frame is transmitted and the receiving baseband processor computed the optimal weights. Then, the optimal weights are applied and 100 data frames are received. This procedure was repeated 25 times at each location and for each frame format by moving the receiver within a small local area (i.e. 900 cm^2).

As an example, Table 11.4 shows the comparison between SISO and RF-MIMO when transmitter and receiver were placed 12 meters apart in an indoor scenario. Using 64-QAM, the RF-MIMO FER is close to 24%, and for SISO it goes up to 86%. From a practical point of view, this means that in SISO we are limited to 18 Mbps, while RF-MIMO allow us to double the rate and transmit at 36 Mbps with similar FER. In terms of power, the gain of RF-MIMO versus SISO is 7 dB: this roughly means that the RF-MIMO increases the coverage by a factor of two.

Table 11.4. Comparison of SISO and RF-MIMO systems

Modulation	System	FER (%)	RSSI (dBm)
BPSK 9 Mbps	SISO	1.3	-58.7
	RF-MIMO	0.0	-52.3
QPSK 18 Mbps	SISO	11.0	-59.0
	RF-MIMO	0.1	-51.3
16-QAM 36 Mbps	SISO	41.8	-58.5
	RF-MIMO	2.3	-51.2
64-QAM 54 Mbps	SISO	85.8	-59.1
	RF-MIMO	23.9	-52.2

Digital baseband and MAC

A test setup that connects two stations with a cable instead of the analogue front-ends has been used to verify the operation and performance of MAC and digital baseband. Each station consists of the following subsystems: a laptop computer running Linux and the WLAN driver software, the MAC board plugged in to the CardBus slot of the laptop and the baseband board connected to MAC via the MIPP cable (Figure 11.12). This way a system assembling the real MAC layer and digital baseband has been tested in the conditions of an ideal radio channel. MIMO effects cannot be investigated with this setup.

Fig. 11.12. Photo of a system assembling the MAC and baseband boards

The monitor right in the background of Figure 11.12 shows the constellation diagram. The video is transferred from the laptop in the back using the 16-QAM signal modulation. The laptop screen in the left displays the received video. One can also recognise the windows of the traffic monitor program (with the yellow and orange bars, which visualise the optimal weight settings) and the terminal programme which controls connection setup and other WLAN parameters.

The primary goal of this test setup is to improve the stability and robustness of the MAC and baseband processors, as well as the WLAN driver software in real-time conditions. Moreover, the signal quality (constellation diagram) and data throughput can be measured for the ideal radio link. The MAC data throughput has been estimated by measuring the time required to copy a large file. The measurement is done at the Linux driver. Thus, it includes the MAC protocol overhead due to frame preambles, acknowledgements and RTS/CTS, the WLAN driver overhead, and other limiting effects like MAC firmware performance limitations. Therefore, the measured throughput is expected to be smaller than the nominal physical layer data rate. The measurement results in Mbps and in % of the nominal PHY data rate are shown in Table 11.5. Since the overhead (e.g. frame preambles, duration of an ACK frame) does not scale as the physical data rate increases, the relative throughput degrades very fast with increasing the data rate, which is normal in WLAN.

Table 11.5. Data throughput of driver, MAC, and baseband

Modulation	BPSK	QPSK	16-QAM	64-QAM
Nominal data rate	6	12	24	48
Throughput without RTS/CTS	4.4 73%	7.2 60%	10.6 44%	13.0 27%
Throughput with RTS/CTS	4.1 68%	6.6 55%	9.2 38%	11.5 24%

The longer the frame is, the smaller is the overhead, and, consequently, the higher is the performance. Table 11.5 also compares the throughputs of the standard IEEE 802.11 mode and the MIMO mode with RTS and CTS. These short frames impose an additional overhead. However, at least in a multiuser scenario, they are required for the receiver to identify the transmitting station and to set the optimal weights for that link before receiving a long frame. The loss can be over-compensated by the higher physical data rate selected when setting the optimal weights.

Complete system

A test setup of the complete system includes the antenna array, analogue front-end board, baseband board, MAC board and WLAN driver. The purpose is to test the system in real-time conditions and to prove its better performance (as well as, the compatibility) in comparison with the standard 802.11a WLAN.

The baseband and analogue front-end boards have been mounted into the housing, which was developed to make the demonstrator laptop-compact and small. A photo of the system test setup is shown in Figure 11.13. The monitor displays a good example of the QPSK constellation diagram. Next to the antenna, the front-end and baseband boards mounted into the black housing are visible. The MAC board is in the foreground right. The lower bar with many stripes shows the transmitted frames and the upper bar with short stripes shows the received frames. Single lines in between indicate the training frames. The bars on top visualize the weight settings (I and Q components) for three antennas.

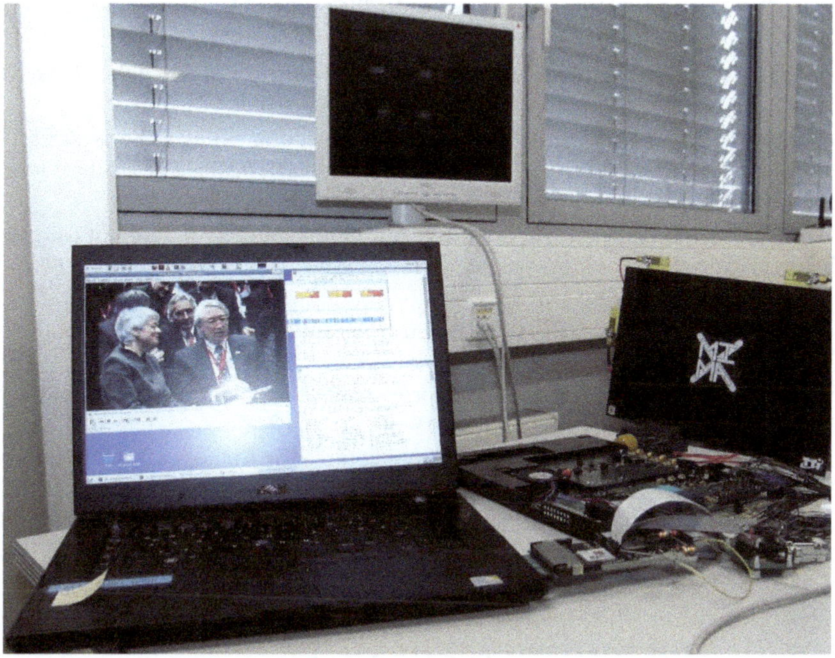

Figure 11.13. Photo of the complete system test setup

This setup has been used in two configurations: one made of two RF-MIMO stations to evaluate a pure RF-MIMO WLAN transceiver with the specific features (in particular, handling of training frames), and the other made of a RF-MIMO station and a commercial 5 GHz WLAN access point to verify the compatibility with the standard 802.11a WLAN. In second configuration, the MIMO features (RTS/CTS and training frames) are disabled since the access point cannot correctly handle them.

References

1. Zhang, H. *et al.* (eds.) (2020). Series on Advanced Electronic Packaging Technology and Key Materials, in *Modeling, Analysis, Design, and Tests for Electronics Packaging beyond Moore*, Woodhead Publishing, 2020, p. ii.
2. Lamy, Y. P. R. *et al.* (2010). RF Characterization and Analytical Modelling of Through Silicon Vias and Coplanar Waveguides for 3D Integration, IEEE Transactions on Advanced Packaging, 33(4), pp. 1072–1079.
3. Nella, A. and Gandhi, A. S. (2017). A survey on microstrip antennas for portable wireless communication system applications, Proc. *2017 International Conference on Advances in Computing, Communications and Informatics (ICACCI)*, pp. 2156–2165.

4. Chen, Z. N. *et al.* (2012). Antennas for WiFi Connectivity, Proceedings of the IEEE, 100(7), pp. 2322–2329.
5. Balanis, C. A. (2016). *Antenna Theory: Analysis and Design*. John Wiley & Sons, 2016.
6. Milligan, T. A. (2005). *Modern Antenna Design*. John Wiley & Sons, 2005.
7. Stutzman, W. L. and Thiele, G. A. (2012). *Antenna Theory and Design*. John Wiley & Sons, 2012.
8. Gago, O. *et al.* (2009). Study of the optimum number of radiating elements for different portable devices operating in a MIMO system, Proc. *2009 European Wireless Conference*, pp. 78–83.
9. Gago, O. *et al.* (2009). MIFA: Modified IFA Radiating Element for Small Handheld Devices, Proc. *ICT-MobileSummit 2009 Conference Proceedings*, p. 7.
10. Best, S. R. (1998). Antenna Properties and their impact on Wireless System Performance. Cushcraft Corporation, 1998.
11. Gholam, F. *et al.* (2009). Simplified Architectures for Analogue Antenna Combining, Proc. *CT-MobileSummit 2009 Conference Proceedings*, p. 9.
12. Joram, N. *et al.* (2009). Fully integrated active CMOS vector modulator for 802.11a compliant diversity transceivers, Proc. *2009 IEEE International Conference on Microwaves, Communications, Antennas and Electronics Systems*, pp. 1–4.
13. Eickhoff, R. *et al.* (2011). Physical layer amendments for MIMO features in 802.11a, Proc. *2011 Future Network Mobile Summit*, pp. 1–8.
14. https://www.ihp-microelectronics.com/en/services/mpw-prototyping/sigec-bicmos-technologies.html
15. Elvira, V. *et al.* (2010). Baseband processor for RF-MIMO WLAN, Proc. *2010 17th IEEE International Conference on Electronics, Circuits and Systems*, pp. 798–801.
16. Tittelbach-Helmrich, K. *et al.* (2010). MAC hardware platform for RF-MIMO WLAN, Proc. *2010 53rd IEEE International Midwest Symposium on Circuits and Systems*, pp. 339–342.
17. Stamenkovic, Z., Tittelbach-Helmrich, K., Wickert, M., Ibanez, J., Ruiz, S. and Dimosthenous, G. (2011). Implementation, integration, and verification of MIMAX WLAN modem, *Proc. 6th Internat. Workshop on Reconfigurable Communication-Centric Systems-on-Chip*, pp. 1–8

Part 4

Verification and Testing

https://doi.org/10.1142/9789811210723_0012

Chapter 12

Verification

Ernesto Sanchez
Politecnico di Torino
Torino, Italy

Introduction

This chapter introduces the main concepts about circuit design verification. In the very first part of the chapter, an introductory part about why it is necessary to verify a digital circuit design is provided. Then, a general verification process is introduced, and some of the most important verification techniques are reported. The chapter also describes the testbench as a verification mechanism, and finally, it reports the concept of coverage metrics.

Digital systems are created by following a series of steps that comprise several intermediate design phases. At the very beginning, the set of specifications is usually defined in cooperation with the customer and the design team; then, at every design phase, an always-increasing detailed model is delivered. Clearly, the lower the abstraction level, the higher the complexity of the resulting model. At the very end of the process, an integrated circuit layout file (usually in GDSII format) is sent to the foundry in order to produce the device. In the following figure, some of the most commonly used design abstraction levels are depicted. It is important to note that in general higher or lower abstraction levels may also exist, however, here only the most relevant ones are described [1], [2].

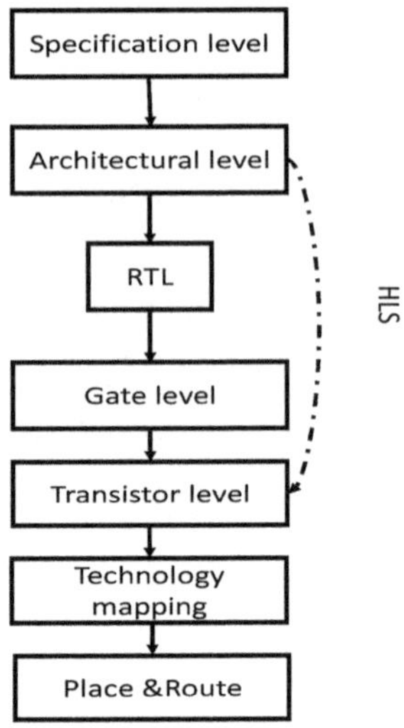

Fig. 12.1. The most common abstraction levels in a digital circuit design cycle.

- *Specification level*
  This is the first step in a design cycle, and the abstraction level is the higher. In this design phase, a natural language description coming from a set of design requirements is created. In most of the cases, the model or device description delivered in this design step is very ambiguous and difficult to verify due to the lack of a formal representation. A possible solution for this lack of standardization is the so called Property Specification Language (PSL), a language that provides the design architect a standard way to specify the design properties using a concise syntax.

- *Architectural level*
  This abstraction level is also called *Algorithmic level*, *Electronic System Level* or even *C-level*. Here, the main details on how to put together the different modules composing the final system are detailed. In this fashion, the model delivered at this level is basically composed of a set of blocks describing the main components of the design. This model is exploited to evaluate the basic operations between of the design and the main interactions among the different components of the system. This is the first abstraction level where a complete simulatable model is built. At the very beginning, at this abstraction level no timing information was included into the delivered model; however, during the last years, loosely timed and approximately timed models have been supported by description languages such as System C and System Verilog [3].

- *Register transfer level (RTL)*
  The most important characteristic of the models delivered at this level is that they include accurate timing information; for the very first time, the models are cycle accurate including all functional details of the design. At this level clocked storage elements become visible, and are mainly synthesized as registers. Clearly, not all the timing details are included here since these models do not include timing information related to the propagation delays. The most common languages used here are *VHDL* and *Verilog*.

- *Gate level*
  At this level, the design is described using logic gates, and every logic gate is again an abstraction of a set of transistors. At this level, the interconnections between all the gates composing the design are detailed. The system complexity is one of the major issues at this level since it is very difficult to simulate large designs due to the high amount of information that the models contain. For example, a 16X16 multiplier that is described in no more than 20 lines using VHDL, may count about 2.5k logic gates. This level is still significantly abstract because there is no information about actual transistors. In most of the cases, at this level the final design is described using the *Electronic Design Interchange Format* (EDIF), Verilog, and other languages.

- *Transistor level*
  The design at this level is described in terms of transistors and their interconnecting wires. Simulation at this level is very costly, for example, supposing that a logic gate may contain in average six transistors, the 16X16 multiplier mentioned before would contain about 15k transistors to analyze. As a matter of facts, designers can usually simulate only some logic gates in order to characterize them, but it is often impractical to simulate whole designs at this level. SPICE is used to describe and simulate the design behavior detailed at the transistor level.

It is important to notice that bringing a design from a very high abstraction level (*Architectural level*) to the very low one (*Transistor level*) is possible by using a different strategy. Actually, it is possible to take advantage of the design process called *High-level Synthesis* (HLS) that produces a synthesized version of the design starting from a very high-level abstraction model. Once the device transistor level has been consolidated, it is possible to proceed to technology mapping and place and route phases. At the end of these phases, the description is used during the fabrication of the device.

Design models could be created resorting to very different approaches, languages and strategies that include manual and automated ones. However, regardless the design approach, any new delivered model is usually prone to *errors*. These errors can be simply defined as logic differences with respect to the actual specifications. In the most of the cases, the designer due to misunderstanding on the specifications, ambiguous specifications, or typographical errors during the coding phase may introduce logic errors.

Identifying and removing logic errors in a design is not a trivial task; in fact, nowadays, the development resources devoted to these tasks amount about 50%-60% of the total cost in the design process. Actually, a series of validation or verification processes are required intending to guarantee that the design model meets the expected specifications. It is clear also that, the earlier a logic error is detected the lower is the cost for correcting it. Different authors and companies have stated that finding bugs before production may cost 100 times less than discovering them when the device is in use by the customers [4]. A famous bug in the FDIV unit in the Pentium® processor was not discovered during the early stages of design generating economic losses of about $500 million associated with the replacement of the involved processors [5].

The terms *validation* and *verification* are usually used in a very ambiguous way. The *Institute of Electrical and Electronics Engineers* in the *IEEE* standard 1012-1998 defines *validation* as "the confirmation by examination and provisions of objective evidence that the particular requirements for a specific intended use are fulfilled"; and *verification* as "the confirmation by examination and provisions of objective evidence that specified requirements have been fulfilled" [6]. In the field of design, this definition is usually synthesized saying that "design validation is concerned with building the right product, while design verification is concerned with building the product right".

In this chapter, we will call *verification* the process that aims at guaranteeing the correct translation of a model delivered at a certain abstraction level to its previous model by discovering errors in the delivered model. The reader should be aware that, despite their exact definitions, in the technical literature the terms *verification* and *validation* are sometimes used interchangeably.

In the following, the chapter describes the full verification process. Then, the verification plan, the verification environment, the device bring-up and device regression are also introduced.

Verification Process

The goal of a hardware design team is to create a device that performs a particular task. Instead, the goal of the verification team is to guarantee that the design accomplished that task successfully, meaning that the design is an accurate representation of the device specifications. The verification process should be run in parallel to the design one, and it usually requires the verification engineers to read the design specifications, create the verification plan and follow it building verification-oriented stimuli sets showing that the model under verification follows the design specifications. Different authors, for example [7], [8] and [9], define the verification process by executing different steps, here, the process is split in four phases as described by Piziali in [8]. The first step consist on writing a *verification plan*, then the *verification environment* should be implemented; at this point, the *device bring-up* is performed, finalizing with the *device regression.*

12.1.1 *Verification Plan*

The verification plan is an executable natural language document that defines the scope of the verification problem and its solution, i.e., it defines what must be verified and how it will be done. The verification problem is quantified using a set of coverage metrics, whereas the solution is composed of a set of specifications to be implemented on the verification environment.

First of all, a *device specification analysis* should be performed in order to extract the most important features of the design. This process may exploit not only the functional specification of the design, but also less formal documents, as well as marketing information and draft notes coming from the engineer team. The information abstraction is done using *top-down* or *bottom-up* strategies.

As an example, the next figure shows the block diagram of a processor based SoC. During the device specification analysis, it is necessary to establish all the possible interactions between the processor core and the other modules included in the system.

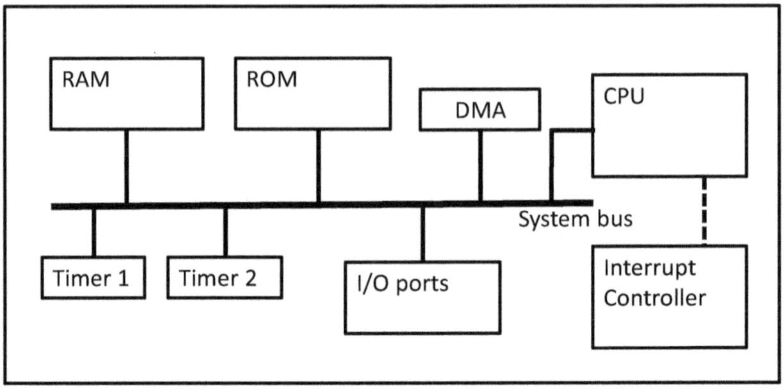

Fig. 12.2. A processor based SoC

In the case of a complex device, the most important features that are required to be extracted after the device specification analysis are:

(1) *Identifying submodules within your device*
(2) *Determine submodules stimulus and response methodology*
(3) *Determine stimulus traffic generation requirements*
(4) *Determine submodules response checking methodology*
(5) *Device interconnection analysis*
(6) *Power consumption requirements*
(7) *Assertion verification*
(8) *Functional coverage*
(9) *Co-verification possibilities*
(10) *Hardware acceleration*
(11) *Virtual platforms*
(12) *Analog/Mixed devices*

The first six elements in the previous list describe the main features to be derived from the device. Points 7-9 gather the metrics that should be used during the verification process. Finally, 10-12 represent some additional possibilities to be exploited to accelerate the development times.

Resorting to Figure 12.1, it is possible to identify the different modules that compose the SoC. Among of them, the Direct Memory Access (DMA) Controller, or DMAC, is designed to allow large blocks of data being transferred between memory and peripherals (or between two memories) without the intervention of the microprocessor. The features supported by DMAs include, among others: different data transfer modes, management of several peripheral channels, programmable arbitration mechanisms to manage multiple data transfers. During the creation of the verification plan, it is necessary to clearly determine how to configure and activate the DMA. For example, it should be important to execute Memory-to-Memory transfers as well as Peripherals-to-Memory. In addition, the expected responses should be established, and the bus behavior controlled in the case of data corruption and collisions. For these purposes, it may be necessary to add a couple of verification monitors that control the bus behavior during the data transfers and the memory content at the end of the process.

The second step to obtain the verification plan consists on the creation of the coverage model starting from the information obtained previously. The idea is to be able to quantify the metrics that must be measured at the end of the process. Then, for any feature extracted previously, it is necessary to define a functional coverage matrix that tracks the results of the verification process. For example, considering again the SoC described in Figure 12.1, for the Memory-to-Memory transfers a coverage matrix similar to the presented in Table 12.2 can be created in order to define the different goals the coverage process must comply with.

Table 12.1. A part of the Coverage Matrix for the verification of a DMA controller.

Memory-to-Memory transfer								
DMA Channel	Transmission mode	Source address	Destination address	Number of elements	Data size [bytes]	Values	Priority	Monitor
	Burst	0x4000	0x8000	256	4	All 0s	Fixed	Mem
Ch. 0	Cycle steal	0x8000	0x4000	256	4	All 1s	Fixed	Mem
	Transparent	0xC000	0x4000	512	8	All 0s	Fixed	Mem
	Burst	0xA000	0x4000	1024	1	All 1s	Round robin	Bus+Mem
...	...	...	...	...	...		...	...

Once the coverage matrixes for all the features of the device under verification have been created, the design engineer usually includes also in the verification plan the expected results in terms of coverage achieved with respect to the high-level code coverage metrics. In general, the code coverage metrics measure how thoroughly a design model has been exercised by a specific set of stimuli; usually, these measurements are expressed as a percentage value [10]. The coverage measures can include apart from the functional and code coverage metrics also assertion coverage.

12.1.1.1 *Stimuli generation*

At this point, it is very important to define how the input set should be created in order to comply with the verification plan just created. It is possible to define *Stimulus* as the input to the design under verification (DUV). A set of stimuli might also include, the device configuration, test patterns, transactions, instructions, assembly programs, exceptions, data packets, etc. The kind of stimulus and its characteristics depend on the circuit under analysis and on the level of abstraction the analysis is performed on.

Depending on the control scheme exploited to generate stimuli, generation methodologies can be defined as either *open loop* or *closed-loop* depending on whether the method uses or not feedback information. While both methodologies directly depend on the circuit description regardless of the abstraction level, open loop methodologies generate stimuli resorting to the previously acquired test experience in terms of design knowledge as well as data gathered in former verification campaigns. The main difference between these methods is the use or not of feedback information during the stimuli generation process.

Considering the open loop mechanisms, the first step the abstraction of valid information about the circuit. This representation of the circuit could be based on very simple *constraints* about the possible values to be assumed by some of the device inputs, or even complex graphs representing all possible interconnections among memory elements belonging the design. Once the circuit representation is acquired, the set of stimuli is generated, resorting for example to some automatic algorithms or a manual strategy.

Closed-loop methodologies, on the other hand, include in the generation processes an *evaluator* or checker module that is able to elaborate the stimulus candidates and return a feedback value representing the stimulus quality. The stimuli generator creates a stimulus candidate that is evaluated, and then it is added or not to the final set of stimuli depending on whether the new stimulus satisfies test quality conditions initially defined in the generation process setup.

The feedback information is usually exploited in one out of three possible aspects of the generation:

- *Threshold*

The feedback information is used by the stimuli generator to determine when the set of stimuli is complete; otherwise, the stimuli generator will continue producing stimuli information.

- *Feedback sifted generation*

The feedback information is exploited to accept or discard new stimuli. For example, only new stimuli able to increase the coverage in a set of selected metrics will be added to the final set.

- *Feedback-driven generation*

The feedback information is used by the stimuli generator to indicate stimuli candidates that better comply with the quality conditions required in the process, and then to optimize them. Evolutionary algorithms usually make use of feedback-driven generation to produce sets of stimuli.

Stimuli generation could be performed following mainly two different methods: manual and automatic. Manual generation is performed by human beings, whereas automatic methods may be based on different approaches, such as random techniques, deterministic or heuristic algorithms.

a) Manual generation

Manual stimuli generation may be performed by verification, test, or even design engineers. Sets of stimuli are manually generated by following the verification strategy defined in the verification plan. Firstly, the verification strategy tackles the main features defined in the functional coverage matrices, and then, both special and corner cases are addressed for verification.

The verification team employs expert engineers that have acquired a significant *know-how* while they were involved for example in circuit design. A verification engineer is able to manually write a set of stimuli tackling special and corner cases, which are really difficult to reach using other generation techniques.

To efficiently devise stimuli generation by hand, a deep knowledge about the device under evaluation is usually required. Let us suppose that a set of stimuli must be manually generated to verify the *handshaking* properties of a device originally devised to perform a communication protocol. It is clear that the verification engineer must be familiar with the handshaking properties of the communication protocol; moreover, some knowledge about the device implementation details might be also useful to competently generate the set of stimuli and more successfully look for possible design errors. In fact, sets of stimuli are really time consuming and expensive to generate by hand.

The main disadvantage of manual generation techniques is that the resulting set of stimuli strongly relies on the engineer's knowledge and might be easily biased by the verification engineer background. Additionally, its cost can be very high, since it requires highly skilled engineers in order to generate effective sets of stimuli.

b) Automatic generation

Methodologies to automatically generate sets of stimuli usually rely on software tools. Some of them resort to an internal representation of the device under verification, generating the sets of stimuli in two phases: first, a special representation of the design is extracted from the available model, in the case that this representation does not exist, yet. This internal representation could be either that produced by the designer, or an *ad hoc* one (possibly extracted from the latter), and based on graphs, state machines, or other representations. Second, the actual set of stimuli is generated according to the previously acquired information. On the contrary, some other automatic generation mechanisms directly use the design under verification to produce the sets of stimuli.

An automatic closed-loop method to generate set of stimuli for digital circuit verification should be characterized by:

- high flexibility regarding the targeted circuit, in order to allow the maximum applicability of the method;
- syntactically correct generation of input information depending on the specific singularities of the targeted design, for example special constraints required in some input variables;
- high versatility with respect to the evaluation system in order to allow tackling different verification problems at different abstraction levels;
- ability to drive the generation process exploiting a feedback measure such as a coverage metric.

Automatic generation of stimuli based on closed-loop mechanisms is really efficient; however, the main drawback regarding these methods is the computational effort involved to produce a good set of stimuli, because it is necessary to evaluate high quantities of candidates to complete the final set. Additionally, automatic generation requires an environment, which in some cases is really difficult and expensive to set up.

c) Deterministic generation techniques

Deterministic approaches generate the stimuli exploiting algorithms that use exact decision mechanisms, and mainly utilize the information extracted from the design under verification during the generation process.

Algorithms that automatically generate test vectors, aiming at testing through simulation, have been called ATPG (*Automatic Test Pattern Generation*). The term ATPG was first introduced for manufacturing testing tools; however, currently this term is also accepted in design validation and verification. Some

ATPG techniques can also be successfully exploited to generate set of stimuli not only for testing, but for verification, too. It is worth noting that most of the ATPG-like algorithms exploited to generate automatically sets of stimuli maximize the coverage of the single stuck-at faults of the circuit.

d) Random generation techniques

Random strategies are some of the most popular methods to generate sets of stimuli because they are simple and require low human intervention; these strategies explore randomly the stimuli search space, looking for acceptable information to complete the set of stimuli; in synthesis, random approaches introduce probabilistic decision mechanisms in the algorithm of generation.

Guaranteeing a thorough verification of the circuit is necessary to generate huge sets of inputs that could require massive computational resources in terms of memory and time. It is important to note that the most of the effort required in this strategy is consumed by the evaluation mechanism exploited to assert the stimuli quality. In fact, even statistical analyses have been proposed to decrease as much as possible the generation of redundant stimuli [11].

Habitually, random strategies are equipped with initialization mechanisms that allow repeatability in the stimuli generation campaigns, for instance most of the random generation tools are provided with the possibility of selecting the initial seed of the random function to produce repeatable results. Even though this fact inhibits the actual randomness properties of a stimuli generator, herein the term *random* is used anyhow, even though a more correct term would be pseudo-random.

Random strategies must be carefully configured before using because these techniques could generate big quantities of information incapable to improve the set of stimuli. In fact, some designs require specific sequences of inputs to be thoroughly verified; the kind of circuits that are difficult to test or verify using random strategies have been called *random resistant* circuits, and in these situations random based generation strategies could fail in reaching the target quality.

A well known random resistant circuit is a pipelined processor; indeed, verifying the correct design of pipelined microprocessors requires sequences of suitable instructions, for example the forwarding and interlock mechanisms in a pipelined processor core are very difficult to verify using random strategies.

In a verification process based on random generation, it is very important to identify stimuli that can be created without any constraints from others that require constraints to ensure that valid stimuli are being created. Other stimuli require additional or modified constrains to hit certain special cases.

The ability to create well defined patterns between different input signals is directly related to the ability to express the constraints able to cause the pattern to be generated.

Let us suppose that it is necessary to generate data communication frames to verify a serial communication device using a random stimuli generator; it is possible that this process becomes an inefficient task if special constraints correlating the *parity bit* with the *data word* at the internal of the data communication frames are not clearly defined.

The main advantage of random strategies is the low human intervention required to perform the generation experiments. Additionally, it is interesting to highlight the fact that generation is not biased by the human understanding of the design behavior.

e) Emerging generation techniques

Emerging methodologies based on new heuristics are being successfully exploited to automatically generate optimal stimuli when it is impractical to find perfect solutions due to the high amounts of resources required in terms of computational time and memory. Evolutionary algorithms are sketched here as an interesting example of these kinds of generation mechanisms.

Evolutionary algorithms (EA) are based on pseudo-random generation able to guide the generation process with a closed-loop methodology. These optimization algorithms mimic the natural selection paradigm, and evolve solutions accordingly.

Roughly speaking, in evolutionary computation there is an initial population of individuals or verification patterns. These individuals are evaluated using a fitness function that provides information about the goodness value assigned to the possible solution. The fittest ones (called *parents*) survive and produce new individuals (or *children*). These new individuals join the population and older individuals may be removed from it. The whole process is performed in genetic cycles called *generations*. New individuals are produced either by mutation or crossover operators. Mutation operators make a random change to the parent; whereas, crossover operators perform recombination of the parents.

Evolutionary algorithms have been successfully exploited to generate sets of stimuli for digital circuits described at RTL and gate level [12] [9], and also for verifying complex circuits such as processor cores [13] [14].

12.1.1.2 *Response checking*

This part of the verification plan should be able to demonstrate if the behavior of the model under verification conforms with the device its specifications. The two possibilities are based on a reference model or distributed data and temporal checks.

The reference model is usually based in a more abstract model with respect to the device under verification. In fact, this model is suitable for architectural verification and not for implementation verification since it does not include detailed information. For example, a gate level model can be verified against a RTL one, however, the one at RTL usually does not include information about the timing of the technology library, so, considering this level of detail, the reference model cannot provide useful information.

On the other side, distribute and temporal checks use data and temporal monitors to capture the device behavior, that should be compared against the expected ones. A verification structure that has been used for these purposes is the scoreboard. The scoreboard should be able to store the data input to the device as well as the expected results. The idea is then to feed the scoreboard as well as the device under verification with the very same input data, then, the scoreboard should be able to compare the obtained results against the expected ones saved previously.

12.1.2 *Verification environment implementation, device bring-up and regression*

Once the verification plan is ready, it must be used as the specification for the implementation of the verification environment. This task can be done from scratch or reusing previous environment already developed in the same organization or exploiting commercial tools.

The first application of the verification environment is to bring the DUV to do its first steps. This process is known as *device bring-up* and can be supported by the test made by the design team during the development phase. Very simple assumptions have to be stated and then, it is necessary to demonstrate that they are true. For example, in the case of a processor core, the simplest assumption is related to the instructions fetch mechanism that should fetch an instruction just after the reset pin is assert and de-assert. If the instruction memory and related buses deliver the instruction as expected, then, this first assumption is considered true. After the first very simple test, the verification environment can start performing the work that is intended to, populating the stimuli set with always new and more advanced input tests.

Device regression is defined as the process that detects the re-introduction of bugs, leading, in this way the device in a previous or less developed state. In order to create a regression test suit, any test must be able to verify the DUV fundamental behavior, highly exercise the design in a few clock cycles, and have discovered some bugs in previous runs. The regression suit should be created in an incremental way, adding new material as soon as it is created.

Verification Techniques

Very different methodologies have been developed to generate verification stimuli. The main possibilities range from *manual verification* techniques to *formal verification* techniques, including random and semi-random approaches. In the following, the verification techniques defined as *formal* or *simulation-based* are described.

12.1.3 *Static or formal*

Techniques called static or formal verification methodologies use exhaustive mathematical approaches to demonstrate that the DUV responses to all possible inputs and all possible reachable states conforming to the device specifications. These methods do not rely on the generation of input information to verify the design. In fact, they may always be seen as theorem-proving within a given logic system [15]. However, in practice, research in this field falls within various sub-categories, such as: *automated theorem proving*, *model checking,* and *equivalence checking*, as briefly described in the following.

12.1.3.1 *Automated theorem proving*

The idea is to represent two models or properties of the DUV as two formulae f and g in an expression logic, then prove $f \Rightarrow g$. While such approach is potentially very general, results in automated theorem proving are limited algorithmically. Most implications in a general context are undecidable, and tackling them requires extensive user involvement. However, in restricted logics such as *Linear-time Temporal Logic* (LTL), proving the theorem $f \Rightarrow g$ is equivalent to testing the unsatisfiability of the formula $f \wedge \neg g$ which is always decidable [11].

12.1.3.2 *Model checking*

This technique can be seen as a special case of proving the theorem $f \Rightarrow g$, where f is a state transition model and g is a logical formula stating a property or specification of f. The computational tree logic (CTL), introduced by Clarke and

Emerson in 1982 [16], is commonly used since it enables to carry out the check in a time linearly dependent on the size of f and g. In some cases, model checking is used to verify the control parts of a circuit, however, it is impractical to verify the circuit data paths using this technique.

12.1.3.3 *Equivalence checking*

This is one of the most popular formal techniques. The main idea behind these techniques is that two models of the DUV are compared. This process is done in two phases: a Boolean net representing the design is obtained from the new model (e.g., the model delivered at gate level); then, to verify the model, an equivalence evaluation is performed between the obtained net and a reference one usually obtained at one of the previous steps of the design. The Boolean nets are usually represented using binary decision diagrams (BDDs). BDDs used in formal verification were introduced in [17].

Equivalence checking is suitable to verify two structurally similar designs that present a one to one mapping of circuit states.

The main issue regarding formal verification techniques corresponds to the computational resources required to complete the verification process, in particular, CPU time and memory occupation become significant for even medium complexity circuits.

12.1.4 *Dynamic or Simulation-based*

Simulation-based methodologies try to completely exercise the current model of the device in order to uncover design errors. Briefly, a simulation-based verification process is composed of three basic elements: input data (also called *set of stimuli*), the model of the device under evaluation (also called *design or device under verification or DUT*); and the *response checker*, which generates the pass/fail information regarding the current process. The checker usually preforms a comparison of the obtained results against the expected ones.

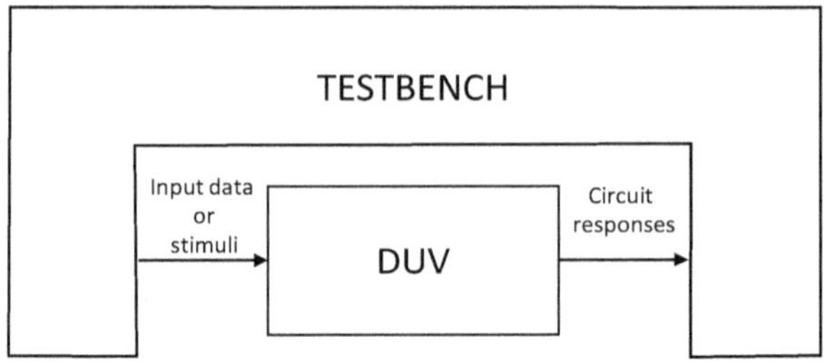

Fig. 12.3. Verification environment at a first glance.

The figure above shows the typical environment used to carry out the verification process. This framework or verification mechanism is usually called a *testbench*. The testbench is the unit inside the verification environment in charge of providing to the DUV the stimuli, collecting the model responses, and checking the obtained results, as will be described in the following section.

A *set of stimuli* is the collection of inputs to the design under verification. It could include for example: configuration information, test patterns, instructions, and communication frames containing protocol errors to excite correction mechanisms [1]. On the other side, *test or verification quality* is the capacity of the set of stimuli to find bugs, errors or faults in the considered model.

Performing exhaustive verification is very costly, for example, exhaustively verifying a 32-bit multiplier by simulation using the verification scheme presented before requires, the set of stimuli, the model under verification, and a response checker. Taking into consideration that the circuit is a combinational block, the set of stimuli that exhaustively verify the circuit has to include all possible combinations of circuit inputs, i.e., 2^{32} X 2^{32} possibilities.

Supposing that the multiplier is described in a hardware description language at RTL, and that the simulation of a single input configuration takes 1 μs (including the comparison process of the circuit outputs *O* with the attended values) the whole verification process may require more than 584 thousand simulation years.

Based on this example, the reader can easily understand why simulation-based verification is rarely exhaustive; actually, the set of stimuli includes a carefully selected sub-set of the possible input configurations, only.

Simulation-based methodologies are strongly dependent on the quality of the set of stimuli used to excite the design. These methodologies are seldom exhaustive and only consider a limited sub-set of possible circuit behaviors. Since the quality of the results first depend on the percentage of applied stimuli with respect to the total possible ones, they almost never achieve 100% confidence of correctness.

Design engineers are usually interested on peculiar stimuli that deserve their attention. Some *special cases* may excite design particularities. Others, instead, may represent interesting cases looking for activating design functionalities in a singular manner. These stimuli are usually called *corner cases*.

In the case of the 32-bit multiplier, a set of special cases may contain some couples of values containing for example one or both of the inputs *A* or *B* set to zero, the multiplication of the largest values as well as the lowest ones, etc.

It is obvious that different sets of stimuli could own different error or bug detection capabilities, even if their length is comparable. Thus, regardless of the generation method, the key question is how good is a set of stimuli produced at every design step? The answer to this question is given by the set of metrics defined on the verification plan, as described previously.

Testbench

In a verification process, the testbench can be thought as the verification mechanism in charge of determining the correctness of the Device Under Verification (DUV). In a general view, a testbench wraps around the DUV, and it could be seen as a hardware tester connected to a physical device that provides the Device Under Test (DUT) with a series of physical stimuli, capturing the DUT responses, and checking the correctness of the results. In the first case instead, the testbench should also provide the DUV with a set of stimuli capturing and checking the device responses. However, the testbench may have also very different goals, for example, it should be able to handle with a system described at different abstraction levels, and it could be able to generate different types of stimuli while tracking the verification process progress.

Thanks to new design technologies, during the last years, testbenches have suffered a very interesting evolution. Actually, these verification mechanisms are able to manipulate very complex projects. A general architecture for a testbench is shown in Fig. 1, as described by Spear and Tumbush in [18] the best strategy for developing a testbench is by implementing a layered scheme.

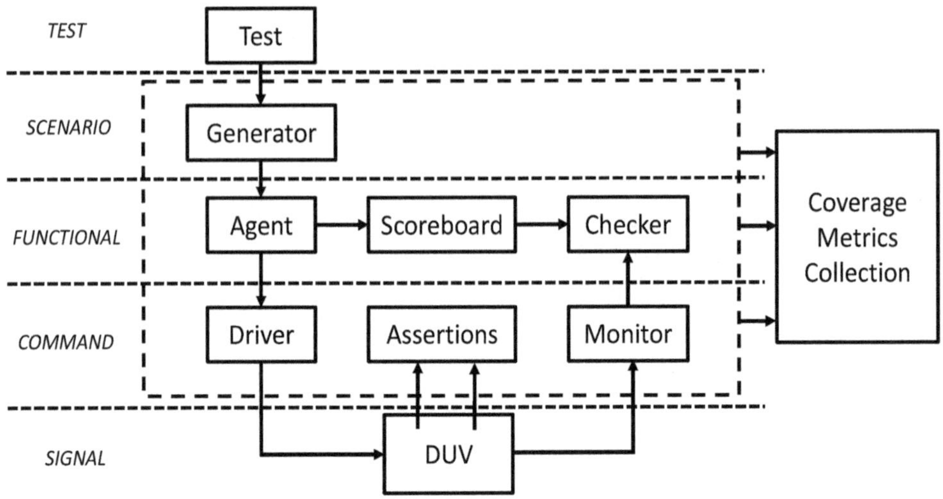

Fig. 12.4. General testbench architecture.

At the bottom of the figure, the Device Under Verification is connected at the *SIGNAL* level to the testbench; the stimuli set that will be delivered to the DUV is provided by the *COMMAND* layer using the *Driver*, and the output signals are collected and sent to the *Monitor* block. At the *COMMAND* layer, it is also possible to gather the results related to the activated *Assertions*, assertions usually monitor internal signals as well as circuit changes. At a higher level, in the *Functional* one, the *Agent* block receives high-level transactions, such as the ones involved in a BUS cycle, and split them in basic commands that are delivered to the lower layer. Contemporary, these commands are sent to the *Scoreboard* that is able to predict the expected results of the transaction. The *Checker* at this level, compares the results coming from the scoreboard against the ones coming from the DUV. A *Generator*, at the *SCENARIO* level, is in charge to provide verification patterns according to different situations or scenarios that are required to be test. The final layer, named the *TEST* one, contains the constraints to create the different sets of stimuli required by the verification plan. At this point, it is necessary to collect the verification results, this process is done by the *Coverage Metrics Collection* block that tracks the evolution of the verification plan during the process.

Coverage Metrics

Guaranteeing the quality of a set of stimuli is a major issue in the verification process. Thus, it is necessary to define a *test criterion* as a condition that describes what constitutes an *adequate* test [19]. Consequently, with respect to verification methodologies, the main idea consists in defining some testing criteria to adequately confirm the quality of the stimuli generation process, as well as to correctly decide when to stop it. Each abstraction level in the design flow could count on appropriate metrics, strictly related to the description of the model, and able to point out when the set of stimuli is satisfactory.

Coverage metrics were firstly defined in software testing as the measure of how thorough a program is exercised by quantifying the capacity of a given input stimuli to activate specific properties of the program code [20].

Thus, borrowing the idea from software testing, a *coverage metric* for hardware verification must be defined to assure the adequacy of the set of stimuli, and the collected information about coverage could be exploited as a useful test criterion.

The goal for a specific coverage metric could be implicitly or explicitly described with respect to the set of stimuli that is able to maximize the metric. For example, the *statement coverage* measures if every statement of the source code is executed and it could be represented by a value in the range between 0 and 100%. It is clear that the metric does not provide any explicit information about the appropriate set of stimuli capable of getting the most out of this metric. On the other hand, a functional metric that evaluates if a series of clearly defined corner cases have been exercised presents explicit information about the set of stimuli in charge of maximizing the coverage. This metric could be also expressed as a 0-100% value. However, it could be better to describe the obtained results using a coverage matrix, as described before.

The higher the coverage values obtained by a given set of stimuli, the higher the confidence in the design it can provide. Intuitively, high coverage values imply high system activation. However, it is worth noting that coverage does not imply that the design conforms to the specifications. A complete coverage on any particular metric could never guarantee a 100% bugs-free design, nor thorough code inspection. Moreover, it is not possible to select one single coverage metric as the most reliable [21], [22]. Therefore, the current verification trend is to combine multiple coverage metrics together to obtain better results [23].

Performing a verification process guided by coverage metrics allows achieving high design verification but limiting the redundant efforts. Coverage metrics act as heuristic measures for quantifying the verification completeness and identifying inadequately exercised design aspects.

In 1974 Brian Kernighan, the creator of the C language, stated that "everyone knows debugging is twice as hard as writing a program in the first place". This statement concerns software design; however, the same implication can be stated in the case of circuit design. Thus, it is really important to clearly define the verification strategy to be adopted.

Usually, the coverage metrics included into the verification plan are selected based on the designers' expertise, the ease of applicability, and a strict evaluation of the computational resources. Depending on the abstraction level of the available DUT model, some coverage data could be irrelevant; therefore, coverage data must be collected as soon as it is practical but not sooner. It is important to collect adequate information: if there are some metrics that do not deserve attention, it is better to avoid spending time on collecting the related information. Collecting meaningless information does not mean performing a real verification progress.

When multiple coverage metrics are used, it might be better to combine the collected values into a single overall coverage metric, using for example a weighted average for every coverage metric. Let us suppose to consider three different metrics CM_1, CM_2 and CM_3; then, a scale factor can be assigned to each coverage value weighting properly the considered metric:

$$Ct = 0.3CM_1 + 0.3CM_2 + 0.4CM_3 \qquad (12.1)$$

in this case, CM_3 is the most relevant metric. Weighting metrics must be done carefully trying to avoid both masking small but essential contributions, and magnifying meaningless information regarding any minor metric.

Unfortunately, no well defined standard set of metrics able to guarantee acceptable design errors detection exist. In fact, choosing a set of metrics to be included into the verification plan is a particularly critical task, which is customarily based on the experience acquired by the verification team.

Some issues should be taken into consideration when choosing coverage metrics. Specifically, it is important to:

- evaluate the computational effort required to measure each coverage metric during the simulation process;
- be able to generate and measure the test stimuli progress avoiding excessive efforts;
- minimize the required modifications to the verification tools.

Additionally, it seems that no metric is superior to another; then a comparative measure of metric goodness needs to be established by intuitive or empiric experiences.

As mentioned before, an exhaustive circuit verification tries to thoroughly evaluate the whole set of possibilities of the stimuli/responses space. However, even for a medium complexity circuit the number of possibilities is quite large.

Aiming at reducing the space of stimuli/responses it is possible to introduce a new concept: the *coverage model*. A coverage model is a sub-set of the stimuli/responses space that will prove, with an acceptable degree of confidence, that the functionality of the design is correct [1].

Let us suppose that you need to verify the circuit model of the programmable 32-bit down counter shown in the figure below; the counter has *Enable*, *Reset* and *CLK* pins. Each time it is reset, it loads a 32-bit word and the count down starts. The additional control bit *E* enables the counter operation. Therefore, the space of stimuli for this simple circuit has 2^{32} x 2^2 possibilities. Thus, assuming that verifying by simulation an input requires in average about 10 μs, about 2 days of simulation are required to exhaustively verify the counter. On the other side, we could consider a coverage model that looks for special cases, only. This means forcing the counter to only load a few values such as: all walking-ones (32), all walking-zeroes (32), all-ones (1), all-zeroes (1), and finally two particular cases: 55555555h (1) and AAAAAAAAh (1); for every configuration, the counter is forced to load it, and then, it is checked whether the counter is able to correctly compute the next value on the outputs, this set of stimuli is composed of 68 patterns. Let us suppose also that for every input word, the counter is programmed twice: one enabling the count down, and one disabling it. The results obtained from these experiments could still leave us with a sufficient confidence about the correctness of the circuit model, and it is not necessary to spend high quantities of time on exhaustively verify the design, since it requires to simulate only 136 input configurations.

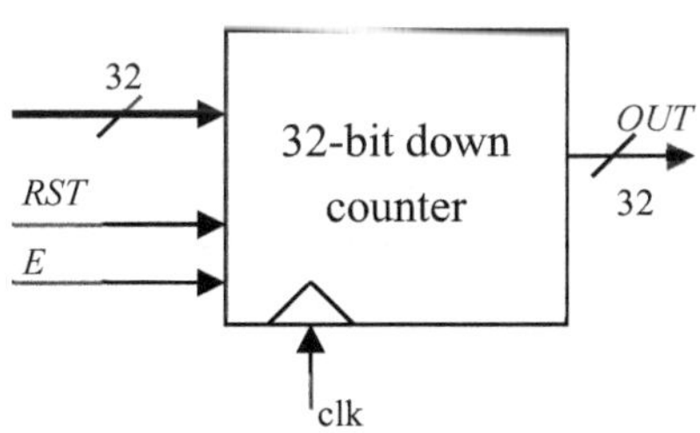

Fig. 12.5. 32-bit down counter.

In the following, an introduction about the metric basic concepts is provided. The classification presented here mainly follows what is proposed in [21] and [24].

12.1.5 *Code coverage metrics*

Code coverage metrics derive from metrics used in software testing. These metrics identify which code structures belonging to the circuit description are exercised by the set of patterns, and whether the control flow graph corresponding to the code description has been thoroughly traversed. The structures exploited by code coverage metrics range from a single line of code to *if-then-else* constructs.

a) Statement coverage (SC)

Statement coverage is the most basic form of code coverage: statement coverage is a measure of the number of executable statements within the model that have been exercised during the simulation. Executable statements are those that have a definite action during runtime, without including comments, compile directives, etc. Statement coverage counts the execution of each statement on a line individually, even if there are multiple statements on that line.

b) Branch coverage (BC)

Branch coverage reports whether Boolean expressions tested in control structures (such as the if-statement and the while-statement) are evaluated to both true and false. The entire Boolean expression is considered one true-or-false predicate regardless of whether it contains logical-and or logical-or operators. Branch coverage is sometimes called decision coverage.

c) Condition coverage (CC)

Condition coverage can be considered as an extension of branch coverage: it reports the true or false outcome of each Boolean sub-expression contained in branch statements, separated by logical-and and logical-or if they occur. Condition coverage measures the sub-expressions independently of each other.

d) Expression coverage (EC)

Expression coverage is the same as condition coverage, but instead of covering branch decisions it covers concurrent signal assignments. It builds a focused truth table based on the inputs to a signal assignment using the same technique as condition coverage.

e) Path coverage (PC)

A circuit description could be embodied by its *control flow graph* that schematizes, using a graph, all possible paths that may be traversed at simulation time. Path coverage refers to the number of exercised paths present in the control flow graph of the circuit during the simulation of a set of stimuli. The number of paths in a circuit description that contains for example a loop structure could easily be very large. Then, it is not feasible trying to cover all paths in a circuit. Instead, a representative set of reduced paths may be chosen.

12.1.6 *Metrics based on the device activity*

These metrics measure the activity of some portions of the design, targeting interconnections, memory elements, and internal networks.

a) Toggle coverage (TC)

Toggle coverage reports the number of storage elements that toggle at least once from 0 to 1 and at least once from 1 to 0 during the execution of a program. At the RTL, registers are targeted and, since RTL registers correspond to memory elements with an acceptable degree of approximation, the toggle coverage is an objective measure of the activity of the design. Indeed, this is a very peculiar metric and can be sensibly used to guarantee high activity in circuits present after the targeted registers.

12.1.7 *Metrics based on FSM*

A *finite state machine* (FSM) is a representation of a sequential circuit by modeling its behavior based on states, transitions, and actions. Thus, a coverage metric that represents the covered states of the circuit's FSM must be able to measure, quantify, and exercise the circuit sequential behavior. The classical representation of the finite state machine of a digital circuit is a connected graph (normally called state graph) corresponding to the states reachable by the circuit.

a) FSM Transition coverage (FSM-TC)

Transition coverage measures the number of distinct edges traversed during the simulation of the set of stimuli.

b) FSM State coverage (FSM-SC)

State coverage counts the number of distinct nodes or states visited during the simulation of the stimuli set.

12.1.8 *Error-based coverage metrics*

These metrics are directly related to specific errors or fault models that are injected in the design and the general goal is trying to understand if the stimuli set is able to discover a discrepancy for any one of them.

a) Mutation coverage (MC)

Design errors are not easily mappable to well defined error models; however, in some cases, design errors due to typographical mistakes produced by the design engineer during the typing process have been successfully modeled by artificial *mutations* of the code.

An interesting way to measure how well design errors are likely to be detected or not by a set of stimuli consists on inserting some artificial errors into the design under verification, and then checking if they are detected by the available set of stimuli. To calculate the test quality of the set of stimuli, the original model (without any change) is simulated in parallel with the model artificially changed, and then if the circuit outputs are different, the injected fault is detected. *Mutation coverage* is a typical example of coverage metric based on this approach.
This coverage metric mimics mutation coverage from software testing. Particularly, this metric models a hypothetical design error by injecting a small local mutation in the design description. The resulting description, called *mutant*, is simulated in parallel with the original design using the same set of stimuli, and the error is detected when a difference is observed while comparing the DUV against the mutant code.

Some possible mutations made at RTL are: replacing arithmetic, logic and relational operators, changing constant values, and replacing variables in operations and assignments.

b) Coverage metrics using faults models of manufacturing testing

This kind of coverage metrics are better described in the chapter dedicated to testing in this book. In a few words, fault models try to bridge the gap between the physical reality of the circuits and the circuit design description. Actually, regarding manufacturing testing, a fault model tries to model physical defects that may appear in the circuit after manufacturing. For example, sticking one of the input or output signals of a logic gate to a predetermined logic value (1 or 0). This fault model is called *single stuck-at* 0 or 1, and it is the most popular fault model used in digital circuit manufacturing test. Usually, it is assumed that only one single fault is present in the circuit model at simulation time.

UVM Verification

In recent years, some of the most important CAD providers in the electronic industry decided to unify the verification process in a unique standard called Universal Verification Methodology (UVM). The UVM standard derives from three industrial standards: Universal Reuse Methodology (URM) by *Cadence*,

Reference Verification Methodology (RVM) by *Synopsys*, and Advanced Verification Methodology by *Mentor*. Since 2017, UVM is a standard approved by the IEEE that facilitate to face a verification process.

According to [1], the main benefits of UVM can be listed in the following:

- It describes a standard testbench structure
- The standard describes a clear verification methodology
- It split the verification problem in two:
  - Defining test scenarios
  - Implementing pin-level protocols

In the following, a brief description of the main components of an UVM testbench and the UVM Cass library are provided.

12.1.9 *Typical UVM Testbench architecture*

The UVM testbench is graphically described in figure 6. The UVM testbench instantiates the *DUV* module and the *UVM Test* class and provides the connections between them. The UVM Test is instantiated dynamically at run time allowing the testbench to be compiled only once but running with different verification sets.

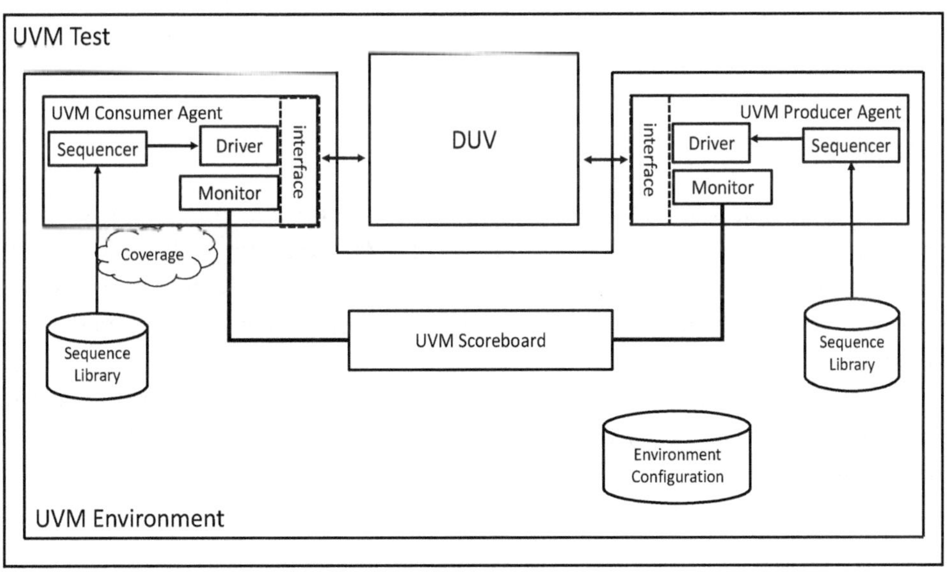

Fig. 12.6. UVM testbench architecture.

12.1.9.1 *UVM Test*

This is the top-level component in the testbench. The UVM Test preforms the following functions:

- Instantiates the top-level environment
- Configures the environment
- Applies the stimuli to the DUV by invoking UVM Sequences

12.1.9.2 *UVM Environment*

This is a hierarchical component that includes the other UVM elements indicating how they are correlated. These modules usually include the UVM Agents, and the Scoreboards.

12.1.9.3 *UVM Scoreboard*

The main function of the UVM Scoreboard is to check the behavior of the DUV during the exercise of the verification process. It usually receives the input signals to be provided to the DUV and exploiting a high-level model of the device (usually named a *predictor*) it produces and compares the expected outputs against the actual ones.

Many different possibilities can be applied in order to create an UVM Scoreboard, since it depends on the type of predictor, the interconnections of the module and the rest of the testbench, and so on.

12.1.9.4 *UVM Agent*

The UVM Agent is a module in charge to intercommunicate with the DUV, in this way, this module includes other sub modules that for example: deals with the DUV interface, creates the stimuli sequences to apply during the verification process, monitors the results in order to compute the coverage metrics involved in the process.

UVM Agents usually includes two working modes: active during the stimuli generation process and passive where it only monitors the DUV interface without controlling it.

12.1.9.5 *UVM Sequencer*

This module works as an arbiter for controlling the transaction flow from multiple stimulus sequences. Specifically, it controls the flow of UVM Sequence Items generated by the different UVM Sequences involved.

12.1.9.6 *UVM Sequence*

A UVM sequence is a module that defines the behavior of a stimuli session. An instance of a UVM Sequence can actuate during a single transaction, or can operate during the whole simulation, or in a specific point of the simulation. Usually, a UVM Sequencer can bind one or more UVM Sequences.

12.1.9.7 *UVM Driver*

The UVM Driver receives an individual UVM Sequence and applies it to the DUV interface. In this way, the driver spans the abstraction levels by converting the transaction-level provided by the UVM Sequence into pin-level stimulus.

12.1.9.8 *UVM Monitor*

The UVM Monitor samples the DUV interface capturing the information in the form of transactions to be sent to other components of the testbench for further analysis. This module acts the abstraction level spans in the different direction of the Driver, since it converts the pin-level activity to transactions.

The UVM Monitor can perform internally some checks of the transactions by for example gathering some coverage metrics information, or can on the contrary, send the appropriate information to the output through the monitor analysis port.

12.1.10 *UVM Class library*

The UVM Class library contains all the necessary building blocks to develop reusable verification components and testbenches. The library is composed of base classes, utilities and macros, a diagram showing some of the main elements in the class library is provided in the Figure 12.7.

The main advantages of using the UVM Class library are:

- The UVM Class contains a set of built-in features that includes printing, copying, and other methods that facilitate the verification process;
- The library correctly implements the UVM concepts defined in the standard. In this mode, using the library increase the project readability as well as guarantee the instantiated modules to be easily reused in different projects.

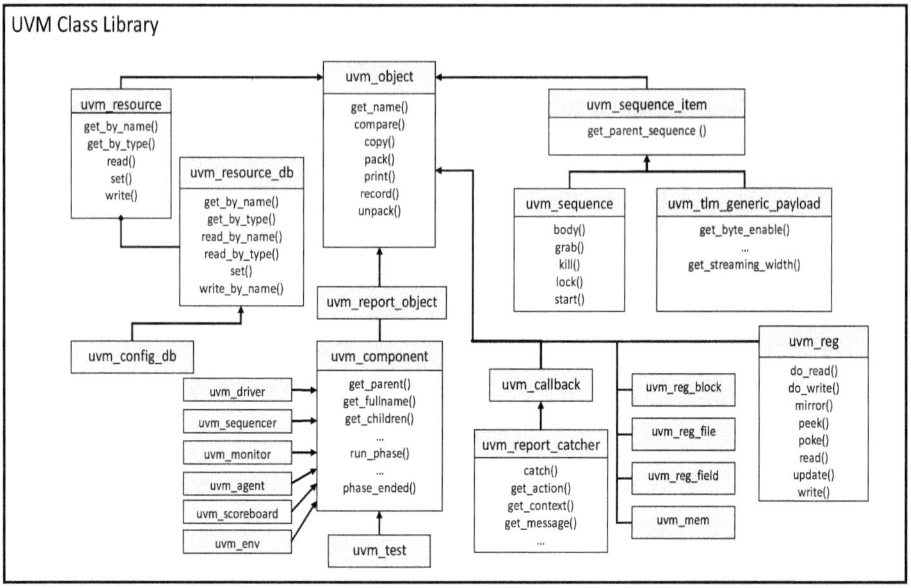

Fig. 12.7. UVM Class library.

Conclusions

In this chapter, a brief introduction of the most important elements involved in a verification process were provided, highlighting in particular, that the design verification is a critical part composing the design cycle of an electronic device. Due to the today's circuit complexity, it is important to define since the very beginning which is the verification process that the final device will pass through. For this reason, a very careful verification plan should be defined and implemented together with the design process. The accurate implementation of the verification mechanisms as well as the correct selection of the coverage metrics that will support the verification process guarantee the success of the whole verification process.

References

1. Sanguinetti, J. (2012). Abstraction and Standardization in Hardware Design. In IEEE Design & Test of Computers, vol. 29, no. 2, pp. 8-13, April 2012. doi: 10.1109/MDT.2012.2184074
2. IEEE Standard for Property Specification Language (PSL)," in IEEE Std 1850-2010 (Revision of IEEE Std 1850-2005), vol., no., pp.1-182, 6 April 2010. doi: 10.1109/IEEESTD.2010.5446004

3. Bergeron, J., Cerny, E., Hunter, A., and Nightingale, A. (2006). Verification Methodology for SystemVerilog, Springer, 2006
4. Synopsys (2015), Delivering Functional Verification Engagements: A Proven, Systematic and Assured Approach, White paper 2015
5. Halfhill, T. R. (1995). The Truth Behind the Pentium Bug. Byte. March 1995
6. IEEE standard for software verification and validation, IEEE Std 1012-1998, 20 Jul 1998
7. Mehta, A. B. (2017). ASIC/SoC Functional Design Verification: A Comprehensive Guide to Technologies and Methodologies (1st ed.). Springer Publishing Company, Incorporated.
8. Piziali Andrew. (2004). Functional Verification Coverage Measurement and Analysis, Springer, 2004.
9. Stamenkovic, Z., Dahmen, H., Glaeser, U., (2000). VHDL design validation by genetic manipulation techniques. IEEE Microelectronics, 2000. Proceedings. 2000 22nd International Conference on, 2000, Vol. 2, pp 735-738
10. Accellera. "Universal Verification Methodology (UVM) 1.2 User Guide", May 2015.
11. Malka, Y., Ziv, A. (1998). Design Reliability Estimation Through Statistical Analysis of Bug Discovery Data. 35th Design Automation Conf., ACM Press, New York, 1998, pp. 644-649.
12. Hahanov, V., Babich, A., Sokolov, A., and Pudov, V. (2002). Deterministic method of genetic algorithms of test generation for digital systems verification. Modern Problems of Radio Engineering, Telecommunications and Computer Science, IEEE Proceedings of the International Conference, 2002, pp 257-258
13. Corno, F., Sanchez, E., Sonza Reorda, M., Squillero, G., (2004). Automatic Test Program Generation - a Case Study. IEEE Design & Test, Special issue on Benchmarking for Design and Test, Volume: 21, Issue 2, 2004, pp 102-109
14. Schiavone, P. D., Sanchez, E., Ruospo, A., Minervini, F., Zaruba, F., Haugou, G., Benini, L. (2018). An Open-Source Verification Framework for Open-Source Cores: A RISC-V Case Study. Proceedings of 26th IFIP/IEEE International Conference on Very Large Scale Integration, VLSI-SoC 2018, pp. 43-48.
15. Kurshan R. (1994). Computer-Aided Verification of Coordinating Processes, Princeton University Press, 1994
16. Clarke, E. M. and Emerson, E. M. (1982). Design and Synthesis of Synchronization Skeletons for Branching Time Temporal Logic, in Proceedings Logic of Programs Workshop, LNCS 131, pp. 52-71, Springer-Verlag, 1982
17. Hu, A. J. (1997). Formal hardware verification with BDDs: an introduction. IEEE Pacific Rim Conference on Communications, Computers and Signal Processing, 1997, pp. 677 - 682 vol 2
18. Spear, C., and Tumbush, G. (2012). Systemverilog for Verification: A Guide to Learning the Testbench Language Features, (3rd ed.). Springer Publishing Company, Incorporated.
19. Zhu, H., Hall, P. A. V., and May, J. (1997). Software Unit Test Coverage and Adequacy. ACM Computing Surveys, Vol. 29, No. 4, December 1997, pp. 366 – 42
20. Goodenough, J. B., and Gerhart, S. L. (1977). Toward a Theory of Testing: Data Selection Criteria, Current trends in programming methodology, vol. 2 R. T. Yeh, Ed. Prentice-Hall. Englewood Cliffs, NJ, 1977, pp. 44 – 79

21. Bledsoe, W. W., and Loveland, D. W. editors, (1984). Automated Theorem Proving: After 25 Years, Contemporary Math 29, American Society, 1984
22. Tasiran, S., Keutzer, K. (2001). Coverage Metrics for Functional Validation of Hardware Designs. IEEE Design & Test of Computers, vol. 18(4), 2001, pp. 36 - 45
23. Chien-Nan, J. L., Chen-Yi, C., Jing-Yang, J., Ming-Chih, L., and Hsing-Ming, J. (2000). A novel approach for functional coverage measurement in HDL Circuits and Systems. ISCAS2000: The 2000 IEEE International Symposium on Circuits and Systems, 2000, pp. 217 - 220
24. Pradhan, D. and Harris, Ian G., (2009). Practical Design Verification, Cambridge University Press

https://doi.org/10.1142/9789811210723_0013

Chapter 13

Digital Testing

Alberto Bosio
Lyon Institute of Nanotechnology, École Centrale de Lyon, Lyon, France

13.1 Introduction

This chapter will introduce the basic knowledge about testing of integrated digital circuits. Figure 13.1. Depicts the main steps of the design, manufacturing and test flow. As already described in Part 4, Chapter 12, before moving to the manufacturing process, the design (i.e., model) has to go through a verification phase aiming at ensuring the absence of bugs. Bugs are human errors introduced in the design as for the case of software development. During the manufacturing process, *defects* can occur in the physical structure of the Integrated Circuit (IC) leading to erroneous behaviors. The role of testing is thus to detect ICs affected by defects and discard them. Moreover, testing is also important in order to collect as much information as possible to be further exploited during the Diagnosis. Diagnosis aims at understanding the root causes of the observed failures, and thus improving the manufacturing process.

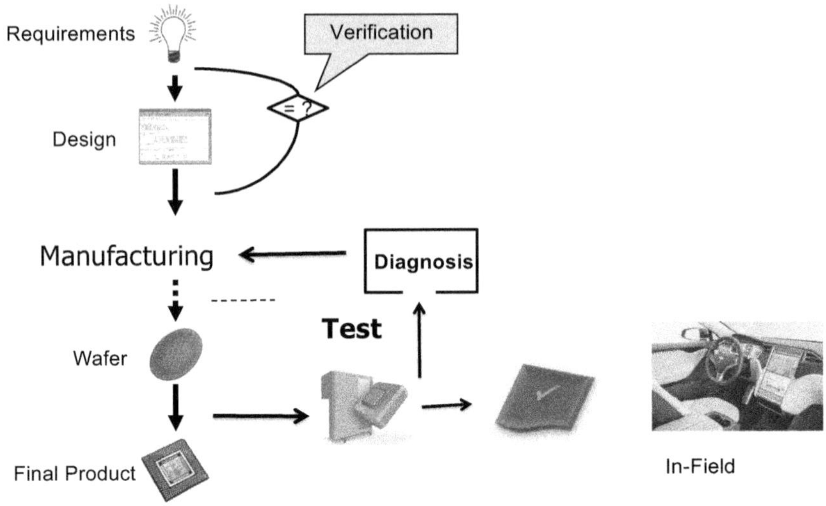

Fig. 13.1. Manufacturing and Test flow

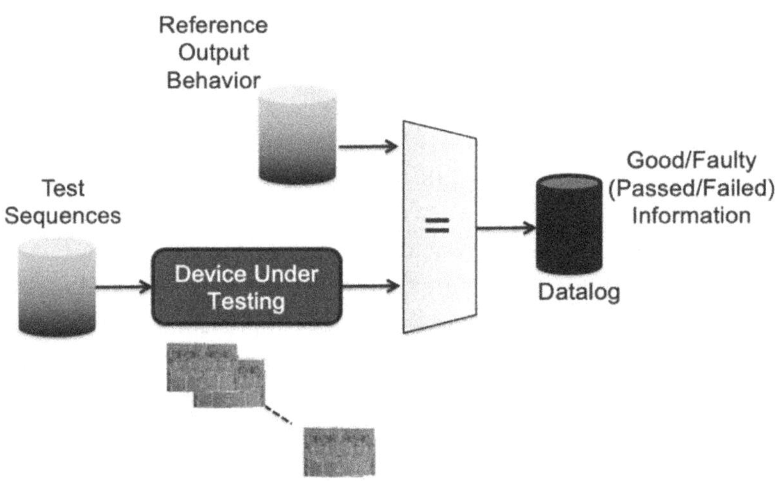

Fig. 13.2. End-of-Production Test

Figure 13.2 sketches the main elements of the *End-of-Production test*. After the manufacturing process, each manufactured device is tested and thus it is called **Device Under Testing (DUT)**. The test consists in applying a sequence of stimuli to the primary inputs of the DUT, measure the primary outputs and compare the values with the pre-computed Reference Output Behavior. If no differences are observed, the DUT is considered as **Good** and it can be shipped out. If a difference is observed, the DUT is considered as **Faulty** and it will thus be rejected. Test information is stored in the so called **Datalog** that will be used during the diagnosis process.

The reader may now have in mind the following questions: Test for what? Which test sequences do I have to apply? What are the parameters to be used to measure the quality of the test? How the test sequence is applied?

The rest of the chapter aims at answering the above questions. It is structured as follows section 13.2 introduces the concept of fault models, section 13.3 details which conditions have to be satisfied in order to test a fault. Section 13.4 talks about the testing issue due to the circuit structure itself and how to alleviate this problem. Finally, section 13.5 concludes the chapter.

13.2 Test for What? Defects and Fault Modelling

In this section, we present the main causes of physical defects and how these defects are usually modeled at transistor-level domain. Then, we discuss about the *faulty behavior* induced by defects and how the *fault models* represent them. Next sections will exploit the concept of fault model to generate a test sequence.

A physical defect can be caused by several phenomena such as metal line broken/deformed, contact or via broken/deformed [1]. Figure 13.3 shows an example of physical defect as an Open (e.g., missing contact).

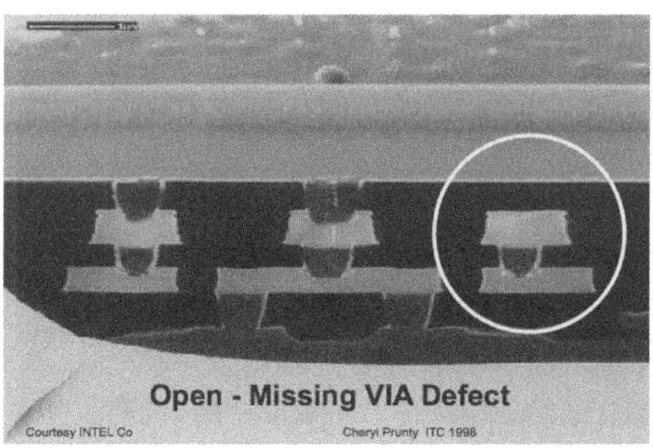

Fig. 13.3. Example of Physical Defect

These phenomena will lead to either an unexpected connection between two or more nets (short) or a missing connection on one net (open) [2] as shown in Figure 13.4. A net, at transistor-level domain, correspond to different elements: polysilicon (the transistor gate terminal), active (the transistor drain and source terminals), metal line (interconnections between transistors), contact (connection trough active and metal level 1) or via (connections through metal levels). Usually, these defects are modeled at transistor-level domain as: (i) an unexpected connection between two nets associated to a specific resistance value (resistive-short), (ii) an unexpected resistance value on a given net (resistive-open) [3]. Depending on the considered resistance value, the effect induced by the defect can vary. Thus, the choice of meaningful resistance values is crucial to have an accurate model of physical defects.

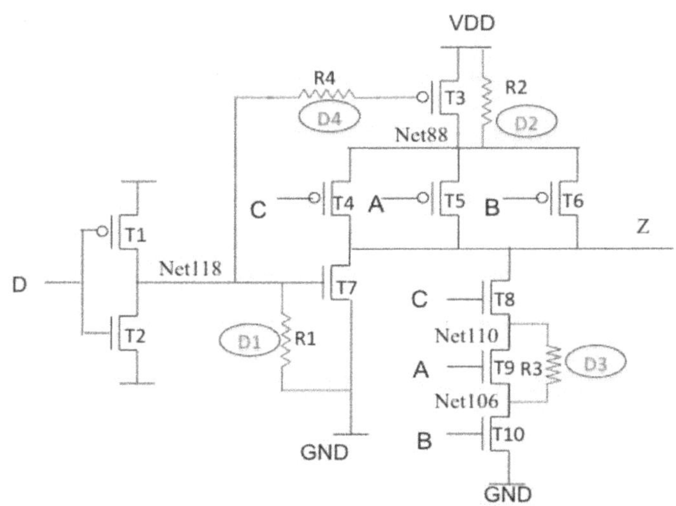

Fig. 13.4. Physical defects modelling at transistor-level domain

Figure 13.4 shows an example of physical defects and the related model at transistor-level domain. The illustrated circuit is composed of four primary inputs (A, B, C and D) and one primary output (Z). In the transistor-level netlist the four defects defined as D1, D2, D3 and D4.

Defect D1 models an unexpected connection between net118 and ground. The behavior of this defect depends on the resistance value R1. If R1 is lower than a threshold value RT, then the Vgs (voltage level between gate and source terminal of transistor T7) is always lower then Vth of T1. Thus, T1 is always switched off (faulty behavior). If R1 is greater than RT, then Vgs depends on the voltage level of Net118 (correct behavior). From a logical point of view the first case (when R1 < RT) is equivalent to have Net118 stuck at logic value '0'. A similar consideration can be done for defect D2. Depending on R2 value, it impacts the voltage level of net88. From a logical point of view, it is equivalent to have net88 stuck at logic value '1'.

Defect D3 models an unexpected connection between net110 and net106. The behavior of this defect depends on the resistance value R3 and also on the applied stimuli (i.e., it can be activated or not). For example, Figure 13.5 gives the circuit when the stimuli "0111" are applied to inputs A,B,C and D. Transistors T8 and T10 are switched-on while T9 is switched-off and output Z is set to logic '1'. However, due to the presence of this defect, the output is now connected to the ground through R3. Thus the defect is activated. Once the defect is activated we analyze the R3 value to determine its behavior. If R3 is lower than Rmin then, the output will be set to logic value '0'. In this case, from logic point of view the Net106 force its logic value '0' to the Net110 and then the output is changed. On the other hand, if R3 is greater than Rmin but lower than Rmax (Rmin< R3 <Rmax) then, output Z will be affected by an undesired transition (from logic '1' to '0') due to the long discharge time. Finally, if R3 is greater than Rmax then output Z will take the correct logic value and the circuit behavior is faulty free. Note that the value of Rmin and Rmax depends on the circuit technology.

Finally, defect D4 affects the net Net118. For certain values of R4 the signal propagation trough gate of transistor T4 is delayed.

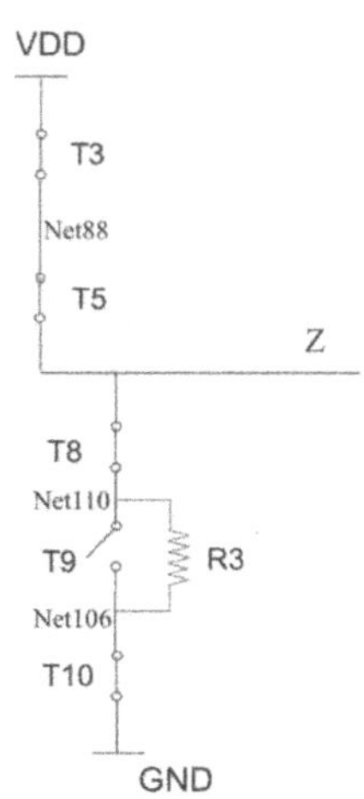

Fig. 13.5. Equivalent circuit when stimuli "0111" are applied

To summarize, the analyzed faulty behavior induced by the four defects of Figure 13.4 are:

- D1, D2: the faulty behavior results in a net always set to a given logic value (either '1' or '0');
- D3: the faulty behavior results in a net set to a given logic value (in the example logic '0') depending on the input configuration or in a signal propagation delay affecting primary output Z;
- D4: the faulty behavior results in net where the signal propagation is affected by a given delay.

Even if we target the transistor-level domain, we exploit the knowledge of the analyzed faulty behavior to be defect independent. So that we avoid to explicitly consider the resistance value.

Usually faulty behaviors induce by defects are represented by means of **Fault Models**. The most used in industry are:

- **Stuck-at Fault Model** [4]: the logic value of a given net appears to be stuck at a constant logic value ('0' or '1'), referred as stuck-at-0 or stuck-at-1 respectively (e.g., defects D1 and D2 in Figure 13.4;
- **Delay Fault Model** [5]: the transition from a given logic value V to the opposite logic value V' is delayed. Two types of delay faults are defined: slow-to-rise transition fault model (slow transition from logic '0' to logic '1') and slow-to-fall transition fault model (slow transition from logic '1' to logic '0').

13.3 Which Test Sequence Do I have to Apply? Test Generation

Let's consider the 32-bit ADDER circuit shown in Figure 13.6. In our example we simply try to apply all the possible input stimuli in order to catch all the faulty behaviors. In this case we need to apply the exhaustive test sequence composed of $2^{(32+32)} = 2^{64}$ input values. Assuming that the circuit can operate at GHz frequency (i.e., a single sum is obtained in the ns range), the whole test sequence application will require more than 1,000 years. This simple example clearly proves that exhaustive test cannot applied, we thus have to find a smarter way to properly test an IC.

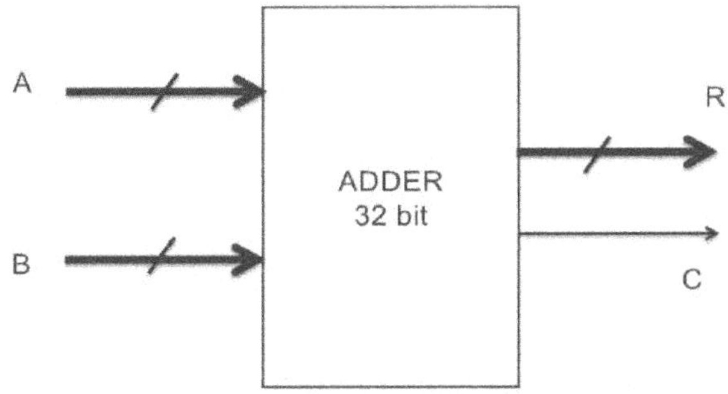

Fig. 13.6. Testing a simple 32-bit ADDER

The smarter way to test requires the use of the fault model concept (defined in Section 13.2). Let's use the AND Gate as depicted in Figure 13.7. As fault model we will use the stuck-at fault: stuck-at-0 (S@0) and stuck-at-1 (S@1). First of all, we will identify the so-called **Fault List (FL)** containing all the faults to be tested. In our case, we can count two faults per each net. The FL will be thus composed of 6 faults: S@0/a, S@1/a, S@0/b, S@1/b, S@0/z, S@1/z. The adopted notation is fault model (S@0, S@1) / circuit net (a, b, z). Once identified the target faults, we can move to the test generation phase. A **Test Vector (TV)** is a set of logic values applied to the primary inputs of our circuit. A TV is able to detect a given fault model if the faulty behavior induced by the fault is excited and the difference is propagated to at least one observable output.

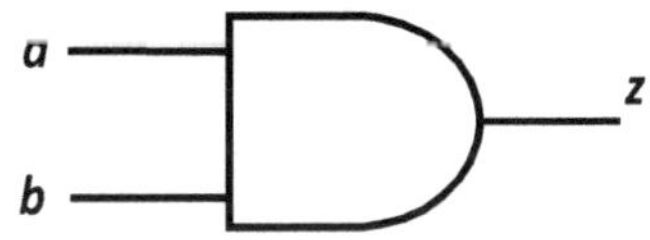

Fig. 13.7. AND Gate

Table 13.1. AND Test Set

TV	a	b	z	Detected Faults
0	0	0	0	S@1/z
1	0	1	0	S@1/a, S@1/z
2	1	0	0	S@1/b, S@1/z
3	1	1	1	S@0/a, S@0/b, S@0/z

Table 13.1 reports the faults detected by each TV. In the table we reported the exhaustive test composed in this case of 4 TVs (22). From the table, we can measure the quality of the test set by computing the **Fault Coverage (FC)**. FC is defined in Equation 13.1 as the ratio among the detected faults (#DF) over the number of target faults contained in the FL (#F).

$$FC = \#DF/\#F \tag{13.1}$$

Without surprise, the exhaustive test set can detect all the target faults, FC = 100\%. However, it is interesting to note that the same fault is detected by more than one TV. S@1/z is detected by three TVs (TV1, TV2 and TV3). We can therefore remove from our test set TV0 since without impacting the FC, our test set is thus reduced from 4 to three TVs. Reducing the size of the test set is crucial since reducing test time always means reducing test cost.

The above example is also interesting because it allows us to introduce the concept of **Equivalent Faults**. Two faults are equivalent if they are tested by the same test vector. In our example S@1/a, S@1/z are equivalent because they are tested by TV1. The importance of equivalent faults stems from the fact that it will reduce the number of target faults. Less target faults implies less effort during the test generation process. Coming back to the example, the other equivalent faults are S@0/a, S@0/b, S@0/z and S@1/b, S@1/z. Exploiting this, we can simply the fault list by only target the following faults: S@1/a, S@1/b and S@0/a. In other words, it is enough testing only for the faults affecting the inputs of the gate. Here, by generating one test per fault we obtain three TVs and thanks to the equivalence we cover all the possible faults (100% of FC).

Figure 2.8 generalizes the concept of fault equivalence. For a Boolean gate (i.e., AND, NAND, OR, NOR and NOT) having *n* inputs and one output, testing the faults affecting the inputs of the gate ensures the detection of the faults affecting the output. The demonstration can be simply done through true table (example of Table 13.1). For a fanout stem, testing the faults affecting each branch ensures the detection of the faults affecting the stem. Again, the demonstration can be easily done through true table.

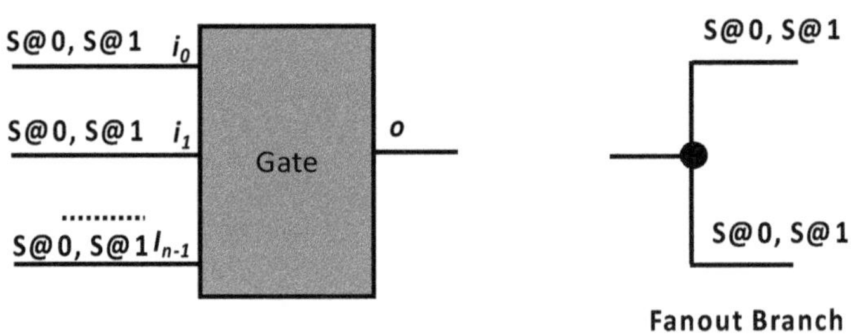

Fig. 13.8. Fault Equivalence

Thanks to the notions of fault modelling, fault equivalence and fault coverage, it is possible to detail the process of the test generation. Figure 13.9 depicts the overall architecture of an **Automatic Test Pattern Generator (ATPG)**. The main input is the "Circuit description" corresponding to a gate-level netlist of the target circuit. The **Fault Manager** is the responsible for identifying the faults to be tested. By using the concept of equivalence it generates the "Reduced Fault List". After that, the **Fault Selector** extract the "Target Fault" that is used by the **TPG Algorithm** in order to generate a "Test Pattern" able to detect it. The generated test pattern is finally used by the **Fault Simulator**, which checks if the generated test pattern detects other faults. The "Detected faults" are fed back to the Fault Selector that will move to the next un-detected fault from the reduced fault list. The main outputs of the TPG are the "Test Pattern" and the "Fault Coverage" used to quantify the quality of the test pattern. The next subsections will detail each block of the ATPG through an example.

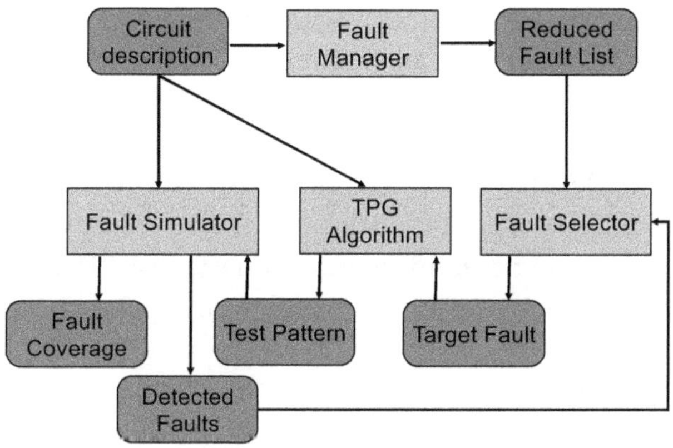

Fig. 13.9. Automatic Test Pattern Generator Architecture

13.3.1 *Fault Manager*

The number of possible fault sites (i.e., where a fault can affect the circuit) in a digital circuit described at gate level is given by Eq. 13.2

$$FS = \#PI + \#PO + \sum_{i=1}^{\#gate}(\#gate_i.inputs + \#gate_i.outputs) \qquad (13.2)$$

In the example of Figure 13.10, we can simply count 2 PIs, 1 PO and three gates with 2 input and 1 output each. It is thus easy to compute FS = 2 + 1 + 9 = 12. Without losing in generality, we will resort to the Stuck-at fault model for all our examples. For each fault site, we have to consider two stuck-at faults: Stuck-at-0 and Stuck-at-1. Therefore, the global number of stuck-at faults is two times the FS leading to 24 for the circuit of Figure 13.10.

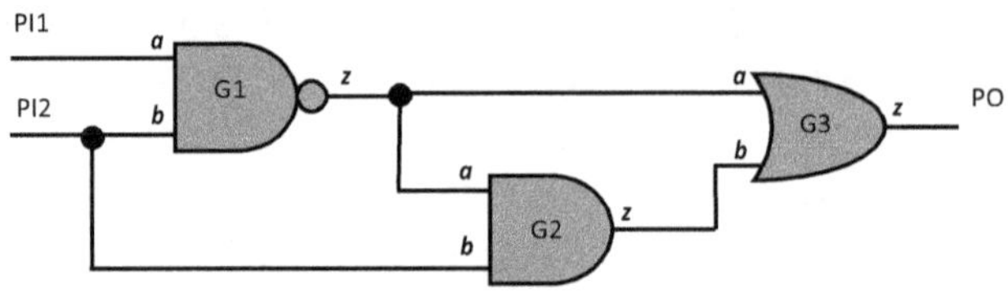

Fig. 13.10. Test Generation Example

It is possible to reduce the number of target faults by using the so-called *Checkpoint Theorem* [6]. In a combinational circuit, primary inputs and fanout branches are called checkpoints.

A test set that detects all single (multiple) stuck-at faults on all checkpoints of a combinational circuit, also detects all single (multiple) stuck-at faults in that circuit. The demonstration of the theorem is based on the concept of fault equivalence [6].

Coming back to our example, we can enumerate the checkpoint as two PI and four fanout branches leading to 12 stuck-at faults. We can thus reduce 50% of the number of target faults (without reducing the fault coverage). The reduced (i.e., collapsed) fault list is composed of the following faults: PI1@0, PI1@1, PI2@0, PI2@1, G1/b@0, G1/b@1, G2/b@0, G2/b@1, G2/a@0, G2/a@1, G3/a@0 and G3/a@1.

13.3.2 *TPG Algorithm*

This section will detail the main steps of a Test Patten Generator (TPG) algorithm. Given a target fault F, the TPG algorithm has to determine the logic values to be applied to the circuit primary inputs ensuring the detection of F. TPG performs the following three steps:

- **Sensitization**: it consists in determining the logic value ('0' or '1') to be applied on the fault location in order to sensitize the faulty behavior of the target fault;
- **Propagation**: it consists in determining the logic values ('0' or '1') to be applied on the inputs of the gates belonging to the *Propagation Path*;
- **Justification**: it consists in determining the logic values ('0' or '1') to be applied on the inputs of the gates belonging to the *Justification Path*.

The above three steps will be detailed by means of the example based on Figure 13.10. First, we consider that the Fault Manager select as target fault F the **G3/a@0** as shown in Figure 13.11.

13.3.2.1 *Sensitization*

The target fault F is a Stuck-at-0, we have thus to set the fault location (G3/a in our example) to logic value '1'. In the general case, given a fault with a polarity $P = \{0 \mid 1\}$ we have to set the fault location to the opposite value P'.

Figure 13.12 reports the formalism used for the sensitization as *V/FV*, where

- *V* corresponds to the logic value required to sensitize the faulty behavior;
- *FV* corresponds to the faulty value.

In our example we have *V* = '1' and *FV* = '0'. The next step has to propagate *FV* to **at least** one primary output in order to make it observable.

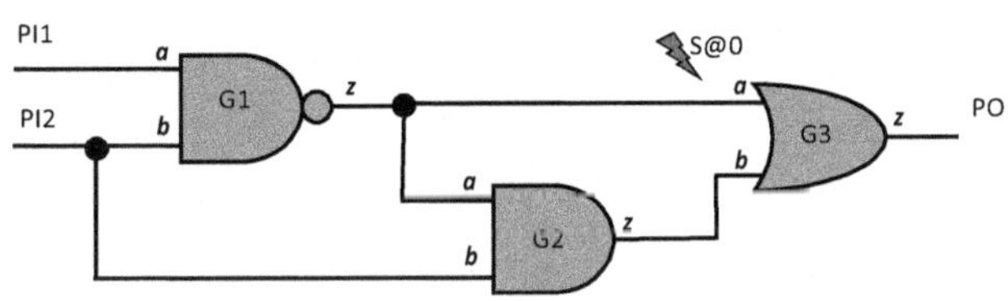

Fig. 13.11. Test Generation Example: Target Fault G3/a@0

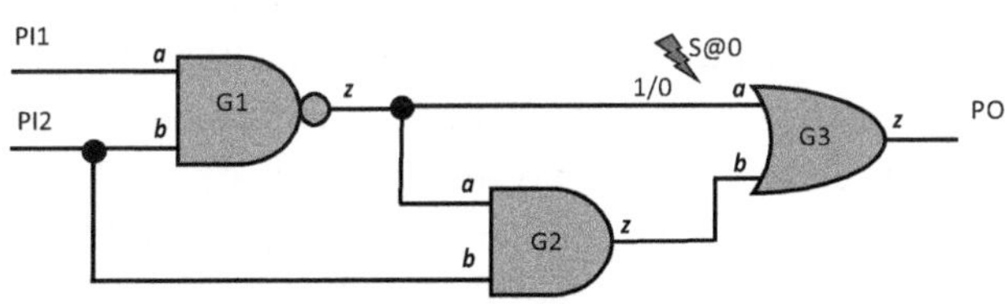

Fig. 13.12. Test Generation Example: Sensitization of G3/a@0

13.3.2.2 *Propagation*

The *FV* has to reach the primary output (PO) of our example. It has therefore to traverse the OR gate G3. The propagation step identify the value '0' to be set at the input b of gate G3 as depicted in Figure 13.13. In the general case, The Propagation exploits a *Forward* propagation: from the faulty net to reach (at least) one of the circuit primary outputs.

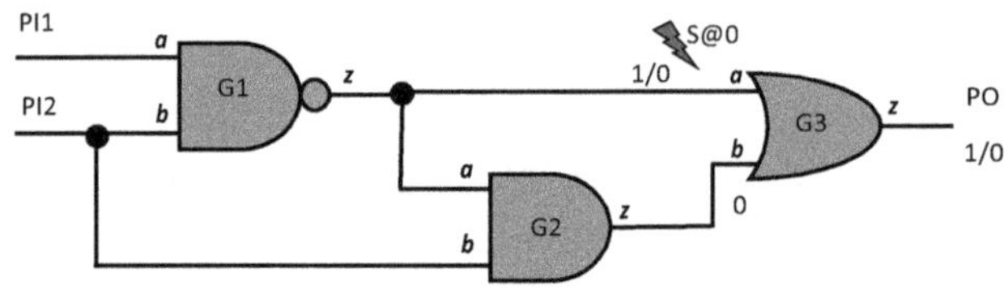

Fig. 13.13. Test Generation Example: Propagation of G3/a@0

13.3.2.3 *Justification*

The last step has to identify a set of logic values to be applied to the circuit primary inputs in order to satisfy the Sensitization and the Propagation values. The Justification exploits a *Backward* propagation: from a given net back to reach the primary inputs. In our example, we have to values to be justified: the logic ‘0’ on G3/b required by the propagation and the logic ‘1’ on G3/a required by the sensitization.

The logic ‘0’ on G3/b depends on the output of G2. In order to have ‘0’ at G2/z we have 3 possible values for G2 inputs: “00”, “01” and “10”. Let is keep the first one “00”. In order to have ‘0’ at G2/b we have to assign ‘0’ at PI2. However, by doing that, we implicitly set G1/z to logic ‘1’ but this is a violation of the logic ‘0’ required on G2/a. The reader can now see that the selection of “00” for G2 cannot be satisfied. We have to discard it and uses another one. This time we keep “10”, this time setting ‘0’ on PI2 satisfies both the values. Please note that the values of PI1 is not important, and we let is at ‘X’ (do not care).

We still have to justify the logic ‘1’ on G3/a. It can be simply seen that the ‘1’ at G1/z already satisfy this condition. We can thus stop and report the test vector a “X0”. To be more precise, “X0” is not really a test vector but it is called a **Test Cube** [6]. Indeed, from “X0” we can derive 2 test vectors: “00” and “10” both of them ensure the detection of the target fault G3/a@0. The Fault Manager update its state from **Undecided** to **Detected**. Figure 13.14 reports the justified values and the obtained test cube.

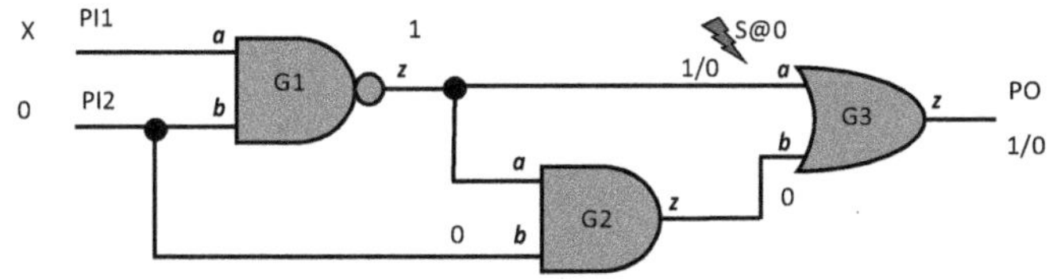

Fig. 13.14. Test Generation Example: Justification of G3/a@0

The reader may have the impression that propagation and justification are not a trivial tasks, and this is true. Indeed, there are many algorithms in the literature implementing these steps. The most popular (and used in the commercial ATPG tools) are the D-Algorithm, [7], and PODEM (an optimized version of the D-Algorithm) [8]. The details of these algorithms are outside the scope of this chapter. All the details can be found in the references [7-9].

13.3.3 *Fault Simulation*

Before moving to the next target fault, the Fault Simulation step is executed. The goal of this step is to simulate the generated test vector and determine if other faults are also detected. In this sense, the Fault Simulation can be considered as a kind of optional steps (i.e., to avoid to generate test vectors not needed). As for the test generation algorithm, there are several fault simulation algorithms existing in the literature such as Parallel [10], Concurrent [11], and Deductive [12] algorithms. Let us resort to the simplest one called Serial Fault Simulation [6] and described by the following pseudo-code:

For each Undecided Fault *F* **do**
  Modify circuit netlist by injecting F
  Simulate faulty netlist with the test vector
  If response differs **then**
    mark *F* as Detected
exit

Coming back to our example, we have one test cube leading to two different test vectors: TV1 = "10" and TV2 = "00". The number of undecided fault is 11 (since we detected one fault). In the rest of this section, we will show a couple of interesting cases. Let us consider TV1 = "10" and three undecided faults: PI1@0,

PI2@1, and G2/b@1. The reader can complete alone the exercise on the remaining undecided faults.

13.3.3.1 *PI1@0*

Figure 13.15 reports the circuit netlist affected by PI1@0. The logic values shown on the figure are known from the test generation step. As reported in the figure, the logic value '1' set to PI1 can sensitize the fault, however the fault effect '0' cannot be propagates to GI/z because of the logic '0' set to PI2. The fault is thus set as undetected and the fault simulation process is stopped.

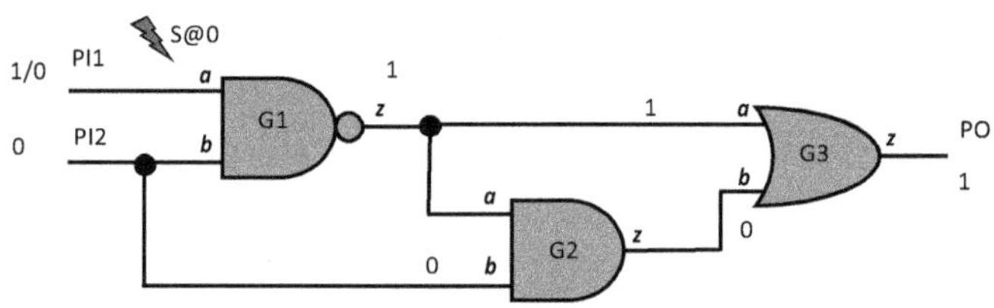

Fig. 13.15. Fault Simulation Example: PI1@0

13.3.3.2 *PI2@1*

Figure 13.16 reports the circuit netlist affected by PI2@1.As reported in the figure, the logic value '0' set to PI2 can sensitize the fault. The interest of this case stem from the fact that the fault effect '1' is propagated through each fan out branch. We can see from the figure that the fault effect is propagates through GI and thus it reaches the input *a* of G3. On the other side, the fault effect cannot be propagated through G2 because both the G2 inputs changes their values in the presence of the fault, thus the output of G2 is still set to logic '0'. Thanks to that, the fault effect is successfully propagated through G2 and reaches the circuit output. The fault is thus set as detected and it will not considered by the fault manager for the next TPG execution.

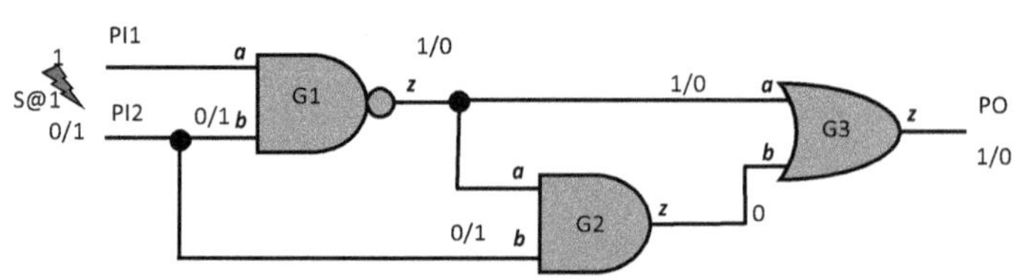

Fig. 13.16. Fault Simulation Example: PI2@1

13.3.3.3 *G2/b@1*

Figure 13.17 reports the circuit netlist affected by G2/b@1. As reported in the figure, the logic value '0' set to PI2 can sensitize the fault. The fault effect can be propagated through G2 but not through G3 because of the logic '1' set to input *a* of G3. The fault is thus set as undetected.

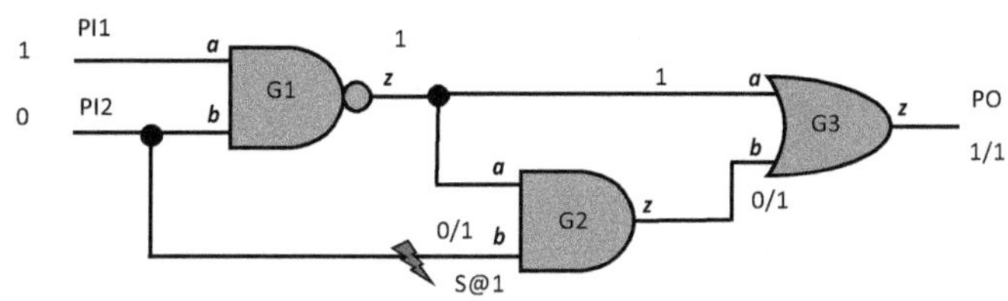

Fig. 13.17. Fault Simulation Example: G2/b@1

13.4 How the Test Sequence is Applied? Design-for-Test

The above section detailed how to generate a test set for a given fault list. The quality of the generated test set can be quantified using the Fault Coverage (Eq. 13.1). In many cases, the FC results may be too low (e.g., lower than 90%). Different reasons may negatively affect the FC. The one we discuss in this section is related to the circuit structure itself and more in particular to the presence of storage elements (i.e., sequential circuits). The classical structure of a sequential circuit is depicted in Figure 13.18.

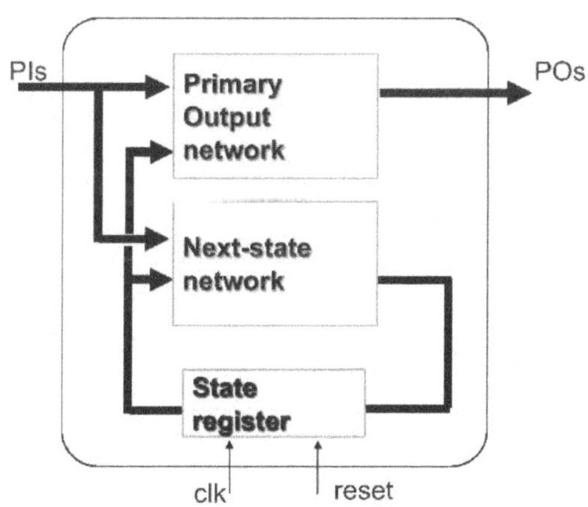

Fig. 13.18. Sequential Circuit Structure

The test set generation for sequential circuit is more complex since a single test is not anymore a "vector" but a "sequence of vectors" aiming at Initializes the circuit to a known state, Activates the fault, and Propagates the fault effect to a primary output [13]. This clearly makes the execution of ATPG algorithm more complex. Let us make a simple example using the circuit s510 from the ISCAS'89 public benchmarks [14]. This circuit has been synthetized using a custom technology leading to a gate-level netlist composed of:

- 21 Primary Inputs;
- 7 Primary Outputs;
- 211 Boolean Gates;
- 6 Flip-Flops;
- 591 Stuck-at Faults.

The reader may note that this circuit is small. We then run a commercial sequential ATPG tool in order to generate the test sequence. The ATPG took about **28 minutes** to generate 20 tests. The Fault Coverage was (only) **71.16%**. Clearly it is simply not possible to run sequential ATPG on real circuits having thousands of FFs.

In this context, the idea is to modify the circuit structure in order to make it easier for the testing, in particular easier for the test generation. It is called **Design-for-Test (DFT)**. The most popular structured DFT approach is so-called scan design. It is implemented by replacing selected storage elements (i.e., Flip-Flops) of a design into multiple shift registers (or scan cells), called scan chains. By connecting the scan output port of a scan cell to the scan input port of the next scan cell, one design can create one or more scan chains.

Depending on how many storage elements selected and replaced, it is classified as a full-scan design (more than 98% elements replaced) or a partial-scan design. Each scan cell has one additional scan input (SI) port and one shared/additional scan output (SO) port. Widely used scan cell designs include: Muxed-D scan,

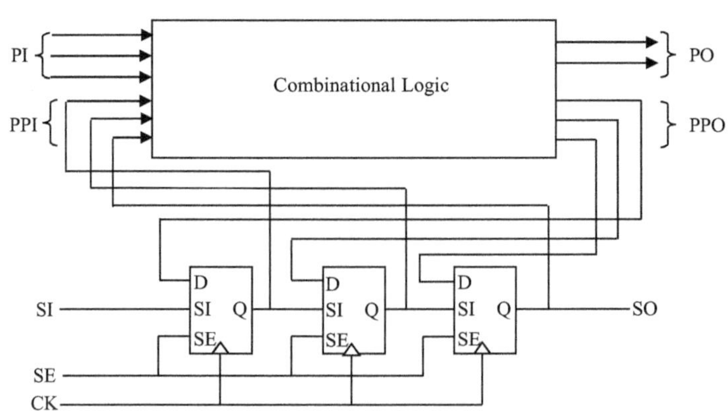

Fig. 13.19. Scan Design

clocked scan and Level-Sensitive Scan Design (LSSD). Muxed-D scan cell is much simpler than others are and does not require any additional clock routing.

Figure 13.19 shows a Muxed-D full-scan circuit. The scan cell is composed of a multiplexer (MUX) and a D flip-flop. An SE (scan-enable) signal is used for MUX port selection. In general, in a full-scan circuit the combination logic has two types of inputs: primary input (PI) and pseudo primary input (PPI), shown on Figure. Similarly, there are two types of outputs: primary output (PO) and pseudo primary output (PPO). PI and PO refer to the external inputs and outputs of the circuit, while PPI and PPO denote the outputs and inputs of scan cells.

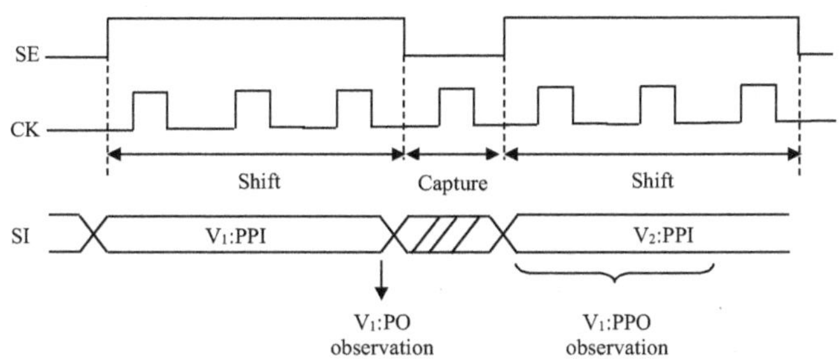

Fig. 13.20. SCAN Test Protocol

Typically, Scan-based testing works first by converting the sequential design into a scan design with three operation modes: normal mode, shift mode and capture mode. In the normal mode, all test signals are turned off, and the circuit operates in the functional configuration. In both shift and capture modes, a test mode signal is often used to turn on all test-related fixes. Figure 13.20 illustrates the test protocol timing diagram of a circuit in test operation for stuck-at faults. To apply vector V1 in the circuit, signal SE is first set to '1'. The circuit is in the shift mode and three clock pulses are applied to the clock CK. V1 includes two parts: [V1: PI] and [V1: PPI]. The portion of [V1: PI] is set directly in parallel from the external inputs of the combinational logic, while [V1: PPI] is shifted into the scan cells in serial and then through the logic circuit. Next, the capture operation is conducted. SE is switched to 0 and one clock pulse to the clock CK in order to capture the test response of the combinational logic to V1. The portion of output [V1: PO] can be observed in the external output PO as long as all the inputs go through the combinational logic. After that, in the second shift operation, the test response [V1: PPO] is shifted out through SO in serially, while scanning in the portion of the second vector [V2: PPI].

Thanks to the scan design, the combinational ATPG can be used. The cost to pay is a more complex circuit (i.e., extra routing for composing the scan chains) and a higher test time application since test vectors have to be serially shifted. Coming back to our example based on the s521 circuit, we implemented a full scan version of the circuit. We ran the same ATPG tool, and this time in less than a **microsecond**, the tool was able to generate 81 test vectors leading to **100%** of Fault Coverage.

13.5 Conclusions

This chapter presented the basics of testing for digital circuits and tried to answer the following questions:

- **Test for what?** Faults that correspond to a model of physical defects occurred during the manufacturing process;
- **Which test sequences do I have to apply?** It depends on the circuit structure and the target faults. We need to resort to an automated tool (the ATPG) to generate a test set;
- **What are the parameters to be used to measure the quality of the test?** Mainly the Fault Coverage (how many faults are detected over the total number of possible faults) and the test length (how many test vectors have been generated to obtain a given FC);
- **How the test sequence is applied?** The test sequence may be applied through dedicated test structures added to the circuit to make easy the testing. We presented the SCAN design as an example (the most popular) of that.

References

1. Mark, D. and Fan, J., Localizing open interconnect defects using targeted routing in fpga's, in 2004 International Conference on Test, pp. 627–634 (2004), doi:10.1109/TEST.2004.1387000.
2. Abuelma'atti, M. T., Modelling of resistive interconnects for computer-aided analysis of vlsi circuits, Electronics Letters 25, 25, pp. 1734–1735 (1989), doi:10.1049/el:19891160.
3. Li, James Chien-Mo and McCluskey, E. J., Diagnosis of resistive-open and stuck-open defects in digital cmos ics, IEEE Transactions on Computer-Aided Design of Integrated Circuits and Systems 24, 11, pp. 1748–1759 (2005), doi:10.1109/TCAD.2005.852457.

4. Abramovici, M., Menon, P. R. and Miller, D. T., Critical path tracing - an alternative to fault simulation, in 20th Design Automation Conference Proceedings, pp. 214–220 (1983), doi:10.1109/DAC.1983.1585651.
5. Waicukauski, J. A., Lindbloom, E., Rosen, B. K. and Iyengar, V. S., Transition fault simulation, IEEE Design Test of Computers 4, 2, pp. 32–38 (1987), doi:10.1109/MDT.1987.295104.
6. Bushnell, M. and Agrawal, V., Essentials of Electronic Testing for Digital, Memory and Mixed-Signal VLSI Circuits, Frontiers in Electronic Testing. Springer US (2006), ISBN 9780306470400.
7. Roth, J. P., Diagnosis of automata failures: A calculus and a method, IBM Journal of Research and Development 10, 4, pp. 278–291 (1966), doi:10.1147/rd.104.0278.
8. Goel, P., An implicit enumeration algorithm to generate tests for combinational logic circuits, IEEE Transactions on Computers C-30, 3, pp. 215–222 (1981), doi:10.1109/TC.1981.1675757.
9. Kirkland, T. and Mercer, M. R., Algorithms for automatic test-pattern generation, IEEE Design Test of Computers 5, 3, pp. 43–55 (1988), doi:10.1109/54.7962.
10. Seshu, S., On an Improved Diagnosis Program, IEEE Transactions on Electronic Computers EC-14, 1, pp. 76–79 (1965), doi:10.1109/PGEC.1965.264063.
11. Saab, D. G., Parallel-concurrent fault simulation, IEEE Transactions on Very Large Scale Integration (VLSI) Systems 1, 3, pp. 356–364 (1993), doi:10.1109/92.238447.
12. Armstrong, D. B., A deductive method for simulating faults in logic circuits, IEEE Transactions on Computers C-21, 5, pp. 464–471 (1972), doi:10.1109/T-C.1972.223542.
13. Auth, E. and Schulz, M. H., A test-pattern-generational gorithm for sequential circuits, IEEE Design Test of Computers 8, 2, pp. 72–85 (1991), doi:10.1109/54.82040.
14. Brglez, F., Bryan, D. and Kozminski, K., Combinational profiles of sequential benchmark circuits, in Circuits and Systems, 1989, IEEE International Symposium on, pp. 1929–1934 vol.3 (1989), doi:10.1109/ISCAS.1989.100747.

https://doi.org/10.1142/9789811210723_0014

Chapter 14

Analog, Mixed-Signal, and RF Circuits Test

Gildas Leger
Consejo Superior de Investigaciones Científicas,
Universidad de Sevilla
Instituto de Microelectrónica de Sevilla
Seville, Spain

14.1 Introduction

The test paradigm for Analog, Mixed-Signal and Radio Frequency circuits (AMS-RF) is fundamentally different from that of digital circuits. That difference resides in the nature of the involved signals. In digital circuits, they are discrete in amplitude and also in time for the most part, while in their analog counterparts, they are continuous both in amplitude and time. Even for sampling circuits, the dynamics are usually too important to be overlooked. As a consequence, the defective behaviors of AMS-RF circuits are much more difficult to distinguish from the normal behavior, particularly if you consider the unavoidable process, voltage, and temperature (PVT) variations.

For digital circuits, the standard test approach is defect-oriented: the circuit under test is considered as good-by-design, and an automated test program generator is used to identify stimuli sequences that will make possible defects visible on monitoring nodes. Circuits with a detected defect are rejected, even if this defect may not impact the functionality for a given application. Indeed, digital circuits of moderate complexity have so many inputs that it is impossible to test for all the possible combinations, even considering only a pure combinational case (which is obviously not the most common case).

For AMS-RF circuits, on the other hand, the standard approach is functional test: the performance metrics specified in the datasheet are measured through standardized test procedures and compared to their acceptance windows. Such an approach can be expensive because the involved measurements require high accuracy equipment and long test times. However, it is usually feasible because the specification list is not too long. A reason why performance testing is almost never completely overlooked (i.e. replaced by a defect-oriented approach) lays in the verification and design closure of AMS-RF circuits. Indeed, due to long simulation times of this type of circuits, it is almost impossible to guarantee that a circuit is good-by-design, which is a necessary condition for defect-oriented testing. Performance-based testing will capture failures linked to spot defects, excessive parametric variations of the fabrication process, but also design errors, model errors or any other systematic error.

That being said, the interest in defect-oriented testing for AMS-RF circuits has not completely vanished and is even regaining some momentum in the test literature. The driver of this trend is quality in safety- or mission-critical applications. If we take volume into consideration, the most important market is definitely the automotive. In this kind of market, the cost of test escape (i.e. a bad circuit that somehow manages to pass the test) can be extremely high and the goal is to reach 0 Defective Parts Per Millions (DPPM). For these extremely demanding markets, it has been shown that some in-field failures could be traced back to circuits that did successfully pass a thorough functional test. Failure analysis further showed that these circuits presented some kind of fabrication defect in the first place. The industry had finally experienced what the researchers in the field of test had warned for decades: Despite the fact that functional test is the golden standard, it is not perfect. Some fabrication defects can weaken the circuit in such a way that it would pass a functional test at time zero, in the production line, but produce a failure in the field of operation due to wear-out mechanisms or simply environmental conditions.

14.2 System-Level Considerations

Before entering the categorical split between defect-oriented and performance-oriented test, let us first describe some system-level DfT techniques that may be used in both approximations.

14.2.1 *A generic approach: Analog test-bus*

In analog systems (even in RF ones), it is usually useful to monitor a potentially large number of DC signals. These can range from internal node operating points in some specific biasing/configuration conditions to the output of some sensors.

The IEEE 1149.4 mixed-signal test access standard was first though for printed circuit board testing, providing a way to measure the impedance of passive components. However, later developments adapted the test bus infrastructure to be able to address measurements inside the chip.

An evolution of such a generic test bus is sketched in [1]. The authors rely on the IEEE P1687 (IJTAG) standard to provide access to embedded instruments that perform the necessary D/A and A/D conversions. The preferred approach consists in encoding analog measurement into timing information which can be made less sensitive to PVT variations. However, direct voltage readout can also be provided resorting to embedded ΣΔ converters. For stimulus generation, the authors rely mainly on ΣΔ encoded bitstreams stored in on-chip recycling registers.

In [2], the authors also propose an embedded instrument for DC readout. The basic idea consists in performing a local A/D conversion in order to avoid possible cross-talks between the tested nodes through the test-bus. For that, they employ simple comparators. One input is connected to the internal node to be probed and the other to a shared reference voltage generated off-chip. By sweeping that reference value, one can identify in parallel the transition point of the different embedded comparators and thus estimate the DC levels. The main drawback of the approach is that the resolution is limited by the comparator accuracy. Some internal nodes may require a very limited resolution such that simple comparators based on Strong-Arm latches would be sufficient. Bu other nodes may require a much higher resolution, making the use of complex pre-amplifiers necessary.

14.2.2 *An approach for transceivers: Loop-back*

In order to test a receiver, one must generate some RF signals with the same characteristics as the signals to be transmitted. Conversely, to test a transmitter, it is necessary to analyze the characteristics of the produced RF signal and verify if it fulfills the expected standard. It is straightforward to observe that the hardware that is necessary to either externally produce or analyze the RF signals is very similar to the one used in a transmitter/receiver. So, why not reusing the transmitter hardware for test purpose? The embedded transmitter may be used to test the receiver and/or the embedded receiver to test the transmitter. That is the concept of loopback. Once the loop is closed, the test can be handled on a digital tester at a very low cost, enabling multi-site.

One of the downsides of loopback testing is that it does not give much diagnostic insight. However, this can be partially improved considering several loop closures.

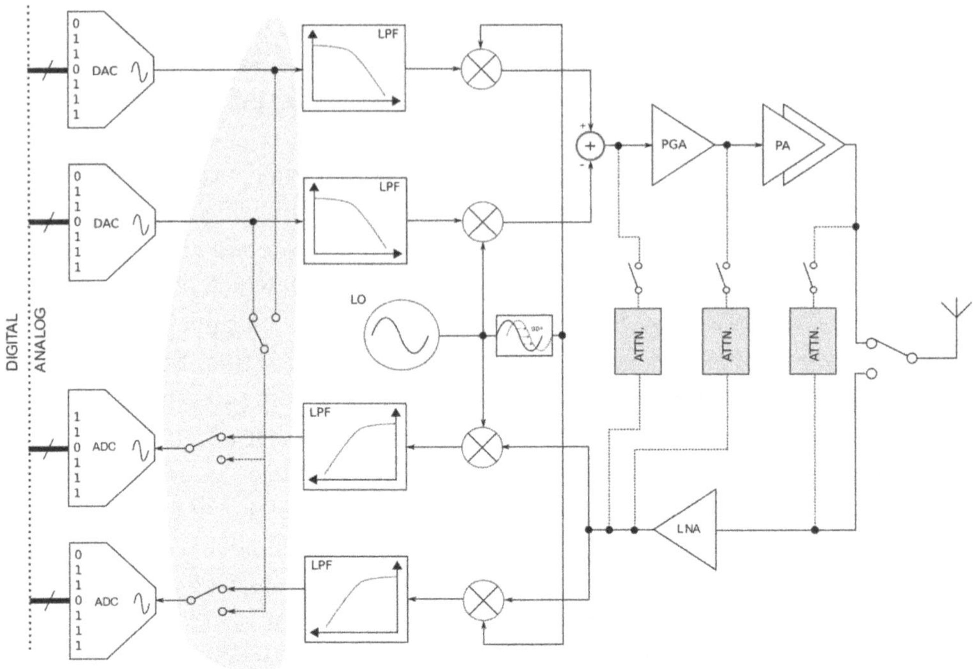

Fig. 14.1. Sketch of a direct-conversion transceiver with several loopback paths

As seen in Figure 14.1, one can consider closing the loop the closer at the antenna, which would hold the potential to detect any fault in the signal path. This implies the use of RF switches and poses some isolation issues that may degrade the normal operation. Another possibility is to exclude the PA (which is the worst aggressor) from the loop and introduce the RF switch at its input. This also relaxes the constraints on the embedded attenuator. It is also possible to close the loop just after the mixers. In the base-band domain, a loopback can also be set between the data converters. The availability of two data converters in the receiver and the transmitter (for I and Q channels) allows to discriminate the origin of the defect if necessary.

In [3], the possibility to simply bypass the LNA and/or the PA with a simple switch is commented. This solution avoids introducing additional multiplexers in the signal path with their corresponding insertion loss. The authors argue that proper sizing of the bypass switch can allow minimal performance penalty in normal mode. Besides, in that paper, some limitations of the loopback architecture for performance measurements are commented. In particular, the noise performance of the necessary attenuator will limit the conditions of any BER test.

As briefly mentioned, the main issue with RF loopback is isolation. In [4], details are given of a practical compact implementation of a 30dB attenuator and an RF switch. It can be concluded that a Π-type implementation is preferable to T-type for the resistor-based attenuator. And with respect to the RF switch, it may be necessary to improve isolation even at the cost of higher insertion loss. Indeed, the experimental measurements showed that the proposed 2-transistor switch caused a -17dB leakage on the off-port, which was 8dB higher than expected. In [5], several attenuator architectures are thoroughly analyzed, which could be used for loopback purpose and provide additional isolation.

While most of the research using loopback schemes rely on defect simulation and/or MonteCarlo simulations for the validation of the approach, the work in [6] is particularly valuable since it validates the loopback approach (plus some other defect-oriented RF BIST) on real production data from more than 1 Million parts of a WLAN transceiver. In loopback mode, a gain test checks the programmable gains of the pre-power amplifier. Similarly, a linearity test sets the receiver (and the loopback attenuator) for a low-power configuration and observes the intermodulation components for a two-tone RF input to determine the IIP3 of the transmitter. Out of the 1 Million parts, 616 were claimed faulty from standard EVM test. 528 of these faults were in the digital sections and 88 in the analog sections. The very basic loopback gain and linearity test was able to detect 9 of these analog faults which could not be detected by the other BIST approaches and thus contributed to reach 100% coverage.

14.3 Performance-Oriented Testing

Let us start with the description of performance-oriented testing. The first comment to be made is that performance-oriented testing can be considered at different abstraction levels.

At system-level, the full transceiver can be considered. In such a case, the main specification to be tested would probably be the Bit-Error-Rate or the Error Vector Magnitude in different operating conditions. At sub-system level, the transmitter and the receiver would be tested separately. The specifications to be measured would be those defined in the corresponding standard (the spectral mask for the transmitter, for instance). Finally, at block-level, each building block would be treated separately: the LNA, the PA, the ADCs and DACs, the mixers, the PLLs, the filters, etc. Here the specifications to be tested would be linked to the system-level design trade-offs. For instance, the requirements in terms of noise (Signal to Noise Ratio, SNR) and distortion (Total Harmonic Distortion) for the baseband ADC of the receiver obviously depend on the architectural choice, the frequency planning and the performance of the surrounding blocks. These block requirements are set during the design process.

The strategy of descending to a lower abstraction level if the functional test is inappropriate at current level (too expensive, inaccurate or simply unfeasible) is known in the test community as a *divide-and-conquer* strategy. Whether such a divide-and-conquer strategy can still be considered performance-oriented is a matter of debate. Indeed, a direct correlation to the system-level performance will not be obtained and the test validation rather resorts to defect-coverage as a quality metric. However, measuring block-level performance still retains some benefits of pure performance-based testing as for instance the ability to detect most non-modeled errors.

14.3.1 *Design-for-testability and built-in self-test*

Design-for-Testability (DfT) is, as its name indicates, any design practice that shall help the testing of the device. We will further consider some cases related to defect-oriented testing, but for performance-oriented testing DfT is mostly related to accessibility and observability. Indeed, if a divide-and-conquer approach has to be considered for a transceiver and the main building blocks are tested independently, then the primary IOs of these blocks should be made available to the test facilities. Making the output of an ADC observable or the input of a DAC accessible is not very challenging. However, multiplexing the input of a baseband filter or an ADC should be considered with care. Needless to say, multiplexing the IOs of an LNA without compromising its performance is not an easy task.

Built-In Self-Test (BIST) approaches take one step beyond the “simple” accessibility and observability improvement. These solutions aim at embedding the test instruments within the chip. For performance-oriented testing, it requires a stimulus generator and a response analyzer. Here again, these test instruments must be embedded without significant degradation of the performance. Thus, BIST and DfT related to AMS-RF circuits imply co-design and the most important corollary is that the test strategy decisions must be taken very early in the design flow, at the architectural design level.

14.3.1.1 *RF blocks*

For testing RF blocks, embedded signal generators are seldom found as such. Indeed, it is extremely difficult to produce a calibrated input power from an

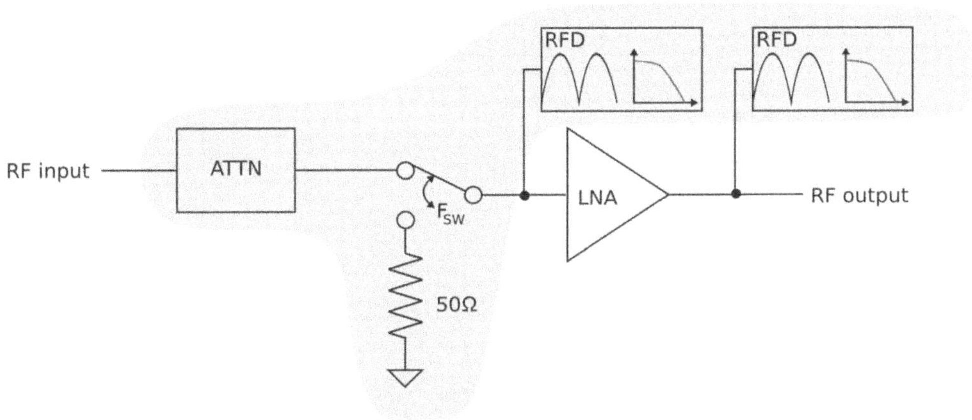

Fig. 14.2. Using RF power detectors to test LNA gain

embedded device. Instead, the embedded signal generator is relaxed and what is actually measured is the transfer function of the block under test.

In [7], for instance, the authors propose to embed power detectors together with an LNA. The RF stimulus generator is a simple switch that generates a square wave to modulate a fixed RF tone. In that paper, this switch is actually used to connect the transceiver in a loopback configuration, as discussed earlier. The power of the modulated RF signal is measured both at the input and the output of the LNA, as shown in Figure 14.2. With such a configuration, it is thus possible to obtain the gain of the LNA. Varying the attenuation of the RF signal then enables retrieving the 1dB compression point.

As mentioned, the introduction of any DfT modification to an RF system (and to a minor extent, any analog system) implies co-design. Indeed, the parasitic loading introduced by the power detector shall be minimized. A possible design of such a detector, with reduced capacitive loading, is detailed in [8]. It achieves a 30dB dynamic range for a fixed biasing current. Though it is not experimentally proven in the paper, the authors argue that the dynamic range can be further improved by programming the biasing current of the detector.

If the complete transceiver is implemented on-chip, the input stimulus generation is usually carried out by the transmitter (or parts of it). However, if the receiver is implemented in a stand-alone fashion, an on-chip stimulus generator holds the potential to greatly reduce test cost, for instance by enabling multi-site testing. In [9], such an RF stimulus generator is proposed, that consists in two low-noise VCOs and an adder to perform single-tone or two-tone tests. The VCOs are implemented using a configurable 5-9 stage Voltage-Controlled Delay Line. That architectural configurability together with the fine-grain tuning of the element

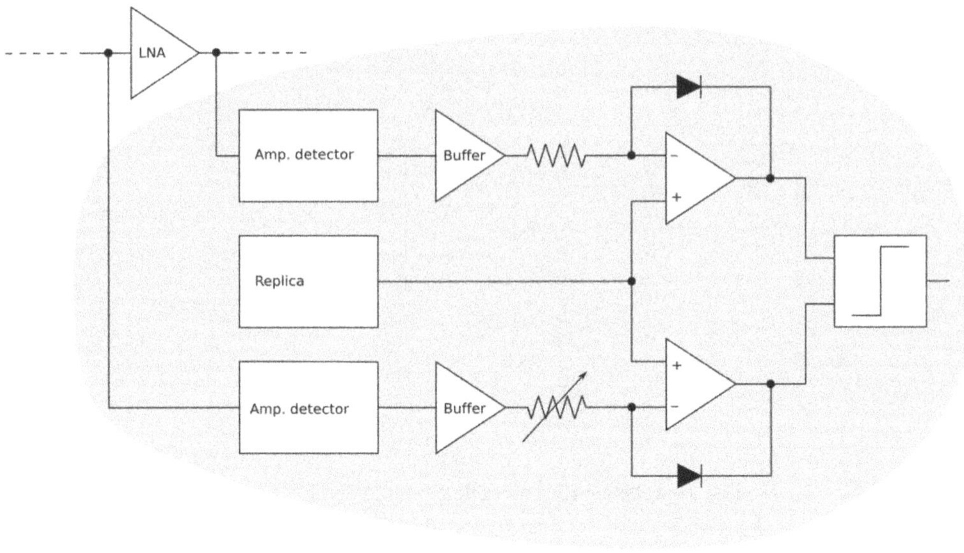

Fig. 14.3. LNA BIST scheme with embedded pass-fail decision. The minimum desired gain for the LNA is set with the resistive attenuator. The scheme can be used to test the circuit at different power levels

delay by the introduction of diode-connected transistors at their output allows for a complete tuning range from roughly 1GHz to 5.6GHz. The choice of ring oscillators instead of LC-tank VCOs favors a reduced area impact but degrades the phase noise. However, that phase noise is contained at -80bBc/Hz for a 1MHz offset (at 2.4GHz). The embedded adder also implements a selective attenuation on the two signals. Together with the wide frequency tuning range, this capability enables to emulate a case with a strong in-band signal and a small (undesired) out-of-band aggressor.

[10] takes a similar approach, but the BIST is completed by a comparator that will directly issue a pass-fail decision. To do so, the authors propose to add an attenuator after the amplitude detector at the output of the LNA. This function is carried out by the tunable resistor in Figure 14.3. Two logarithmic amplifiers help relaxing the dynamic range requirements for that comparator and also perform a subtraction operation with a replica detector to get rid of non-meaningful DC component. Indeed, the amplitude detector itself is a simple 1-transistor amplifier where the transistor is biased in the subthreshold region and AC-coupled to the RF signal. This amplifier is followed by a passive low-pass filter that removes unwanted RF power. The output is thus a DC value which is the sum two components: one fixed and another

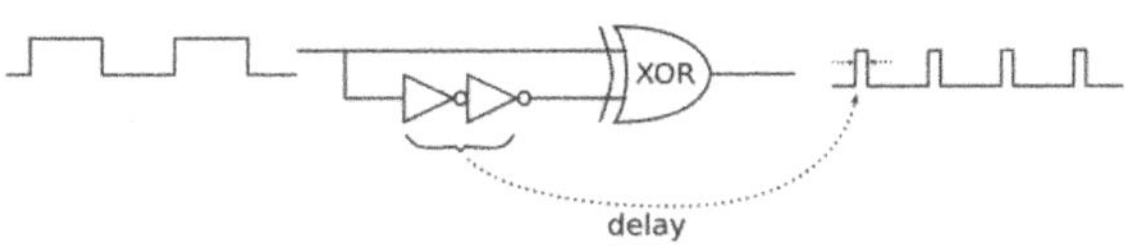

Fig. 14.4. Simple RF signal generator based in a train of pulses. The width of the pulse defines the bandwidth of the signal around the clock frequency

that is proportional to the square of the RF signal amplitude. The fixed component is what is removed by the replica circuit.

One example of RF signal generator for receiver testing can be found in [11]. This very simple block, depicted in Fig. 14.4, generates a train of pulses from a clock using a simple XOR gate and a delay. The spectral transform of such a signal is also a train of pulses but the non-zero width of pulses effectively implements a cardinal sinus (sinc) envelope on the spectrum. As a result, the train of pulses is only flat within a reduced bandwidth. That being said, it can be properly set to address the receiver band of interest. The signal generator is also complemented by a programmable attenuator to control the amount of RF power to be processed by the receiver. Setting the generator's clock to a particular frequency, we can exactly know in which FFT bin of the receiver baseband should the fundamental tone appear. However, due to the distortion induced by the receiver, that FFT bin will also be contributed by harmonic content (intermodulation components between the different pulses of the train). Considering a third order polynomial model of the receiver, the authors compute the expected harmonic power in the monitored bin. This power is amplitude dependent and thus, testing the device for different power levels and performing a regression allows retrieving the coefficient of the polynomial model and compute some specifications like the IIP3.

Some dedicated BIST also exist for another critical block in transceiver, namely, the phase-locked loop (PLL). A PLL is mostly a digital circuit, at least in the sense that its main IOs are digital signals. However, there is a characteristic that is analog in nature and utterly critical for proper RF operation: the clock jitter. [12] presents a DfT module for measuring clock jitter. This is not a full-BIST solution because the module needs an additional input: a clock signal whose frequency is coherent with the PLL reference clock but slightly lower. Basically, the proposed approach performs under-sampling of the PLL output clock using that additional clock. The resolution of the jitter measurement is equal to the period difference between the two clocks, which can be set very small (typically below 5ps). The transitions of the clock signal under test can thus be precisely located onto an equivalent period. Such an equivalent period is taken as the difference between the median of rising edge glitches. This allows to separate Low-Frequency jitter from High-Frequency jitter. The HF jitter can be characterized by the counting statistics of the position of the glitches with respect to these median edges. Conversely, the LF jitter can be characterized by comparing the glitch-filtered clock to a constant period reference. More details can be found in [13].

While the above mentioned module is agnostic to the PLL/DLL architecture, the work in [14] presents a BIST scheme for charge-pump PLLs. In that setup, the PLL loop is re-used to convert it into a time-to-digital converter. The only extra hardware is a modified PFD and charge pump, and a counter. Basically, the PLL is configured in normal mode until it locks properly. At that point, the frequency is supposed to be properly tracked and the phase is what is actually read by the PFD. The jitter thus appears as variations of the pulse duration at the output of the PFD. So, once the PLL is locked, the loop is opened-up and reconverted into a time-to-digital converter. The phase (i.e. time) difference between the reference clock and the divided VCO output is converted into a frequency shift which can be simply measured by a counter. Repeating this operating several times gives the statistics of the pulse duration at the output of the PFD after locking, from which the jitter can be deduced. The paper reports a jitter measurement for an integrated prototype as small as 5.5ps RMS with a resolution of 1.2ps and an error of 15% with respect to conventional measurement methods.

14.3.1.2 *Baseband circuits*

For baseband circuits, the test literature is much more extensive, particularly for data converters. Many papers can be found that deal with signal generation. Linear ramps can be generated by simply integrating a fixed current into a capacitor. The practical implementation of such ramp generators can vary significantly depending on the target linearity. The interested reader can refer to [15] for a detailed study. Such linear ramps can be used to perform simple histogram testing for ADCs: for a ramp input, the Integral Non-Linearity (INL) for a given code is directly proportional to the deviation from the average code count. The digital hardware necessary for the response analysis is thus very limited. That being said, the linearity of the on-chip ramp generator must be superior to that of the ADC under test. This may be quite challenging to ensure in some cases.

For a monotonous and continuous input signal – as would produce, for example, the simplest implementation of a ramp generator – a technique called double-histogram allows separating the non-linearity contribution of the signal from that of the ADC [16][17]. The basic idea is that two histograms (i.e. code counts) are acquired: one with the original test stimulus and another for the same stimulus plus a reduced offset. The correlations between the code counts for the two histograms can be used to retrieve the INL of the ADC.

However, for data converters in RF transceivers, the most important performance parameters are probably the spectral ones that can account for the dynamic behavior (THD, SNDR, ENOB). Measuring these parameters for an ADC requires a sine-wave generator.

The most straightforward approach is to use an on-chip oscillator. Many analog oscillator designs can be found in the literature and, most of the time, they rely on a high-Q bandpass filter with a non-linear feedback (typically a comparator followed by a one-bit DAC). However, it can be shown that, in order to test a 10b ADC, a quality factor above 70 would be necessary for the bandpass filter, which can be quite challenging to achieve [18]. In order to relax the linearity requirements (and the filter quality), techniques based on harmonic cancellation have been proposed. They consist in summing properly time-delayed and possibly scaled versions of an oscillation waveform in such a way that some of the critical harmonics are cancelled out while the fundamental is not affected. In [19], a discrete implementation managed to reach -82dB THD for a slow sine wave (at 30kHz), while in [18], a pseudo-differential integrated oscillator with harmonic cancellation reached a THD level of -72dB for a 10MHz sine-wave.

Besides on-chip oscillators, another approach for signal generation relies on stripped-down versions of DACs. In [20], the authors propose to use a simplified version of a switched-capacitor DAC such that only the capacitors required to generate a sine-wave at rational fractions of the sampling frequency are effectively implemented. In the same line, another approach consists in relying on $\Sigma\Delta$ modulation [21]. Using a simple 1-bit DAC, which is inherently linear, and a low-pass filter, it is possible to encode a $\Sigma\Delta$ bitstream corresponding to a pure sine-wave in a recycling register. Here, the filtering requirements are quite different from the conventional oscillator approach: What has to be removed are not the harmonics but rather the quantization noise that is pushed out of the band of interest by the $\Sigma\Delta$ coding. The downside of this approach is that, for a low-pass modulator, the sampling frequency must be significantly higher than the encoded sine-wave frequency. In turn, band-pass $\Sigma\Delta$ coding schemes would relax the oversampling ratio at the cost of a more complex bandpass filter. Notice that this approach is much simplified if the ADC to be tested is a $\Sigma\Delta$ modulator itself. In such case, no reconstruction filter is actually necessary if the $\Sigma\Delta$ encoded test stimulus is properly attenuated or shaped at a higher order than the modulator under test [22].

For all the signal generators that have been overviewed, it is important to consider the adaptation to the input impedance of the data converter. Indeed, many

ADCs based on switched capacitors (like for instance most pipeline converters, Successive Approximation Register converters (SAR)) do not have a high fixed input impedance. Hence, the signal generator must be able to properly drive the sampling capacitor and withstand a potentially large kickback noise. A successful BIST implementation may thus require the implementation of a high bandwidth buffer [23].

14.3.2 *Machine-learning indirect test*

Machine-learning indirect test can be considered a performance-oriented test, but it is significantly different from the direct measurement approach that we have considered so-far.

The unavoidable variations of some physical parameters of the fabrication process—for instance, the thickness of the gate oxide varies across a wafer—induce changes in the electrical parameters of otherwise identical transistors. During the design phase, this variability is acknowledged and taken into account through process corners and/or Monte-Carlo simulations. However, the fierce competition in the semiconductor market pushes toward minimizing the guard-bands and state-of-the-art products – particularly in RF circuits which tend to very sensitive to process variations – seldom have a 100% parametric yield. In other words, the tails of the distribution of the fabrication process parameters usually cause some circuit to fail the specifications.

The rationale behind indirect test is that, under normal process variations, the measurements carried out on a circuit are correlated because they all rely on the same root causes (the process variations), driven by a stochastic process. Making use of these correlations, it should thus be possible to substitute costly specification measurements by other simpler measurements and achieve the same coverage.

The first version of such an indirect test is the so-called blind model-based [24], which is essentially a linear regressor. It represents the complex specification measurements as a linear combination of some simpler measurements.

In the early 2000, the idea was extended and renamed as Alternate Test [25], to leverage the power of modern machine-learning algorithms. The conceptual ground, though, remains identical: For a small but representative set of circuits, both the specification and the simple measurements are measured. This set is called the training set. Then a model is learnt for this training set to map the simple measurements onto the specification measurements. This training must be carried out minimizing the generalization error, that is, the error committed by the model to predict the result on new data. Finally, in the final stage in the production line, only the simple measurements are acquired (saving most of the test cost) and the model is used to predict the performance.

Models

First, the choice of the tool to be used may appear intimidating to a test engineer, since machine-learning has experienced an exponential development in the past few years. However, the observation of the test literature seems to indicate that the model choice is not the main bottleneck of the approach. Obviously, some algorithms may perform better than others in a given situation but still, any modern algorithm shall clearly outperform linear modeling.

That being said, there are some important trade-offs that one should be aware of. The first one is related to the type of task to be performed: a classification or a regression? Typically, a classifier produces a binary pass-fail decision and the specification acceptance threshold is embedded into the training process. However, smarter approaches may include another category to account for uncertainty. This allows two-tier approaches, where the circuits labelled as uncertain undergo additional testing. Alternatively, the algorithm can be aimed at regression and will not produce a pass/fail decision but an indirect estimation of the circuit performance. This approach gives more information on the circuit and would enable performance binning or process drift monitoring. On the other hand, it requires one model to be trained per performance measurement. Classifiers, on the contrary, can embed several specifications into a single pass-fail boundary but greatly reduce the diagnostic capabilities. With respect to accuracy, training a regressor will aim at minimizing the generalization error equally over the full exercised range of performance without concentrating the effort in the vicinity of the specification threshold.

Another acknowledged machine-learning trade-off is that of bias versus variance. The bias of a model is the systematic part of the prediction error that it will produce, while the variance part of the prediction error relates to the variability in the training data. For instance, a linear regression usually provides a model with strong bias and low variance: if you change the training data, the linear part of the trend in the data will probably remain unaltered, so the variance is low. However, if the underlying relation between the data has any non-linearity, it will always produce an error. Conversely, a model based on nearest neighbor has high variance and low bias: it will adjust very well to any non-linearity in the data, but at the same time the prediction greatly depends on the training data. High variance models are thus more sensitive to outliers and noise in the data.

These two examples are archetypical but there exist many types of models that try to strike the right balance between bias and variance. Generally speaking, model training is based on optimizing a number of parameters to minimize an estimate of the generalization error. Typically, the higher the number of free parameters, the lower the bias and the higher the variance. For instance, in a perceptron neural network, the bias/variance trade-off is governed by the number of hidden layers and neurons. In the field of electronic testing, the most used regressors are Neural-Networks and Multivariate Adaptive Regression Splines for regressors. Neural Networks are also widely used for classification as well as Support Vector Machines.

If the choice of an adequate model still seems a daunting task, it has been shown that heterogeneous ensemble learning can be of help [26]. It consists of training a variety of model types and combine their output by weighted averaging (for regression) of voting (for classification). In that way, the training algorithms that take care of the averaging also performs an embedded model selection which guarantees that the best model type(s) for the problem at hand will be used. The model selection and output weighting are based on estimates of the model generalization error based on cross-validation. The training set is split in N equal parts and the model(s) are trained on (N-1) parts and then the prediction error is computed on the Nth part that has not been used for training. This process can be carried out N times, so that the generalization error is estimated for a large number of samples. Taking that approach to the limit, that is to say considering N as equal to the number of samples, is known as leave-one-out cross-validation.

Input space

Despite these considerations on machine-learning algorithms, the most important ingredient for a successful machine-learning indirect test is the definition of the input space. It is important to have a set of information-rich features that hold the potential to actually represent the target specifications. A feature can be relevant if it brings additional information to an existing feature set but it can also be redundant if the information that it brings is already available in the set. Obviously, these two properties are context dependent: they will depend on the actual set of feature but also on the model type.

Feature selection algorithms form an extensive chapter of the machine-learning bibliography. In the field of electronic testing, the question has drawn some interest since the process of proposing feature candidates – which must be associated to some simple measurements – is still mainly ad-hoc and guided by expert knowledge. In most approaches, the features that are considered are dc voltages corresponding to the operating point of some internal nodes of the circuit under test or the output of current probes. For the case of RF circuits, we have already mentioned that most embedded test equipment shall be co-designed with the circuits themselves in order to guarantee that additional loading of internal nodes do not impair normal operation. An example of such a co-design is the envelope detector associated to an LNA in [26]. However, another clever approach consists in using embedded process monitors [27]. Such process monitors can be replicas of some parts of the circuit which will not be used during normal operation, such that unwanted loading is completely avoided. Obviously, the downside is that local mismatch effects will not be captured but these are usually not dominant compared to global process variations in most RF blocks.

To enhance the information contained in all these features, it is also possible to perform the measurements in a variety of operating conditions (at different supply voltage, operating frequencies, temperature, etc.). In the end, one can come up with a large number of feature candidates. While this may not seem problematic, it actually is. An excess of features will most of the time reduce the accuracy of the trained model through overfitting. And what is more important, each feature corresponds to an additional measurement that has to be performed with the corresponding test cost. Feature selection is thus of much interest. In [28], several approaches were reviewed that included filtering as well as wrapper approaches. Filtering approaches aim at ranking the relevant features a-priori, without training any model. They are fast but their accuracy is usually limited to the first (few) features. Wrapper approaches are much slower but much more accurate, since they consist in a combinatorial search in the space of the available features, where the generalization error of the true learning algorithm is minimized. A hybrid approach is thus proposed based on an incremental filtering steps over training residues, which manages to keep computational cost low while achieving a good accuracy.

In the last few years, some research has also addressed the issue of feature generation: How to guide the designer/test engineer to propose new features, if the available ones are insufficient? One possible idea is to perform stepwise addition of Monte-Carlo process parameters (which are available through the simulation process) as if they were test features. Since we know that they correspond to the root causes of the performance degradation, if we can identify which Monte-Carlo variable(s) brings the most additional information to the existing feature set, we can then trace back its physical meaning and propose some additional tests that may capture its variations. In [29], it could be shown that the information on passive components (R, L and C) was missing in a feature set consisting of internal DC operating point voltages.

Shortcomings

The main limitation of machine-learning indirect test is the same that was reported in the early days for model-based testing: non-model errors. Indeed, the approach relies on the existence of stable correlations between the simple measurements and the specifications.

The definition of the test is based on simulation, and simulations may not accurately represent the actual behavior of the circuit (the models from the design kit may not be mature enough, RF package and parasitic models may be deficient). Though machine-learning indirect test approaches assume that training will be made on real measurements, the test itself is selected and approved based on simulation results. There is no a-priori guarantee of test quality. That being said, this challenge has been acknowledged and some mitigation strategies have been proposed. As quoted above, if the uncertainty in the pass/fail decision is too large it is possible to adopt a two-tier approach where circuits will only be classified by the machine-learning indirect test if the confidence in the classification is high enough [25]. Otherwise, the circuits tagged as uncertain can undergo some additional testing. If a regressor is used instead of a classifier, then the estimate of the generalization can be used for the same purpose. Adaptive testing has also been proposed as a way to guarantee that the parameters of the fabrication process do not exhibit too much drift [31]. This imposes a periodic complete re-testing of some circuits such that the test error can be monitored. If it is too high, then some re-training could be done to recover an acceptable accuracy level. Not much work has been done in the line of re-using simulation data. However, an interesting concept was presented by [32] which consists in calibrating the parameters of a model to the experimental data. So, the model was effectively trained using simulation data but tweaked by observational measurements using Bayesian model fusion.

Finally, the most important bottleneck of machine-learning indirect test is the occurrence of spot defects. The premise of indirect test is that there exists an underlying stochastic process that explains the variations of both the performance and the features (i.e. the simple measurements). Spot defects do not lay in the category of stochastic process. The circuit behavior in the presence of a given spot defect cannot, generally speaking, be predicted by the behavior observed for a different spot defect. In order to be able to predict such behavior, the only solution would be to have observed all the possible defects, which is impossible. A prior step necessary for indirect test is thus the construction of a so-called defect filter. [33] proposes such a defect filter as the detection of outliers from the joint probability density function of the same features later used for the indirect test. However, the a-priori validation of such an approach is not straightforward. Actually, it can be observed that the defect filters proposed for indirect test are no other thing than defect-oriented approaches, which is the subject of the next section.

14.4 Defect-Oriented Testing

The goal of defect-oriented testing is to detect fabrication defects which typically are spot defects related to small particles depositing on any layer during the fabrication process. The circuit is thus considered as good-by-design and what is searched for during the test process is any deviation from the normal behavior, not necessarily in a performance measurement but in any kind of measurement.

A good defect-oriented strategy should thus propose a number of measurements such that the normal process variations are contained while the sensitivity to spot defects is high.

14.4.1 *Simulation*

Defect-oriented strategies must be designed and validated before fabrication, so they necessarily rely on simulations. Monte-Carlo simulation is mandatory to properly define the normal (acceptance) range for the proposed measurements, but this can be routinely done with existing EDA tools. Defect simulation is a different issue and commercial EDA tools are just appearing [34].

14.4.1.1 *Defect models*

There is currently an IEEE group [35] working towards the standardization of defect-oriented simulation and reporting for AMS-RF circuit. This working group has already taken some decisions:

- Defects are considered at circuit level. Circuit level is common to all AMS-RF circuits. Trying to reach higher levels of abstraction would imply translating defects into block-level faults, with the consequent loss of generality.
- Only single defect injection is considered. In modern fabrication process, the defectivity level is kept extremely low once the process is mature. The probability of two defects is thus much smaller than the probability of a single defect. The current state-of-the-art for defect simulation is far from reaching the accuracy level at which multiple defects should be considered.
- The focus is put on detecting day-0 defects (latent defects are not targeted). Defects related to (or activated by) aging are not considered in the defect-oriented methodology for now.
- Both catastrophic and (strong) parametric deviations are considered. Notice that the term catastrophic, here, refers to the local defect and not

necessarily to its consequence. For instance, a catastrophic short would be an undesired connection of a few Ohms between two nets. If such a defect occurs between the source and drain of a small trimming transistor, its effect on the circuit may be negligible. But still, it is a catastrophic defect. A parametric defect is an undue variation in a model parameter. For instance, the sheet resistance of a poly resistor may be affected by some defect (like a blob under the poly layer). The resistance would thus be altered in a much larger extent than nominal process variations. These parametric variations in a defect simulation are typically handled considering an increase (or decrease) of a factor 2 with respect to the nominal model parameter.

The defects that are typically considered are opens and shorts, undue AC coupling (intra-layer or inter-layer), stuck-on and stuck-off transistors, leakage in PN junctions or through MOS oxide and extreme variations of some process parameters (for instance, in passive components).

The modelling of shorts is almost straightforward since a resistor can be included between two nodes of the netlist. However, even in that simple case, determining a suitable value for the short resistor is linked to the root cause of the short. Ideally, the foundry should provide some input in the form of a distribution. Otherwise, ad.-hoc decisions must be taken. A typical value is in the range of 100Ω. Opens are much more difficult to model. Indeed, transient simulations may be conveniently addressed by the inclusion of a large open resistor (in the range of $1M\Omega$) in the affected net. However, in the case of MOS transistors with gate opens, such a simplistic model would not alter the DC operating point [36]. Furthermore, an open in a metal line is very layout dependent: If a node consists of a metal line with several branches, the effect may be very dependent on which branch is actually affected.

AC coupling can be handled by the introduction of an extra capacitor between the affected nodes. The leakage defects can be simply modeled by a large resistor and/or a small current source (in parallel with the leaking oxide or PN junction). The effects of stuck-on and stuck-off transistors are usually similar to some open and short defects so that they can be considered as redundant in most defect simulations.

14.4.1.2 *Defect simulation*

Once the different types of defects have been modeled, a defect simulation campaign can be carried out. The first step would consist in extracting defect candidates. One of the ways to do it is to resort to inductive fault analysis (IFA) and simulate spot defects in the different process masks. This will generate random disks of extra or missing material in the layout, from which we should be able to deduce the occurrence of shorts and opens. Equivalent shorts and opens should also be identified to avoid duplicated simulations. Furthermore, the likelihood of occurrence of these opens or shorts must be a good approximation of reality. However, this approach implies that the layout of the circuit is available at early stages of the design, which in turn supposes that the test validation will not lead to the inclusion of additional DfT structures. An issue with the IFA approach is that it is not possible to know a-priori the total number of defect candidates.

On way to circumvent this issue is to use the results of a parasitic extraction as a proxy to compute open and short likelihoods. Indeed, it appears the probability of a short between two nets (in the same layer) is proportional to the parasitic capacitance between these two nets. Imagine two metal lines running in parallel: The longer they run in parallel, the higher the probability of a short. Similarly, the closer they are to each other, the higher the probability of a short, due to the higher number of particles with sufficient radius. These two trends are correctly accounted for by the parasitic capacitance. In a similar fashion, the open probability can be taken as proportional to the parasitic resistance of a metal line: A wide metal line is less susceptible to be opened (since the large spot defect is necessary) and is also less resistive. Conversely, a long metal line is more susceptible to an open and is also more resistive. With these proxies, one can extract a list of defect candidates associated to their relative likelihoods.

Finally, if layout is not available, a typical approach consists in performing the test defect simulation at schematic level and resort only to shorts and opens in the different devices. For instance, a defective MOS transistor can have a DS, a GS or a GD short, or a G, D or S open. In that case, the likelihood of occurrence is usually taken as proportional to the transistor size. This is definitely a coarse approximation but there is no more information available at schematic level. Here again, a list of defect candidates together with their relative likelihood is obtained.

Once defect candidates are available, the defect simulation can actually take place: the defect is inserted in the netlist and the test setup is simulated. Unfortunately, for moderate to large circuits, exhaustive simulation is impossible. In order to get a meaningful estimate of the defect coverage (i.e. the proportion of defects that are effectively detected by the test), a statistical approach is necessary. It has been shown that the best way to proceed to avoid biases in the estimate and to maximize the convergence rate is to perform likelihood weighted random sampling [37-38]. At first sight, it may seem a good approach to simulate the defects from the most likely to the less likely. However, this is not true because there may exist a correlation between defect likelihood and detectability. Consider, for instance, that most likely defects could be linked to large analog transistors, while the less likely ones may be linked to smaller digital transistors. These two populations (large "analog" transistors and small "digital" transistors) may though have a very similar statistical weight if volume compensates for likelihood. In other words, one cannot overlook small less likely defects if they are much more numerous than big likely ones.

In order to speed-up defect simulations, it is possible to resort to high-level models of the parts that are not affected by the defect. However, some care should be taken to ensure that the defect does not produce any kickback effect on the surrounding blocks. For instance, it may appear that the biasing section of an amplifier can be modeled with some voltage and current sources when defect-free. However, a defect in the amplifier itself may bring the impedance seen by the biasing section out of its normal range, generating a non-modeled interaction. It is also possible to achieve computational cost reduction by performing early stopping if the defect can be tagged as catastrophic.

Despite all the efforts to produce a realistic estimate of the defect coverage, it is still not possible to guarantee its absolute accuracy. However, defect simulation is still a valuable tool to compare different test strategies and motivate the inclusion of some DfT capabilities. Coming back to the parallel with the digital case, the coverage metrics obtained from a simple stuck-at fault models are probably not accurate either. But they have fostered the emergence of a lot of tools and test practices that have greatly increase IC quality.

14.4.2 *Generic defect-oriented approaches*

14.4.2.1 *IDDQ*

The observation of quiescent power supply current is a well-established technique for digital circuits. The idea is that digital circuits with static inputs shall only consume some static current related to leakage. In the presence of shorts and opens, the current level may be significantly altered for some input patterns.

The IDDQ test method has been adapted to analogue circuits, though it has clearly found a much more limited success. Indeed, quiescent power supply current in analogue circuits is usually not zero and can vary significantly due to normal process and mismatch variations. This broadens the acceptability region of the current signature and thus reduces the sensitivity to defects.

In order to mitigate this issue, it was proposed in [39] to monitor the quiescent current for a ramping supply in order to improve diagnosis capability. The main idea is that ramping the power supply will drive the different analog transistors across different biasing regions, leading to an IDDQ profile that should be somehow characteristic of the topology. In essence, this approach is similar to applying several patterns to the input: it increases the amount of information.

This defect-oriented approach was adapted to RF circuit testing. Noticeably, in [40] it is shown how the combination of several such current monitoring tests achieve acceptable coverage levels. [41] also present an LNA with a co-designed IDDQ sensor that does not significantly affect performance.

To get rid of the issue of process-related current variations, we can take advantage of the defect-filter methodology as proposed in [42]. This defect filter computes the joint probability density function of several signatures over normal process parameter variation. This PDF is then used to assess the existence of outliers (i.e. possible spot defects). The benefit of this approach is that it accounts for the correlations that exist in the data, while a traditional hyperbox approach – setting independent pass-fail limits on each signature -- does not. Including conventional process monitors in the set of signatures may thus remove the process dependency, as if it were a sort of "differential" IDDQ. Figure 14.5 shows an illustration of this process. Imagine that signature 1 is an IDDQ sensor of the circuit and signature 2 a process monitor. These two signatures are clearly correlated. Taking independent pass/fail limits on the two signatures would not detect the outliers because their projections on both the x and y axis are within the normal variation range. However, building the joint PDF allows to see that these outliers fall out of the expected correlation.

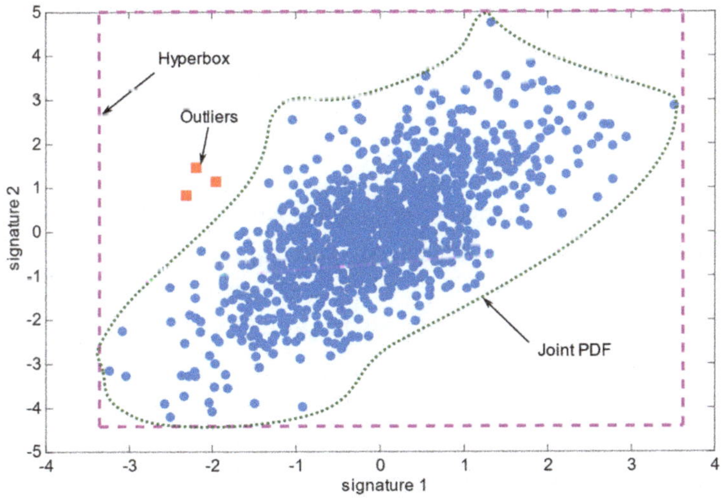

Fig. 14.5. Example of the construction of a defect filter in a two-dimension signature space. The circles correspond to valid circuits obtained from the MonteCarlo simulation of expected process and mismatch variations, while square markers correspond to spot defects. The dashed line is an hyperbox defect filter (with independent pass-fail limits). The dotted line is a defect filter based on the estimation of the joint probability density function. It appears that the spot defects are quoted as outliers by the latter but not by the former.

14.4.2.2 *DC probes*

As mentioned in the system-level section, monitoring internal DC levels is a common practice, either using a test bus or a more conventional multiplexer. Unless these internal DC probes come from the output of embedded sensors like the RF power detectors commented above, it is likely that internal DC operating points suffer from a large intrinsic variability related to process. Just like in the case of IDDQ sensors, this variability can be accounted for if process monitors are included in the chip. DC probes, current monitors and RF sensors in several supply conditions were successfully used for machine-learning indirect test in [43].

A defect-oriented framework relying essentially on DC probes has recently drawn some attention. [44] proposes to introduce controllability features inside a circuit in the form of pull-up or pull-down transistors. This idea is that these transistors can change the topology of the circuit in test mode and sensitize some defects. The signatures are based on internal DC probes. An optimizer based on a genetic algorithm is used to maximize the defect coverage while keeping the number of probes and configurations as low as possible. A Pareto front of the optimum solutions is produced.

This approach is very generic and holds great promise for high defect coverage but suffers from two important drawbacks. The first is related to the computational cost since the optimizer requires an extensive defect simulation campaign. The second is particularly relevant for RF blocks: the inclusion of the observability and controllability structures may degrade performance. If co-design is mandatory, it means that a re-design is necessary within the optimization loop itself.

14.4.2.3 *OBT*

Another well-known defect-oriented approach is Oscillation-Based Test [45]. The idea is to reconfigure the block under test into an oscillator such that the oscillation frequency is sensitive to the defects in the circuit. The main benefit of this approach is that the oscillation frequency can be easily acquired by a simple counter, leading to a full-BIST approach.

Measuring the amplitude of the oscillation has been shown to improve the defect coverage but implies the inclusion of more hardware, even if compact solutions based on $\Sigma\Delta$ modulation have been proposed [46].

Though OBT methodology has been recently proposed for an LNA [47], it is usually considered as a good approach for filters, since these can easily be reconfigured into oscillators by the addition of an adequate non-linear feedback loop [48].

References

1. Sunter, S. and Roy, A. (2011). A Mixed-Signal Test Bus and Analog BIST with "Unlimited" Time and Voltage Resolution, *Proc. Sixteenth IEEE European Test Symposium*, ETS, pp. 81–86.
2. Liu, N. *et al.* (2018). Concurrent Sampling with Local Digitization — An Alternative to Analog Test Bus, *Proc. IEEE International Symposium on Circuits and Systems, ISCAS*, Florence, pp. 1–5.
3. Dabrowski, J. J. and Ramzan, R. M. (2010). Built-in Loopback Test for IC RF Transceivers, IEEE Trans. Very Large Scale Integr. VLSI Syst., 18(6), pp. 933–946.
4. Yoon, J.-S. and Eisenstadt, W. R. (2005). Embedded loopback test for RF ICs, Instrum. Meas. IEEE Trans. On, 54(5), pp. 1715–1720.
5. Dogan, H. *et al.* (2008). Analysis and Design of RF CMOS Attenuators, IEEE J. Solid-State Circuits, 43(10), pp. 2269–2283.
6. Webster, D. *et al.* (2011). Replacing Error Vector Magnitude Test with RF and Analog BISTs, IEEE Des. Test Comput., 28(6), pp. 66–75.
7. Valdes-Garcia, A. *et al.* (2007). Built-in self test of RF transceiver SoCs: From signal chain to RF synthesizers, Proc. *IEEE Radio Frequency Integrated Circuits Symposium*, RFIC, pp. 335–338.
8. Valdes-Garcia, A. *et al.* (2008). A Broadband CMOS Amplitude Detector for On-Chip RF Measurements, IEEE Trans. Instrum. Meas., 57(7), pp. 1470–1477.
9. Ramzan, R. *et al.* (2010). On-Chip Stimulus Generator for Gain, Linearity, and Blocking Profile Test of Wideband RF Front Ends, IEEE Trans. Instrum. Meas., 59(11), pp. 2870–2876.
10. Yen-Chih Huang *et al.* (2008). A Build-in Self-Test Technique for RF Low-Noise Amplifiers, IEEE Trans. Microw. Theory Tech., 56(5), pp. 1035–1042.
11. Zeidan, M. A. *et al.* (2012). Phase-Aware Multitone Digital Signal Based Test for RF Receivers, IEEE Trans. Circuits Syst. Regul. Pap., 59(9), pp. 2097–2110.
12. Sunter, S. and Roy, A. (2007). Purely digital BIST for any PLL or DLL, *Proc. 12th IEEE European Test Symposium, ETS*, pp. 185–192.
13. Sunter, S. and Roy, A. (2004). On-chip digital jitter measurement, from megahertz to gigahertz, IEEE Des. Test Comput., 21(4), pp. 314–321.
14. Hsu, J.-C. and Su, C. (2008). BIST for Measuring Clock Jitter of Charge-Pump Phase-Locked Loops, IEEE Trans. Instrum. Meas., 57(2), pp. 276–285.
15. Provost, B. and Sanchez-Sinencio, E. (2003). On-chip ramp generators for mixed-signal BIST and ADC self-test, IEEE J. Solid-State Circuits, 38(2), pp. 263–273.
16. Holcer, R. *et al.* (2003). DNL ADC testing by the exponential shaped voltage, IEEE Trans. Instrum. Meas., 52(3), pp. 946–949.
17. Jalón, M. A. and Peralías, E. (2010). ADC non-linearity low-cost test through a simplified double-histogram method, J. Electron. Test., 26(1), pp. 47–58.

18. Elsayed, M. M. and Sanchez-Sinencio, E. (2010). A Low THD, Low Power, High Output-Swing Time-Mode-Based Tunable Oscillator Via Digital Harmonic-Cancellation Technique, IEEE J. Solid-State Circuits, 45(5), pp. 1061–1071.
19. Karthik Vasan, B. *et al.* (2013). Low-Distortion Sine Wave Generation Using a Novel Harmonic Cancellation Technique, IEEE Trans. Circuits Syst. Regul. Pap., 60(5), pp. 1122–1134.
20. Barragán, ManuelJ. *et al.* (2011). Analog Sinewave Signal Generators for Mixed-Signal Built-in Test Applications, J. Electron. Test., 27(3), pp. 305–320.
21. Dufort, B. and Roberts, G. W. (1999). On-chip analog signal generation for mixed-signal built-in self-test, Solid-State Circuits IEEE J. Of, 34(3), pp. 318–330.
22. Hong, H.-C. (2007). A Design-for-digital-testabilty circuit structure for σ-δ modulators, IEEE, 15 (12), pp. 1341–1350.
23. Gines, A. J. *et al.* (2016). Linearity test of high-speed high-performance ADCs using a self-testable on-chip generator, Proc. *21th IEEE European Test Symposium, ETS*, pp. 1–6.
24. Stenbakken, G. N. and Souders, T. M. (1994). Developing linear error models for analog devices, IEEE Trans. Instrum. Meas., 43(2), pp. 157–163.
25. Variyam, P. N. *et al.* (2002). Prediction of analog performance parameters using fast transient testing, IEEE Trans. Comput.-Aided Des. Integr. Circuits Syst., 21(3), pp. 349–361.
26. Barragán, M. J. *et al.* (2011). Alternate Test of LNAs Through Ensemble Learning of On-Chip Digital Envelope Signatures, J. Electron. Test., 27(3), pp. 277–288.
27. Zjajo, A. *et al.* (2007). BIST Method for Die-Level Process Parameter Variation Monitoring in Analog/Mixed-Signal Integrated Circuits, *Proc. Design, Automation & Test in Europe Conference & Exhibition, DATE*, pp. 1–6.
28. Leger, G. and Barragan, M. J. (2016). Brownian distance correlation-directed search: A fast feature selection technique for alternate test, Integr. VLSI J., 55, pp. 401–414.
29. Barragan, M. J. and Leger, G. (2015). A Procedure for Alternate Test Feature Design and Selection, IEEE Des. Test, 32(1), pp. 18–25.
30. Kupp, N. *et al.* (2009). On Boosting the Accuracy of Non-RF to RF Correlation-Based Specification Test Compaction, J. Electron. Test., 25(6), pp. 309–321.
31. Biswas, S. and Blanton, R. D. (2009). Maintaining Accuracy of Test Compaction through Adaptive Re-learning, *Proc. 27th IEEE VLSI Test Symposium*, Santa Cruz, CA, USA, pp. 257–263.
32. Liaperdos, J. *et al.* (2015). Fast deployment of alternate analog test using Bayesian model fusion, *Proc. Design, Automation Test in Europe Conference Exhibition, DATE*, pp. 1030–1035.
33. Stratigopoulos, H. G. *et al.* (2010). Defect filter for alternate RF test, Proc. *15th IEEE European Test Symposium, ETS*, pp. 265–270.
34. Zivkovic, V. and Schaldenbrand, A. (2019). Requirements for Industrial Analog Fault Simulator, Proc. *Proceedings of International Conference on Synthesis, Modeling, Analysis and Simulation Methods and Applications to Circuit Design, SMACD*, p. 4.

35. IEEE P2427 (2018). IEEE P2427 - Standard for Analog Defect Modeling and Coverage (working group), 2018. https://standards.ieee.org/project/2427.html (accessed Oct. 01, 2019).
36. Esen, B. *et al.* (2016). Effective DC fault models and testing approach for open defects in analog circuits, Proc. *IEEE International Test Conference, ITC*, pp. 1–9.
37. Sunter S. et al. (2014). Practical random sampling of potential defects for analog fault simulation, Proc. International Test Conference, pp. 1–10.
38. Gutierrez V., Gines A. and Leger G. (2019). Assessing AMS-RF test quality by defect simulation, IEEE Trans. Device and Materials Reliability, 19(1), pp. 55-63.
39. Somayajula, S. A. S. *et al.* (1996). Analog fault diagnosis based on ramping power supply current signature clusters, IEEE Trans. Circuits Syst. II Analog Digit. Signal Process., 43(10), pp. 703–712.
40. Zjajo, A. and Pineda de Gyvez, J. (2005). Evaluation of Signature-Based Testing of RF/Analog Circuits, *Proc. European Test Symposium, ETS*, Tallinn, Estonia, pp. 62–67.
41. Cimino, M. *et al.* (2008). Design of a 0.9 V 2.45 GHz Self-Testable and Reliability-Enhanced CMOS LNA, IEEE J. Solid-State Circuits, 43(5), pp. 1187–1194.
42. Stratigopoulos, H.-G. *et al.* (2009). Defect Filter for Alternate RF Test, Proc. *14th IEEE European Test Symposium*, Sevilla, Spain, pp. 101–106.
43. Barragan, M. J. *et al.* (2011). Improving the Accuracy of RF Alternate Test Using Multi-VDD Conditions: Application to Envelope-Based Test of LNAs, *Proc. 20th Asian Test Symposium (ATS)*, pp. 359–364.
44. Coyette, A. *et al.* (2015). Automatic generation of autonomous built-in observability structures for analog circuits, *Proc. 20th IEEE European Test Symposium, ETS*, pp. 1–6.
45. Arabi, K. and Kaminska, B. (1997). Testing analog and mixed-signal integrated circuits using oscillation-test method, IEEE Trans. Comput.-Aided Des. Integr. Circuits Syst., 16(7), pp. 745–753.
46. Vazquez, D. *et al.* (2002). On-Chip Evaluation of Oscillation-Based-Test Output Signals for Switched-Capacitor Circuits, Int. J. Analog Integr. Circuits Signal Process, 33(2), pp. 201–211.
47. Nel, H. P. *et al.* (2018). Built-In Oscillation-Based Self-Testing of a 2.4 GHz LNA in 0.35μm CMOS, *Proc. 25th IEEE International Conference on Electronics, Circuits and Systems, ICECS*, Bordeaux, pp. 837–840.
48. Goyal, A. and Swaminathan, M. (2008). A low-cost approach for testing embedded rf passive circuits based on oscillation principle, Proc. *Electrical Design of Advanced Packaging and Systems Symposium*, Seoul, Korea (South), pp. 194–197.

https://doi.org/10.1142/9789811210723_0015

Chapter 15

Embedded Memory Test

Hassen Aziza
Aix-Marseille Univ, Univ Toulon, CNRS, IM2NP
38 rue Joliot Curie, F-13451, Marseille, France

Alberto Bosio
University of Lyon - Lyon Institute of Nanotechnology
Ecole Centrale de Lyon, 36 av. Guy de Collongue, F-69134 Ecully, France

15.1 Introduction

Previous chapters reviewed general reliability issues and fault models which are generic among various semiconductor technologies. However, there are a number of failure modes and mechanisms which are specific to memories. In addition, future SoCs are expected to embed memories of increasing sizes. As a result, the overall SoC yield will be dominated by the memory yield. In this section we introduce some general ideas in memory testing. The section will discuss test and reliability issues related to SRAM memory. Then, NVMs are addressed with a focus on embedded Flash (eFlash) memories.

15.2 SRAM Memory Test

Among different types of memories, SRAMs are still dominant in several applications, especially those that require a high access performance (e.g. upper memory hierarchy, such as on-chip L1 caches). Nowadays, SRAMs are designed to exploit the technology limits in order to reach the highest storage capacity and the highest access speed. As a consequence, SRAMs are statistically more likely to be affected by manufacturing defects compared to other IP cores that may be embedded in SOCs, which makes such devices the main yield detractors of systems where they are embedded. It is therefore imperative to develop efficient test and diagnosis methods for embedded SRAMs in order to ensure quality of the final product. Functional test is the most popular type of test applied in SRAMs. It consists of executing sequences of operations (read and write) in the SRAM under test in order to highlight possible faulty behaviors that may occur.

In the next sections, we first present a formal definition for faulty behaviors, as well as a formalism that may be used to represent them as functional faults. Secondly, we present the main concepts behind March tests, which is the most popular functional test used in SRAM testing. In particular, we present an in-depth analysis on the most relevant works on SRAM testing, with special focus on state-of-the-art March test algorithms and stress aware test flows.

15.2.1 *Describing faulty behaviors*

A faulty behavior corresponds to an observed memory behavior that deviates from the expected one. Faulty behaviors are normally caused by physical flaws in chips, called defects. A defect often consists of impurities in silicon, such as dust particles. Faulty behaviors in memories are usually modeled as functional faults, such that functional tests can be used to detect such faults. This mapping of defects into faulty behaviors, and thereafter into functional faults is called fault modeling. A functional fault is defined as the deviation of the observed memory behavior from the expected behavior, when a given set of operations (read and write) is executed in the memory. Therefore, two basic ingredients can be identified to any functional fault:

- A list of operations executed in the memory;
- The faulty behavior observed when such list of operations is executed.

Any list of operations performed on the memory is called an *operation sequence* (OS). An operation sequence that results in a difference between the observed and the expected memory behavior is called a *sensitizing operation sequence* (SOS). In the following subsection, a formalism used to represent functional faults is introduced.

15.2.2 *Representation of functional faults*

Throughout the 1980s and during the first half of the 1990s, the only functional parameter considered relevant to represent a functional fault was the logic value stored in the core-cell where the fault is observed [1]. Recently, another functional parameter has also been considered to be relevant: the logic value available at the memory output when the operation sensitizing the fault is a read operation [2].

A functional fault can be formally represented by a *fault primitive* (FP), denoted by notation $< S/F/R >$ [1]. *S* refers to a logic value or the SOS that sensitizes the fault, *F* is the logic value stored in core-cell where the fault is sensitized, $F \in \{0, 1\}$, and R is the logic value on the memory output when the sensitizing operation is a read, $R \in \{0, 1, -\}$. Notation '-' is used for R when a write operation sensitizes the fault. The notion of FPs makes it possible to give a formal definition of a *functional fault model* (FFM), as understood for memory devices: FFM is a non-empty set of FPs [1].

For example, let us consider two well-known FFMs: *transition faults* (TFs) and *stuck-at-faults* (SAFs), defined in [1]. Such FFMs are composed by the FPs presented in Table 15.1.

Table 15.1. Fault primitive of stuck-at-faults and transition faults

FFM	Fault Primitives = < S/F/R >
SF	< 0 / 1 /- >, < 1 / 0 /- >
TF	< 0w1 / 0 /- >, < 1w0 / 1 /- >

In the case of a SF, FPs < 0 / 1 /- > and < 1 / 0 /- > represent faulty behaviors where a faulty core-cell always store logic '1' and logic '0', respectively, regardless of operations performed in such core-cell. In the case of a TF, FPs < $0w_1$ / 0 /- > and < $1w_0$ / 1 /- > represent faulty behaviors where a faulty core-cell fails to undergo transitions 0→1 and 1→0, respectively.

15.2.3 *Classification of faults*

FPs can be classified according to #C, the number of different core-cells accessed during an SOS, and according to #O, the number of different operations performed in an SOS. Depending on #C, FPs can be divided into the following classes:

- If #C = 1, then the FP sensitized by the corresponding SOS is called a *single-cell* FP.
- If #C > 1, then the FP sensitized by the corresponding SOS is called a *coupling* FP. If #C = 2, then it is described as 2-coupling FP or 2-cell FP.
- If #C = 3 then, it is described as 3-coupling FP or 3-cell FP, etc.

Depending on #O, FPs can be divided into the following classes:

- If #O = 1, then the FP sensitized by the corresponding SOS is called a *static* FP.
- If #O > 1, then the FP sensitized by the corresponding SOS is called a *dynamic* FP. If #O = 2, then it is described as 2-operation dynamic FP. I
- f #O = 3, then it is described as 3-operation dynamic FP, etc.

Figure 15.1 shows a taxonomy of the space of FPs. It is important to note that the two ways to classify FPs are independent, since their definition is based on independent factors of the SOS. As a result, a single-cell FP can be static, or dynamic with any number of operations. The same applies to coupling FPs.

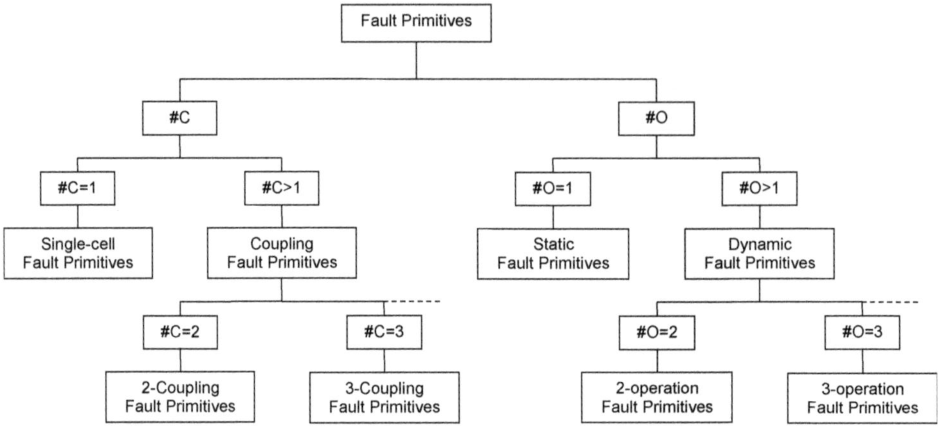

Fig. 15.1. Taxonomy of fault primitives

Since an FFM is defined as a set of FPs, it is expected that FFMs will inherit the properties of FPs, as follows:

- If an FFM is defined as a collection of single-cell static FPs, then the FFM is a single-cell static fault model. SAFs and TFs, described in detail in the end of Section 4.2.1, are examples of such FFMs.
- If an FFM is composed by a set of two-operation FPs, then the FFM may be either a single-cell dynamic FFM or a coupling dynamic FFM. For example, *dynamic read destructive fault* (dRDF) [3] is a single-cell dynamic FFM and *un-restored destructive write fault* (URDWF) [4] is a 2-coupling dynamic FFM.
- If an FFM consists of FPs classified into inconsistent classes, single-cell and two-cell FPs for example, it is described as a single-cell and a two-cell fault model.

15.2.4 *Test solutions and algorithms*

Traditional memory tests include many well-known functional test algorithms, such as *Scan*, *GALPAT* and *Walking 1/0* [1]. The main drawback of such algorithms is the high complexity, generally not linear *w.r.t.* the memory size (complexity may in the order of *O(N2)*, where *N* is the total number of addresses of the SRAM). This makes them not applicable to large memories that are commonly designed nowadays, with storage capacity higher than 1 Gbit. To overcome the complexity issues of the test solutions discussed above, the class of March tests has been introduced and defined in [1].

The formal definition of a March test is the following: *A March test M is a functional test algorithm with a finite number of March elements (MEs) M = { M1, M2, ..., Mk }, where each ME Mi = Di(O1, O2,..., On), 1 ≤ i ≤ k, consists of an addressing order Di ∈ { ↕, ↑, ↓ } and a finite number of read and write operations. In a ME, each operation Oj, 1 ≤ i ≤ n, may consist of writing logic '0' or logic '1' into core-cells (represented as w0 and w1, respectively), or reading logic '0' or logic '1' from core-cells (represented as r0 and r1, respectively). All operations of a ME are executed sequentially in every core-cell of the memory under test. Considering a given memory address, all operations of a ME are executed in the core-cells accessed through such address before proceeding to core-cell accessed through the next address. The way one moves from a given memory address to another is called addressing order. There are two addressing orders: an increasing addressing order (i.e. from address 0 to address N - 1), denoted as ↑; and decreasing addressing order (i.e. from address N - 1 to address 0), denoted as ↓. When addressing order is irrelevant, notation ↕ is used.*

The complexity of a March test is in the order of *O(N)*, i.e. linear with respect to the number of addresses (size) of the memory under test. More precisely, the complexity of a March test is defined as the sum of the number of operations in each ME, multiplied by the number of addresses N of the memory. As an example, the structure of the classical March test MATS+ [1] is shown below:

$$\textbf{MATS+:}\ \{\ \underset{ME1}{\updownarrow(w_0)};\quad \underset{ME2}{\uparrow(r_0,w_1)};\quad \underset{ME3}{\downarrow(r_1,w_0)}\ \}$$

It is composed of three MEs (ME1, ME2 and ME3). The complexity of March MATS+ is 5N, since it is composed of five operations in total. In terms of fault coverage, March MATS+ is able to detect only static faults, in particular stuck-at-faults and address faults [1]. The definition of March tests also provides six degrees of freedom (DOFs) [5, 6] that make such algorithms more flexible. All six DOFs are summarized as follows:

- *DOF I*: The address sequences can be freely chosen as long as all addresses occur exactly once and the sequence is reversible.
- *DOF II*: The address sequence for initialization can be freely chosen as long as all addresses occur at least once.
- *DOF III*: If the March test is built symmetrically (detects, for example, both SA0 and SA1 faults), the data written to the cells can be exchanged completely.

- *DOF IV*: The data within a read/write operation does not necessarily has to be equivalent for all memory addresses as long as the detection probabilities for basic faults are not affected.
- *DOF V*: The input data is not defined during read operations.
- *DOF VI*: The output data is not defined during write operations.

The main advantage provided by the use of DOFs is that a March test may have higher fault coverage without increasing its complexity (*i.e.* without adding operations) and without changing the capability of the March test to detect the initial targeted faults. For example, the well-known March C- [1], which has complexity 10*N*, is effective to detect some static faults, such SAFs, TFs and 2-coupling faults. This March test normally covers 0% of dynamic faults. In [3], however, authors propose the execution of March C- using the particular addressing sequence *word line after word line*, which is possible due to the first DOFs of March tests. The goal was to detect *dynamic read destructive faults* (dRDFs) [3]. This modification requires no extra operations in March C- and does not change the capability of March C- to detect the initial targeted faults (SAFs, TFs and 2-coupling faults).

As mentioned above, memory testing is based on the use of algorithms, especially March test algorithms, which can provide high fault coverage with low execution time. The class of static faults has been deeply analyzed, especially in [1], and several meaningful test solutions have been developed. Unfortunately, in SRAMs designed with recent technologies, dynamic faults appeared as the predominant root cause of errors. Since test solutions targeting static faults normally do not cover dynamic faults, a high effort has been devoted to develop new and dedicated test solutions to detect dynamic faults. In [7], an exhaustive set of single-cell and two-cell two-operation FPs is generated. Based on this complete set, authors extracted a subset of 12 single-cell two-operation faults and a subset of 24 two-cell two-operation faults, considering that those are the most realistic ones based on ad-hoc assumptions. In [8], the authors present two March test algorithms, called RAW1 (13N) and RAW (26N). March RAW1 covers the set of realistic faults in the class of single-cell two-operation dynamic faults, while March RAW covers the set of realistic faults in both classes of single-cell and two-cell two-operation dynamic faults. In [9], new March test algorithms have been proposed, called AB1 (11N) and AB (22N). March AB1 and AB have the same fault coverage as March RAW1 and RAW, respectively, but with lower complexity.

In all March tests presented above, the aim was to cover the largest set of dynamic faults as possible. Such test solutions can cover even a set of faults with very low probability of occurring. This leads to obtain very effective test solutions, but at the expense of a high complexity, which results in a high test cost.

In order to minimize the effort devoted to develop test solutions and produce more efficient, optimized and dedicated memory tests, it is possible to rely on information provided by preliminary electrical analysis of the SRAM in presence of realistic defects that may affect specific circuit blocks.

15.3 Flash Memory Test

15.3.1 *Flash memory test flow*

Flash memories are now embedded in a large number of specific applications requiring a broad range of memory sizes: a few bytes will be necessary for sensor calibration or ID devices, while high-end applications such as networking, set-top boxes, games, will require several Mbits of embedded memory. The testing strategy is driven by the specific characteristics of Flash memories such as long programming time, analog like parameters extraction, high voltage requirements, but also by the constraints of the application in terms of maximum pin count of the chip, estimated flash testing cost versus overall testing cost of the chip, production quantities etc. Moreover, compared to SRAM memories, there are a number of failure modes and mechanisms which are specific to Flash memories such as data retention, data margin loss due to disturbs, endurance and cycling [10].

For standalone Flash memory, parallel access makes controllability and observability optimum and test and characterization procedures faster and easier. Unfortunately, this approach results in an unacceptable pin count. Using a Serial Test Interface (STI) dramatically reduces the pin count to an acceptable value, making the use of a STI a cost-effective solution for embedded Flash testing. Even if the characterization of some dynamic parameters such as the access time is made more difficult by the serial access, the testing time penalty is not so important due to the intrinsic programming time of the memory. Furthermore, the STI provides flexibility to apply test patterns, which can be useful in case of memory debug. Flash embedded in SOCs has a very limited pin count, or high security constraints, which is the case for smart card chips where the test of the Flash memory should be entirely generated on chip. One effective solution that has been implemented uses a ROM memory to store test patterns that are applied to the memory by the the MCU [11]. Memory outputs are then compared to expected ones. If this strategy is optimal for production testing, debug of the memory is very difficult due to the low flexibility and very limited access. Although most of the aspects developed in this section may be valid for other Flash technologies, this chapter refers to the mainstream eFlash technology: NOR architecture erased by Fowler-Nordheim and programmed by Channel Hot Electrons.

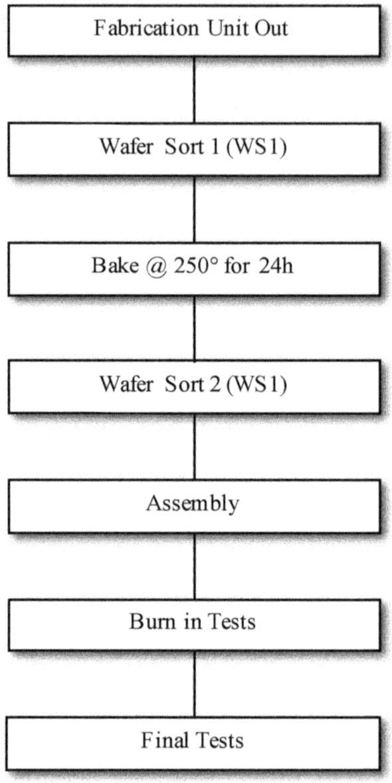

Fig. 15.2. Generic Flash memory production test flow

Flash memory wafer sort and screen tests are performed to guarantee performance within specifications and screen dies that do not meet the specifications. For high-density memory, a 100% test for all possible operating combinations is almost impossible due to the extremely long test time required (10μs to 1ms write time). Figure 15.2 presents a generic production test flow for Flash memory [12]. Depending on the specific memory type, the maturity of the product, and the manufacturer, this generic test flow can be adapted. The test flow consists of WS1 (Wafer Sort), Bake tests, WS2, Assembly tests, Burn-in tests and Final tests. The most important test is WS1, which includes parametric tests, check for shorts/opens, test for input/output leakage currents, static and dynamic standby currents and functional tests. Other test stages are a subset of WS1. Bake tests are intended to look for charge loss and to repair the leaky bits. WS2 are post-bake tests used to check device parametric parameters prior to burn-in tests. Burn-in tests are performed at high temperature during a specific period to look for biased charge gain and charge loss and for adjacent bit charge transfer (i.e. disturbs). Final tests monitor speed degrades and parametric shifts. Functional tests as well as burn-in tests will be developed in the next sections.

15.3.2 *Functional tests*

Due to the intrinsic very low speed of the programming operation, eFlash testing strategy is very different from other types of memory. March like algorithms are not suitable, and a very limited number of patterns are used to test the memory in order to keep the testing cost acceptable. In the literature, we often find the most known 7-patterns sequence performing the following programming (erase and write) and read operations [13]:

- *Write all 0s and Read all 0s* – This pattern ensures the operability of the Flash memory page mode. Moreover, this pattern preconditions the memory for the next pattern (i.e. GalPat0).
- *Galloping Pattern (GalPat0)* – All the memory array is set to '0' by the previous pattern. GalPat0 detects read disturbances between byte or word by reading '00' alternately in a byte or a word sharing the same column or row.
- *Checkerboard (CHK)* – A succession of '0' and '1' is written to obtain an opposite state between two adjacent core cells in the memory array. Then, a second pattern corresponding to the complement of the previous one is programmed (inverse CHK). The CHK pattern ensures that every cell can be written to a '1' or '0' but also that the adjacent cells, which are written with an opposite state do not affect the state of the cell under test.
- *Columns Pattern* – This pattern writes alternating columns composed of a byte or a word with '00' and 'FF'.
- *Diagonal Read of Alternating Data (DRAD)* – In this pattern only the bytes or word belonging to the memory array diagonal are read after write. This configuration is extreme as for each read cycle the address decoders change the value. Moreover, this pattern is often used for AC parametric tests as it corresponds to the worst case in terms of read access time.
- *Bit unique test* – This pattern verifies that every bit within a byte is unique and ensures that two bits are not shorted.
- *March element* – The March element is used to test address uniqueness and multiple selections. This pattern writes a March algorithm element in the entire Flash memory chip. If the March LR algorithm is considered, this March element pattern could be executed using the element: read '1',

write '0', read '0' and write '1' with ascending addresses ($\Uparrow(r1,w0,r0,w1)$) [14]. However, as Flash granularity is word-based, the March element is performed with a data background 'FF' or '00' (for an 8-bits word) instead of '1' and '0' respectively.

To illustrate functional tests, the CHK algorithm is considered. Figure 15.3a presents the CHK pattern where '1' and '0' are written into alternate memory locations. The algorithm includes different stages: write CHK with up addressing order, read CHK with up addressing order, write inverse CHK with up addressing order and read inverse CHK with up addressing order. The algorithm divides the cells into two groups such that every neighboring cell is in a different group. Figure 15.3b presents the corresponding memory array V_T distribution. Two different cell populations are clearly identified referred to as '0' and '1'. V_T distribution extraction is a critical step in Flash memory test as it allows to monitor marginal cell populations.

1	0	1	0	1	0	1	0	1	0
0	1	0	1	0	1	0	1	0	1
1	0	1	0	1	0	1	0	1	0
0	1	0	1	0	1	0	1	0	1
1	0	1	0	1	0	1	0	1	0
0	1	0	1	0	1	0	1	0	1
1	0	1	0	1	0	1	0	1	0
0	1	0	1	0	1	0	1	0	1
1	0	1	0	1	0	1	0	1	0
0	1	0	1	0	1	0	1	0	1

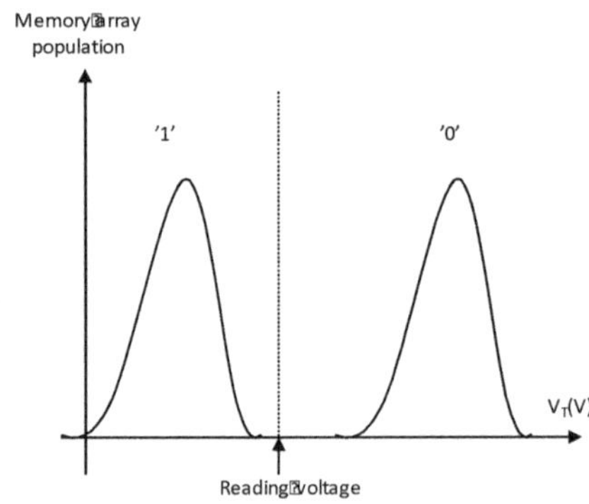

(a) (b)

Fig. 15.3. a) CHK algorithm and corresponding b) VT distribution

15.3.3 *Reliability tests*

Reliability is the probability that the product remains effective in performing its designated function for a specific period of time. In eFlash, **data retention** (i.e. ability to maintain the programmed state for an acceptable period of time) and **endurance** (i.e. number of write/erase cycles without degradation) are the most critical reliability concerns. **Burn-in**, which is conducted at elevated temperature and voltage, is an example of reliability tests [15]. By exposing a device to accelerated burn-in stress (for a shorter time), one can with some specified confidence level predict device lifetime under normal operating conditions. Many of these extreme operation condition tests are described in detail in the Joint

Electron Device Engineering Council (JEDEC) publication JESD22 and can be found at the website *www.jedec.org*. Commonly used tests include High-Temperature Operating Life Test (HTOL), Low-Temperature Operation Life Test (LTOL), Data Retention Storage Life (DRSL), Endurance Cycling, etc.

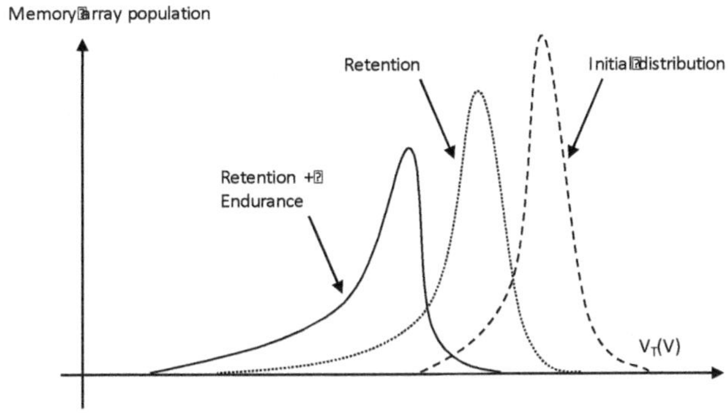

Fig. 15.4. Marginal cell population after retention and endurance tests

Figure 15.4 shows schematically the impact of retention and cycling tests on an initial V_T distribution of Flash memory cell. Retention tests are meant to accelerate the loss of charge off the floating gate, resulting in a shift of the V_T distribution. Long-term (>10 years) data retention test is usually accelerated by temperature to a period of days or weeks. It has been discovered that for advanced Flash devices with very thin tunnel oxide (< 90Å), the charge-loss activation energy could be lower after endurance [16]. Indeed, endurance tests lead to the generation of the defects responsible for charge leakage and read disturb, resulting in a post-retention V_T distribution shift. Both retention and endurance pose difficult test and verification dilemma as the process of testing endurance by cycling implies that the testing would consume much of the useful lifetime of the product [12].

15.3.4 *Repair strategies*

As the size and density of Flash memory grows rapidly, keeping a high level of yield and reliability becomes more and more difficult. Both **redundancy** and **Error correcting codes (ECC)** techniques have been widely used for enhancing the yield and reliability of eFlash.

Redundancy should be considered as a way to improve yield. The use of redundancy can help in allowing a production yield much higher than natural yield even when the process has reached its mature stage [12]. Redundancy is also particularly useful during production ramp-up when the yield can be very low due to the learning curve in a new process flow. Column redundancy is performed in

Flash memories by adding one or more column to each sector. Redundant columns is intended to replace the failing columns during EWS1. Row redundancy is performed by adding one or more word lines to each sector for failing word lines replacement. Experience over the last 20 years has resulted in the formulation of well-established row and column redundancy design techniques for Flash memories [17]. Criteria for choosing one solution or another (row or column redundancy) are based on the defect model associated with the particular array architecture and process. The downside is that redundancy is difficult to implement, it affects the total area and access time, and increases testing time and complexity. So decision making to add redundancy will be a trade-off between memory size, production quantities and testing cost. In addition, redundancy is not useful to fix errors due to endurance or data retention because it can only be used one time during EWS1.

ECC can be used in Flash memories to increase endurance and improve retention. After a certain number of erase/program operations, a Flash memory can have defective cells due to oxide degradation or charge trapping. ECC circuit is meant to automatically repair the amount of failing bits. The apparent endurance of the chip is extended to a greater number of cycling operations. The ECC is the most effective way to protect data and correct errors, so all modern flash controllers (external and embedded) have built-in ECC engine. ECC operation mode is the following: when data is written to the Flash memory, the controller generates special checksum and write it to the spare area of a page. Then, when data is read, the controller tests data integrity of every page by re-computing the checksum and correct the data if necessary (provided the number of bit discrepancies is within the limits of the ECC potency). There are several types of ECC codes, but only BCH type is widely used in Flash devices [18]. This algorithm is universal and implemented in on-chip ECC to correct errors as soon as they happen in the application. Unfortunately, area penalty of ECC can be huge, and this solution should be considered very carefully due to its big impact on cost.

15.4 Test Challenges of Emerging Memories

PCM, STT-MRAM and RRAM emerging memories have shown promising characteristic to gain a position inside the memory hierarchy, and even enable new computing paradigms (an introduction to emerging memories is provided in Part 4). However, as any other emerging technology these devices are affected by concerns to be resolved before they could become a mainstream [19]. This section reviews the main reliability and testability challenges of aforementioned emerging non-volatile memories.

15.4.1 *Test and reliability of PCMs*

Due to its special materials, device structure, and operation mechanisms, PCM has its specific failure modes. A comprehensive introduction of PCM failure modes and fault models can be found in [20]. PCM errors are characterized into four types: program interference faults, read faults, and write '1' and write '0' faults, and basic March algorithms are utilized to classify and detect them. Alternatively, novel March algorithms have been explored in recent years to cover the specific failure modes of PCM. In [20], a March algorithm called March-PCM is introduced to detect PCM specific faults in addition to common faults. In [21], the March-PCMP is presented to detect additional faults named weak transition faults and write destructive faults caused by the parasitic capacitance and resistance defects.

At the early stages of phase-change memory research, it was suggested that the cycling endurance could be as high as 10^{12} [22]. However, various implementation efforts have resulted in agreement that the empirical endurance limit of typical phase-change memory cells is the range of 10^6-10^9, which is much better than Flash memory, but less than the desired limit of 10^{12} or even 10^{15} needed for working memory applications. The endurance limitation is attributed to the high operating temperature (melting temperature typically at ~900 K) that PCM cells experience during the write operation. To raise the temperature above the melting temperature through joule heating, the current density needs to be quite high, typically ~10 MA/cm^2 [23]. This current density causes atomic migration and is the root cause of cycling endurance failures, whether they are stuck-SET (caused by elemental segregation) or stuck-RESET (caused by void formation). Thus, the cycling endurance of PCM became the last hurdle to overcome to enable its adoption in the larger market for persistent memory products.

Regarding retention, it has been demonstrated that data retention capability is greater than 10 years and that retention is independent of cycling (while, for example, floating gate devices show a reduction in retention as they are cycled). One of the main issue is the RESET sate retention at high temperatures. This issue has been investigated widely in the last decade [23] and results from the fact that the amorphous state is meta-stable with respect to the stable crystalline phase. Any energy applied to this state (primarily thermal or electrical) can accelerate the crystallization process.

15.4.2 *Test and reliability of STT-MRAMs*

Although STT-MRAM switching has proven sub-nanosecond potential, the operation is intrinsically stochastic and some desired data may fail to be stored correctly on the MTJs [24]. Soft errors (i.e. wrong signal) and hard errors (i.e. device damage) affect the core cell. Soft errors are associated with the limited thermal stability of STT-MRAM memories, which affects read operations where the read current can erroneously change the stored data. Hard errors are associated with the oxide barrier breakdown due to the high current density that occurs during switching. STT-MRAM also suffers from electrical defects. In [25], the types of resistive, capacitive and coupling defects are identified. Their manifestation as read and write faults as well as fault activation patterns are analyzed. Apart from read and write faults, the technology can suffer from bit-flip caused by thermal noise. However, such failures are negligible in STT-MRAM due to the high-energy barrier between the two resistance states.

STT-MRAM reliability, like the reliability of any other memory circuit, is a topic that contains a number of focus areas including data integrity issues related to the MTJ itself, but also packaging related issues and CMOS subsystem reliability. STT-MRAM has virtually unlimited program/erase/read endurance, meaning it is possible to write/read/erase the memory array virtually forever without degrading the MTJ's ability to change or maintain its bit state. STT-MRAM available datasheets typically quote about 10^{15} cycles for their endurance specification [26]. This is effectively unlimited endurance as it takes over 300 years to write just 100,000 addresses even 10^{12} times (assuming a 100 ns write-cycle time).

Regarding retention, failure mechanisms is different from traditional charge-storage- based memories. In floating-gate memories like Flash, thermal excitation of the stored charge in the floating gate will cause it to leak off the gate, eventually changing the threshold voltage of the transistor enough to cause the wrong bit state (i.e., '1' instead of '0' or vice versa) to be read out. This process is dependent on time and temperature, and independent of whether or not the device is powered. In DRAM, a volatile memory, charge is stored in small capacitors, and the charge leaks out almost instantly when powered down. However, with spin-based memories, the bit state changes not when charges leak, but when the polarization of the free magnetic layer in the MTJ flips, thus altering the resistance of the MTJ. In other words, in magnetic memory, magnetic polarization is stored rather than electrical charge. To date, Everspin, which is the only company to have successfully commercialized STT-MRAM, specifies their MTJs to have a data retention capability of 20 years at 125°C [26].

Despite exceptional reliability characteristics, it has been reported in the literature that MRAM devices are susceptible to upset and damage from Stray Magnetic Fields (SMF) [27]. Common sources of SMF include nearby electric current, permanent magnets, and objects containing iron, nickel, or cobalt. SMFs can cause soft bit flips as well as physical damage to the device. SMFs can also alter the polarization of the free magnetic layer causing data corruption to a cell.

15.4.3 *Test and reliability of RRAMs*

RRAM's main challenge is the variability of its switching parameters due to the stochastic nature of the ion migration/filament creation [28, 29]. Such a concern imposes the necessity to design adaptive sensing circuits to identify correctly the RRAM resistance value. Thus, programming algorithms and write assist circuits play a crucial role on performance and yield optimization of resistive switching memories. Program-verify algorithms have been developed to ensure distinct resistance distributions. These algorithms adjust the write signal (pulse number, width or amplitude) until the cell resistance reaches the target resistance state. However, in this scheme, the writing is interrupted periodically to allow read operations and determine whether the switching has occurred. This interruption can increase the write-time significantly. An alternative to avoid the use of read operations is to implement a write-termination scheme where the write current pulse is cut off right after the resistance transition completes. This technique prevents wasted write energy but introduces an important area overhead [30]. Moreover, due to its specific physical construction and switching mechanism, RRAM can be affected by unique faults due to the FORMING operation. Indeed, prior to performing SET or RESET operations, RRAM technology needs a FORMING step. The FORMING step, achieved one time in the RRAM device life is a voltage-induced resistance switching from an initial virgin state with a very high resistance to a conductive state. High voltage is typically needed during FORMING since it generates the prerequisite number of defect/ions within the thin dielectric layer to construct a Conductive Filament (CF). However, FORMING involves high currents (~ 100 μA) and the high voltage can initiate thermal damage if not controlled. These special faults have caused test engineers to modify the traditional March test algorithms and have signified the importance of utilizing new Design for Testability (DFT) techniques to cover all fault types [31, 32].

RRAM retention and endurance characteristics suffer from the intrinsic variability of the technology. During endurance cycling, defect generation occurs in the resistive layer due to electric field stress, especially during RESET operation. When defect density becomes critical, a conductive path is formed within the dielectric, leading to oxide breakdown and endurance collapse. During retention, filament dissolution in the resistive layer follows a stochastic behavior, leading to dispersion of memory retention time. To improve retention, one can reduce the energy provided to the system during programming modes, reducing ion migration rate and improving RRAM stability during bake. In particular, it has been demonstrated that SET programming conditions affect HRS retention in RRAM technologies. If the programming energy is too strong, defects are created in the resistive layer, enhancing the ion motion during bake and affecting the memory stability.

References

1. Van de Goor, A.J. (1998). *Testing Semiconductor Memories: Theory and Practice*, COMTEX Publishing, Gouda, The Netherlands.
2. Van de Goor, A.J. and Simonse, J.E. (1999). Defining SRAM Resistive Defects and Their Simulation Stimuli, *Proc. of Asian Test Symposium*, pp. 33-40.
3. Bosio, A. et al. (2009), Resistive-Open Defects in Core-Cells, *Advanced Test Methods for SRAMs Effective Solutions for Dynamic Fault Detection in Nanoscale Technologies*, ISBN: 978-1-4419-0937-4.
4. Bosio, A. et al. (2009), Resistive-Open Defects in Pre-charge Circuits, *Advanced Test Methods for SRAMs Effective Solutions for Dynamic Fault Detection in Nanoscale Technologies*, ISBN: 978-1-4419-0937-4.
5. Nicolaidis, M. (1996). Theory of Transparent BIST for RAMs, *IEEE Transactions on Computers*, vol. 45, N° 10, pp. 1141-1155.
6. Niggemeyer, D., Redeker, M. and Otterstedt, J. (1998). Integration of Non-classical Faults in Standard March Tests, *Records of the Int. Workshop on Memory Technology, Design and Testing.*
7. Van de Goor, A.J. and Al-Ars, Z. (2000). Functional memory faults: a formal notation and a taxonomy, *Proc. Of IEEE VLSI Test Symposium*, pp. 281-289.
8. Hamdioui, S. et al. (2002). Testing Static and Dynamic Faults in Random Access Memories, *Proc. of IEEE VLSI Test Symposium*, pp. 395-400.
9. Benso, et al. (2005). March AB, March AB1: new March tests for unlinked dynamic memory faults, *Proc. of IEEE International Test Conference*, 8 pp.-841.
10. Pavan, P., Bez, R., Olivo, P., & Zanoni, E. (1997). Flash memory cells-an overview. *Proceedings of the IEEE*, 85(8), 1248-1271. https://doi.org/10.1109/5.622505
11. Godard, B., Daga, J., Torres, L. and Sassatelli, G. (2007). Architecture for Highly Reliable Embedded Flash Memories, *IEEE Design and Diagnostics of Electronic Circuits and Systems*, pp. 1-6. https://doi.org/10.1109/DDECS.2007.4295257

12. Brewer, J. and Gill, M. (2008). *Nonvolatile Memory Technologies with Emphasis on Flash: A Comprehensive Guide to Understanding and Using Flash Memory Devices*, John Wiley & Sons. https://ieeexplore.ieee.org/servlet/opac?bknumber=5201541
13. Ginez, O. (2007). Fault Modeling and Testing of Flash Memories, *Doctoral dissertation.*
14. Ginez, O., Daga, J.M., Girard, P., Landrault, C., Pravossoudovitch, S. and Virazel, A. (2006). Embedded flash testing: Overview and perspectives. *In International Conference on Design and Test of Integrated Systems in Nanoscale Technology*, pp. 210-215. https://doi.org/10.1109/DTIS.2006.1708721
15. Plantier, J., Aziza, H., Portal, J.M., Reliaud, C., Regnier, A. and Ogier, J.L. (2009). Retention test and electrical stress correlation to anticipate EEPROM tunnel oxide reliability issues. *In 2009 International Semiconductor Device Research Symposium*, pp. 1-2. https://doi.org/10.1109/ISDRS.2009.5378306
16. Aziza, H., Portal, J.M., Plantier, J., Reliaud, C., Regnier, A. and Ogier, J.L. (2012). Non volatile memory reliability evaluation based on oxide defect generation rate during stress and retention test. *Solid-state electronics*, pp.151-155. https://doi.org/10.1016/j.sse.2012.05.069.
17. Mihara, M. et al. (1994). Row-redundancy scheme for high-density flash memory, *Proceedings of IEEE International Solid-State Circuits Conference*, pp. 150-151. https://doi.org/10.1109/ISSCC.1994.344692
18. Huang, Q., Lin, S. and Abdel-Ghaffar, K.A. (2011). Error-correcting codes for flash coding. *IEEE transactions on information theory*, 57(9), pp.6097-6108. https://doi.org/10.1109/ITA.2011.5743580
19. Hamdioui, S., Pouyan, P., Li, H., Wang, Y., Raychowdhur, A. and Yoon, I. (2017). Test and reliability of emerging non-volatile memories. *In 2017 IEEE 26th Asian Test Symposium*, pp. 175-183). https://doi.org/10.1109/ATS.2017.42
20. Mohammad, M.G. (2011). Fault model and test procedure for phase change memory. *IET computers & digital techniques*, 5(4), pp.263-270. https://doi.org/10.1049/iet-cdt.2010.0083
21. Pan, X., Cui, X., Zha, J., Lin, X. and Lee, C.L. (2012). Modeling and test for parasitic resistance and capacitance defects in PCM. *In 2012 12th Annual Non-Volatile Memory Technology Symposium Proceedings*, pp. 73-76. https://doi.org/10.1109/NVMTS.2013.6632866.
22. Fong, S.W., Neumann, C.M. and Wong, H.S.P. (2017). Phase-change memory - Towards a storage-class memory. *IEEE Transactions on Electron Devices*, 64(11), pp.4374-4385.
23. Kim, S., Burr, G., Kim, W., & Nam, S. (2019). Phase-change memory cycling endurance. *MRS Bulletin*, 44(9), 710-714. https://doi.org/10.1557/mrs.2019.205

24. Zhao, W. S., Zhang, Y., Devolder, T., Klein, J. O., Ravelosona, D., Chappert, C., & Mazoyer, P. (2012), Failure and reliability analysis of STT-MRAM, *Microelectronics Reliability*, 52(9-10), 1848-1852. https://doi.org/10.1016/j.microrel.2012.06.035
25. Chintaluri, A., Naeimi, H., Natarajan, S. and Raychowdhury, A. (2016). Analysis of defects and variations in embedded spin transfer torque (stt) mram arrays. *IEEE Journal on Emerging and Selected Topics in Circuits and Systems*, 6(3), pp.319-329. https://doi.org/10.1109/JETCAS.2016.2547779.
26. https://www.everspin.com/spin-transfer-torque-mram-technology.
27. Jenkins, S., Meo, A., Elliott, L.E., Piotrowski, S.K., Bapna, M., Chantrell, R.W., Majetich, S.A. and Evans, R.F. (2019). Magnetic stray fields in nanoscale magnetic tunnel junctions. *Journal of Physics D: Applied Physics*, 53(4), p.044001.
28. Chen, C.Y., Shih, H.C., Wu, C.W., Lin, C.H., Chiu, P.F., Sheu, S.S. and Chen, F.T. (2014). RRAM defect modeling and failure analysis based on March test and a novel squeeze-search scheme. *IEEE Transactions on Computers*, 64(1), pp.180-190. https://doi.org/10.1109/TC.2014.12.
29. Aziza, H., Bocquet, M., Portal, J. M., Moreau, M., & Muller, C. (2013). A novel test structure for OxRRAM process variability evaluation. *Microelectronics Reliability*, 53(9-11), 1208-1212. https://doi.org/10.1016/j.microrel.2013.07.012.
30. Aziza, H., Hajri, B., Mansour, M., Chehab, A., & Perez, A. (2018). A lightweight write-assist scheme for reduced RRAM variability and power. Microelectronics Reliability, 88, 6-10. https://doi.org/10.1016/j.microrel.2018.07.065.
31. Chen, C.Y., Shih, H.C., Wu, C.W., Lin, C.H., Chiu, P.F., Sheu, S.S. and Chen, F.T. (2014). RRAM defect modeling and failure analysis based on March test and a novel squeeze-search scheme. *IEEE Transactions on Computers*, 64(1), pp.180-190. https://doi.org/10.1109/TC.2014.12.
32. Hamdioui, S., Taouil, M. and Haron, N.Z. (2013). Testing open defects in memristor-based memories. *IEEE Transactions on Computers*, 64(1), pp.247-259. https://doi.org/10.1109/TC.2013.206.

https://doi.org/10.1142/9789811210723_bmatter

Index

Lightning Source UK Ltd.
Milton Keynes UK
UKHW050839281120
374278UK00003B/54

9 789811 210716